T0315225

Marine Algae

Biodiversity, Taxonomy, Environmental Assessment, and Biotechnology

Marine Algae

Biodiversity, Taxonomy, Environmental Assessment, and Biotechnology

Editors

Leonel Pereira and João M. Neto
Department of Life Sciences
IMAR-CMA and MARE
(Marine and Environmental Sciences Centre)
University of Coimbra
Coimbra
Portugal

CRC Press
Taylor & Francis Group
Boca Raton London New York

CRC Press is an imprint of the
Taylor & Francis Group, an **informa** business

A SCIENCE PUBLISHERS BOOK

CRC Press
Taylor & Francis Group
6000 Broken Sound Parkway NW, Suite 300
Boca Raton, FL 33487-2742

First issued in paperback 2020

ISBN-13: 978-1-4665-8167-8 (hbk)
ISBN-13: 978-0-367-73975-1 (pbk)

Library of Congress Cataloging-in-Publication Data

Marine algae : biodiversity, taxonomy, environmental assessment, and biotechnology / editors: Leonel Pereira, João M. Neto.
    pages cm
Includes bibliographical references and index.
ISBN 978-1-4665-8167-8 (hardcover : alk. paper) 1. Marine algae--Classification. 2. Marine algae--Biotechnology. 3. Indicators (Biology) I. Pereira, Leonel. II. Neto, João M.

QK570.2.M37 2015
579.8'177--dc23   2014025997

Visit the Taylor & Francis Web site at
http://www.taylorandfrancis.com

and the CRC Press Web site at
http://www.crcpress.com

Preface

This is a book consisting of 11 chapters covering three thematic areas of great impact in modern societies. Based on the main web site of algae (www.algaebase.org), developed in Chapter 11, it includes a revision of the taxonomy used on algae studies, as well as general aspects of biology and the methodologies used in this sector of marine biology (Chapter 1). The second thematic area comprises five chapters (Chapter 2 to Chapter 5) focused on the use of algae as potential environmental sentinels; the threats that algae may represent when dispersed around the world due to the uncontrolled commercial trades' activity; and their use for a sustainable modern world. Following the conservational concerns presently implemented in most Western economies and some emerging countries, this information is of vital importance for a proper management of aquatic environments, and the sustainable management of their natural resources. The third area is centered on the use of different strands of algae and its potential use in the industrial sector: food (human and animal feed), pharmaceutical, cosmetics, and agricultural fertilizers (Chapter 6 to Chapter 10).

This book is intended to find a wide market of potential users, from the academic field, research institutions and industry, to government agencies responsible for the implementation of integrated management of natural resources and environmental quality assessment of aquatic systems. Two added values of the book are: i) the wide experience the authors of different chapters possess in different marine biology research areas; and ii) the combination of the potential uses of algae in modern society (industry) with a sustainable use of natural resources of aquatic ecosystems.

A special acknowledgement is addressed to our colleague Dr. Joana Patrício by her great contribution and productive discussions had initially to structure and select the contents of the book.

Leonel Pereira
João M. Neto

Contents

Marine Algae: General Aspects (Biology, Systematics, Field and Laboratory Techniques)

Tomás Gallardo

1 Introduction

Today, algae are not a taxonomic category. However, the term is very useful for grouping both prokaryotic organisms, in which cell organelles are not delimited by membranes, and eukaryotic organisms, in which they are. Considering biochemical criteria, their ecological affinities and common photosynthesis with oxygen production, in this chapter we will focus on both photosynthetic bacteria with chlorophyll *a*, division Cyanophyta, and the different divisions of eukaryotic algae.

Algae are simple organisms. Many are unicellular, while others are multicellular and more complex, but they all have rudimentary conducting tissues. They also exhibit a wide range of variation from a morphological and reproductive point of view. Algae are biochemically and physiologically very similar to the rest of plants: they essentially have the same metabolic pathways, possess chlorophyll, and produce similar proteins and carbohydrates. Some algae, such as euglenophytes, dinophytes and ochrophytes, have lost their photosynthetic capacity and live as saprophytes or parasites. However, there are also representatives of other groups, such as green algae, in which more than a hundred heterotrophic species have been described. An essential characteristic which distinguishes algae from other photosynthetic plants is their lack of an embryo and multicellular

Dep. Biología Vegetal, Universidad Complutense, 28040 Madrid, Spain.
Email: tgallar@ucm.es

envelope around the sporangia and gametangia (except for freshwater green algae, charophytes). Algae are different from fungi in that they lack photosynthetic capacity.

Algae have been estimated to include anywhere from 30,000 to more than one million species, most of which are marine algae (Guiry 2012). The most accurate estimate obtained from Algabase (Guiry and Guiry 2013) cites over 70,000 species, of which about 44,000 have probably been published. It is still not well known how many species comprise some groups. For diatoms, some phycologists estimate a number of over 200,000 species. Algae are ubiquitous and live in virtually all media. Although they are mainly related to aquatic habitats, they can also develop on the ground or on snow and ice, as these living organisms tolerate the most extreme temperatures. In aquatic ecosystems, they are the most important primary producers, the base of the food chain.

The classification of algae has experienced great changes over the last thirty years, and today there is no general scheme accepted by all phycologists. There are several systematic proposals, ranging between 5 and 16 divisions. Different treatments are found in Bold and Wynne (1978), South and Whittick (1987), Dawes (1998), Margulis et al. (1989), Hoek et al. (1995), Johri et al. (2004), Barsanti and Gualtieri (2006), Lee (2008) and Graham et al. (2009). In this text, the adopted system is summarized in Table 1. It partly follows the recommendations of Yoon et al. (2006) for red algae, those of Leliaert et al. (2012) for green algae, and those of Riisberg et al. (2009) and Yoon et al. (2009) for ochrophytes, as well as data compiled by Algabase (Guiry and Guiry 2013).

2 General Aspects

Algae are unicellular or multicellular organisms which, with the exception of the cyanophytes, have cellular organelles surrounded by membranes. All autotrophic algae have chlorophyll a and the accessory pigment β-carotene. Sexual reproduction by means of specialized cells involves alternating nuclear phases and a zygote that never develops a multicellular embryo. In general, the cells of eukaryotic algae are surrounded by a wall produced by the Golgi apparatus. The wall in most of them has a fibrillate appearance, because it consists of cellulose, often containing polysaccharides formed by amorphous mucilage. Their cells have numerous organelles, among which the mitochondria, chloroplasts and nucleus are the only organelles surrounded by a double membrane (Fig. 1a). Invaginations of the inner membrane of mitochondria, called mitochondrial crests, can have two different shapes. They are laminar in algae with phycobiliproteins and in those with both chlorophyll a and b (Table 2), whereas they are tubular in the rest of the groups (Roy et al. 2011).

Table 1. Classification scheme of different algal groups.

Kingdom	Phylum	Subphylum	Class
Prokaryota eubacteria	Cyanophyta		Cyanophyceae
Eukaryota	Glaucophyta		Glaucophyceae
	Rhodophyta	Cyanidiophytina	Cyanidiophyceae
		Eurhodophytina	Compsopogonophyceae
			Porphyridophyceae
			Rhodellophyceae
			Stylonematophyceae
			Bangiophyceae
			Florideophyceae
	Cryptophyta		Cryptophyceae
	Dinophyta		Dinophyceae
	Haptophyta		Haptophyceae
	Ochrophyta	Khakista	Bacillariophyceae
			Bolidophyceae
		Phaeista	Chrysophyceae
			Synurophyceae
			Eustigmatophyceae
			Raphidophyceae
			Dictyochophyceae
			Pelagophyceae
			Pinguiophyceae
			Phaeothamniophyceae
			Chrysomerophyceae
			Xanthophyceae
			Phaeophyceae
	Euglenophyta		Euglenophyceae
	Chlorarachniophyta		Chlorarachniphyceae
	Chlorophyta	Prasinophytina	Prasinophyceae
		Tetraphytina	Chlorophyceae
			Chlorodendrophyceae
			Trebouxyophyceae
			Ulvophyceae
			Dasycladophyceae
	Charophyta (Streptophyta *p. p.*)		Coleochaetophyceae
			Conjugatophyceae
			Mesotigmatophyceae
			Klebsormidiophyceae
			Charophyceae

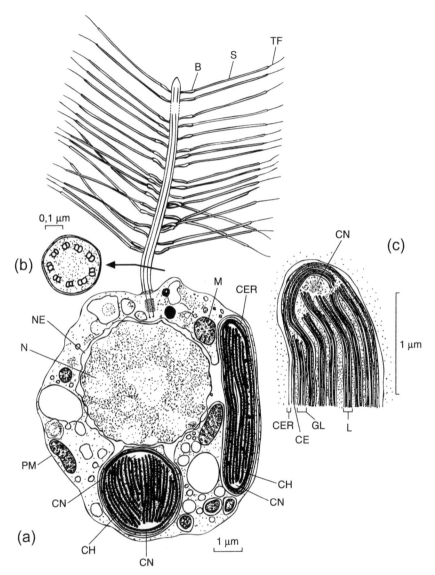

Figure 1. Ultrastructure of the flagellate male gamete of a central diatom. (a) Longitudinal section through whole cell. (b) Cross section through the flagellum; note the absence of the central pairs of the microtubules. (c) Detail of the chloroplasts with girdle lamella. B: base of mastigoneme. CE: chloroplast envelope. CER: chloroplast endoplasmic reticulum. CH: chloroplast. CN: chloroplasts nucleoid. GL: girdle lamella. L: lamella composed of a stack of three thylacoids. M: mitochondrion. N: nucleus. NE: nuclear envelope. PM: plasma membrane. S: tubular part of mastigoneme. TF: terminal fiber of mastigoneme (After Hoek et al. 1995).

Table 2. The main pigments of the algal phyla.

Phylum	Chlorophylls	Phycobilins	Carotenoids	Xanthophylls
Cyanophyta	*a* and *a, b*	Allophycocyanin	β-Carotene	Myxoxanthin
		c-Phycoerythrin		Zeaxanthin
		c-Phycocyanin		
Glaucophyta	*a*	Allophycocyanin	β-Carotene	Zeaxanthin
		c-Phycocyanin		
Rhodophyta	*a, d*	Allophycocyanin	α-, β-Carotene	Lutein
		r-Phycoerythrin		
		r-Phycocyanin		
Cryptophyta	*a, c*	Phycoerythrin	α-, β-, ε-Carotene	Alloxanthin
		r-Phycocyanin		
Dinophyta	*a, b, c*	Absent	β-Carotene	Diadinoxanthin
				Peridinin
				Fucoxanthin
				Dinoxanthin
Haptophyta	*a, c*	Absent	α-, β-Carotene	Fucoxanthin
Ochrophyta	*a, c_1, c_2, c_3*	Absent	α-, β-, ε-Carotene	Fucoxanthin
				Violaxanthin
				Diadinoxanthin
				Heteroxanthin
				Vaucheriaxanthin
Euglenophyta	*a, b*	Absent	β-, γ-Carotene	Diadinoxanthin
Chlorarachniophyta	*a, b*	Absent	β-Carotene/ absent	Lutein
				Violaxanthin
				Neoxanthin
				Siphonaxanthin
Chlorophyta	*a, b*	Absent	α-, β-, γ-Carotene	Lutein
				Prasinoxanthin
Charophyta (Streptophyta *p.p.*)	*a, b*	Absent	α-, β-, γ-Carotene	Lutein

The pigments responsible for photosynthesis are located in a membrane system in the form of flat vesicles called thylakoids, where carbon dioxide fixation occurs. Thylakoids are free in plastid stroma, isolated or in groups of two or more thylakoids, called lamellae. In red algae, thylakoids are not grouped, and they are associated with granules, the phycobilisomes, where phycobiliproteins (mainly phycoerythrin and phycocyanin) are contained. In the remaining groups of algae, thylakoids are gathered in groups. In golden brown algae, thylakoids form packs of three, which are surrounded by a band of three thylakoids or a girdle lamella (Fig. 1c). In some green algae, clusters of thylakoids are interconnected by other thylakoids forming compact stacks known as grana, such as in land plants. Chlorophylls and carotenoids are associated with thylakoids. Carotenoids, as previously mentioned for phycobiliproteins, constitute auxiliary pigments, and there are two types: free oxygen or hydrocarbon carotenes and their oxygenated derivatives called xanthophylls.

In certain groups of algae, the chloroplast is surrounded by one or two additional membranes. When there are two additional membranes, the innermost membrane represents the plasma membrane of the alga that was phagocytized, while the outer membrane often has attached ribosomes and is considered to have originated from the endoplasmic reticulum. In these cases, the outer membrane also surrounds the nucleus, and microtubules and vesicles with storage products can be found in between the two membranes, leading us to think that these chloroplasts may have a endosymbiotic origin, so-called secondary endosymbiosis (Fig. 2). When ribosomes are only present on a third membrane, as in dinoflagellates, it is interpreted that the host plasma membrane was destroyed upon phagocytosis. Species of Cryptophyta present a different situation, because their chloroplasts have important remnants of genome present in phagocytosed alga, the nucleomorph. In Chlorarachniophyta and Glaucophyta, the chloroplast has four membranes and the outermost membrane lacks associated ribosomes, suggesting that it was originated by a digestive vacuole.

Chloroplasts contain circular DNA without histones and 70S ribosomes. They often exhibit electron-dense areas, called pyrenoids, consisting of polypeptides with enzymatic properties. They have been associated with carbon dioxide fixation, since reserve products such as starch tend to accumulate around them. Another structure that can be found in many unicellular algae is the stigma or orange or red colored eyespot, consisting of packed carotenoids. Eyespot is considered to be related to phototaxis and is associated with photosensitive proteins. Eyespot seems to be a shading device to the true photoreceptor. In some groups of algae, such as euglenophytes and dinoflagellates, the stigma is located outside the chloroplast. The nucleus is surrounded by a double membrane, called the nuclear membrane, which contains DNA, proteins, small amounts of RNA and the bulk substance or nucleoplasm. The nuclear membrane, derived from the endoplasmic reticulum of the cell, is perforated by numerous pores. DNA is organized into chromosomes which are not visible during interphase, as in most plants and animals with the exception of euglenophytes, dinoflagellates and cryptophytes, in which DNA is condensed in chromosomes during interphase. The number of chromosomes in algae varies greatly from 2 to over 80.

Many algae, or their reproductive cells, are motile by flagella. The flagellum is an axoneme, consisting of nine pairs of microtubules that encircle

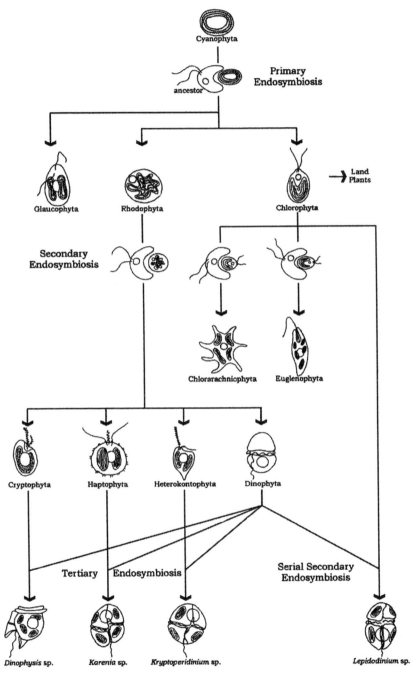

Figure 2. Algal evolution and endosymbiotic events (After Barsanti and Gualttieri 2006).

two central microtubules and are surrounded by the cell plasma membrane (Fig. 1b). The flagellum at the insertion point with the cell body undergoes changes in its structure, so that the two central microtubules form a plate, while the nine peripheral pairs are transformed into triplets. The basal body or basal corpuscle of the flagellum is located inside the cell and has the same structure as the centriole. In many cases, basal corpuscles act as centrioles. In the basal region, groups of microtubules, known as microtubule roots, go through the protoplast (Fig. 27). Filaments or hairs, known by the name of mastigonemes, can be seen on the surface of the flagellum (Fig. 1a). There are two types of mastigonemes: fibrous mastigonemes, formed by solid fibers of glycoproteins, and tubular mastigonemes, formed by proteins and glycoproteins. The latter is differentiated into three parts: the base, which connects the mastigoneme to the flagellar membrane without penetrating it, an intermediate zone or microtubular duct, and a thin apical part. In addition to hairs, different types of scales may appear on the flagellum surface. Flagellated cells vary greatly in terms of shape, arrangement and number of flagella. When flagella are the same length, the cell is called isokont, while if they are of different lengths, the cell is anisokont. If one of the flagella is smooth and the other has mastigonemes, the cell is heterokont. When the flagella are smooth, they are called acronematic, while flagella with mastigonemes are called pleuronematics.

3 Anatomy and Reproduction of Algae

There is considerable variability in the organization of algae, from unicellular organisms only a few microns in size to thalli of complex structure, such as the large phaeophycean kelp *Macrocystis*, which can reach 100 meters in length. Many unicellular algae are motile by flagella (Fig. 3a), whereas others like the so-called coccoid forms are immobile (Fig. 3b). In some groups, like in ochrophytes, there are amoeboid or rhizopodial species. Unicellular forms can live isolated or grouped in colonies (Figs. 3c, d). The colonial algae known as coenobium consist of a fixed number of cells whose number does not increase during its life (Fig. 3e).

Multicellular simple or branched filaments are common in algae (Fig. 3f), some of which are heterotrichous with differentiated prostrate and erect parts (Fig. 10G). Compact multicellular, cylindrical, band-like or foliaceous thalli show great diversity. Most compact forms are branched filaments in origin, and at least in its early stages of development, the thallus often consists of a single filament: uniaxial thalli. In other cases, the thallus has a central portion with several filaments: multiaxial thalli. The axial filaments can produce many lateral branches that are cohered forming a structure with a parenchymatic aspect, the pseudoparenchyma, which is typical in most macroalgae (Fig. 6). In brown algae (Phaeophyceae),

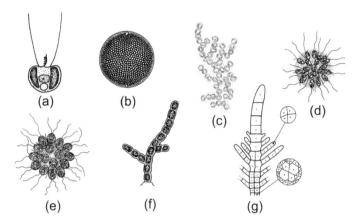

Figure 3. Diversity of algal body type. (a) Motile unicellular (*Chrysochromulina* sp., Haptophyceae). (b) Coccoid unicellular (*Cerataulus smithii*, Bacillariophyceae). (c) Colony of coccoid cells (*Sphaeridiothrix compressa*, Chrysophyceae). (d) Motile colony (*Uroglena volvox*, Chrysophyceae). (e) Coenobium (*Gonium pectorale*, Chlorophyceae). (f) Branched filament (*Asterocystis smaragdigna*, Bangiophyceae). (g) Parenchymatous thallus (*Sphacelaria plumula*, Phaeophyceae).

branched filaments of one or several axial filaments are cohered by mucilages, forming pseudoparenchymatic thalli called haplostichous. In other cases, phaeophyceans have parenchymatous thalli formed by cells that undergo division into two or more planes, originating polystichous thalli (Fig. 3g). Some parenchymatous algae have simple laminar thalli, consisting of cells similarly arranged in a single layer, as in some species of the genus *Porphyra* (Bangiophyceae), or in two layers, as in *Ulva* (Ulvophyceae). These thalli are caused by the division of its cells into two planes. Parenchymatous organization reaches its greatest complexity in Phaeophyceae, as in *Nereocystis* or *Laminaria* (Fig. 21), which already have conducting cells with functions similar to those of phloem. Some Chlorophyta have a siphonocladal organization with thalli formed by simple or branched filaments which have several multinucleated cells. Siphonal organization is typical of some xanthophyceans and chlorophyceans in which thalli are formed by a large, multinucleated cell without septa. Its range of variation includes unicellular forms, like *Acetabularia* (Fig. 33) or pseudoparenchymatous as *Penicillus* (Fig. 31).

The growth of multicellular thalli can occur by the division of any cell in the thallus, diffuse growth, or vegetative cell division restricted to certain parts of the alga, known as localized growth. When growth capacity is restricted to only one apical cell as in *Sphacelaria* (Phaeophyceae) (Fig. 3g) or a few apical cells as in *Padina* (Phaeophyceae), growth is called apical. Intercalary growth occurs when cells with the ability to divide are located in one specific area of the thallus, as in *Desmarestia* (Phaeophyceae), or in several, as in *Ectocarpus* (Phaeophyceae). Branching results from the rotation of the spindle's longitudinal axis. Depending on the angle of rotation of the spindle, thalli produce dichotomous or lateral branches. The postgenic junction of lateral branches, as in many red algae, result in pseudoparenchimatous, compact thalli whose appearance can be terete, fleshy or foliose (Fig. 10).

Algae also show great diversity in their propagation mechanisms. In unicellular algae, the formation of new individuals often occurs through the process known as binary fission or bipartition, where the parent cell produces two new identical individuals. In simple multicellular thalli, asexual or vegetative reproduction occurs by means of fragmentation. Some algae, like the brown alga *Sphacelaria* (Pheophyceae), produce propagules which are specialized structures of vegetative propagation. Algae also produce a variety of spores for asexual reproduction: zoospores if they are able to move, aplanospores if they lose their ability to move, and hipnospores when the cell is surrounded by a thick wall which allows spores to act as resistance bodies. Spores frequently occur within vegetative cells, but they are often generated in the sporangia, which are specialized differentiated cells.

Sexual reproduction is known in most algal groups, with the exception of some of the unicellular or colonial species. When fertilization occurs by the fusion of morphologically indistinguishable gametes called isogametes, it is known as isogamy. In unicellular algae, vegetative haploid individuals can act as isogametes. If one of the gametes is smaller than the other, anisogametes, then the process is known as anisogamy. If one of them is large and immobile, oosphere, and the other is smaller and mobile, sperm, fertilization is known as oogamy. Gametes originate in vegetative cells or in special structures called gametangia. The female gametangia in algae that reproduce sexually by oogamia are called oogonia. Monoecious species produce male and female gametes in the same individual, whereas dioecious species produce male and female gametes in separate individuals.

There are several types of life cycles in the sexual reproduction of algae, depending on when meiosis occurs. The three basic types are: a) haplontic reproductive cycle: meiosis takes place in the first division of the zygote, zygotic meiosis, and produces genetically haploid individuals. As it involves a single generation, is called monogenetic; b) diplontic and monogenetic reproductive cycle: meiosis occurs during gametogenesis, gametic meiosis, as in animals, and there are only diploid individuals; c) haplodiplontic reproductive cycle: two different types of individuals alternate, one haploid gametophyte that produces gametes and other diploid sporophyte that produces spores. As two different generations occur, the cycle is called digenetic and meiosis occurs during sporogenesis, sporic meiosis. This process, which occurs in two phases, is known as alternation of generations (Fig. 4). When gametophytes and sporophytes are morphologically identical, they are called isomorphic, but if they are different, they are heteromorphic. In the latter life cycle there are wide variations within different groups of algae. Finally, in some algae, as in the trebouxiophycean *Prasiola*, meiosis occurs in some vegetative cells of the thallus, somatic meiosis, and the life cycle is known as somatic. Reproductive cycles sometimes present alterations in their development, such as the appearance of new individuals from unfertilized gametes or parthenogenesis. Red algae often have a cycle with alternation of three generations, or a trigenetic life cycle (Fig. 11). Haplontic cycles are more common in unicellular and colonial algae, while macroscopic algae usually have haplodiplontic cycles. Isomorphic alternation of generations is common in morphologically simple algae, whereas heteromorphic alternation of generations occurs in more complex algae. Diplontic cycles have only been described for some groups of algae, such as diatoms (Fig. 18), some siphonous ulvophycean, and brown algae of the order Fucales (Fig. 25). Numerous chemicals have been discovered to act as sex hormones or pheromones during the fertilization process, favouring attraction, adhesion and fusion of sex cells. In brown and green algae, male gametes are attracted to volatile hydrocarbons produced by female gametes.

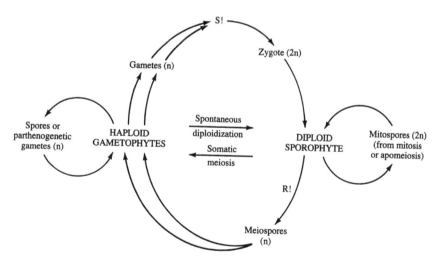

Figure 4. Haplodiplontic life cycle with sporic meiosis and alternation of generations. Asexual reproduction is possible in the gametophyte by spores or parthenogenic gametes, and in the sporophyte by diploid mitospores originated from mitosis or apomeiosis. S!: syngamy. R!: meiosis.

4 Fossil Records of Algae

Fossils attributed to photosynthetic organisms have been dated at 3,000 million years; traces similar to current cyanobacteria are known from 2,500 million years ago, and acritarchs, possibly eukaryotic photosynthetic organisms, have been described from strata of about 2,500 million years. Red and green algae fossils resembling existing species have been dated at about 600 million years ago, by the end of Precambrian. These algae are well preserved due to their calcium carbonate deposits. The green algae of the order Dasycladales are the most abundant macroscopic fossil algae from early Cambrian, in which the number of fossil genera is larger than the number of living genera. Coccoliths (haptophytes) with calcified walls, and diatoms with silicified skeletons, are found in Mesozoic sediments. Marine diatoms formed large Cretaceous deposits known as diatomaceous earth, reaching thicknesses of several hundred meters. Fossil dinoflagellates are known from the Silurian, 450 million years ago. Hundreds of species of dinoflagellates and diatoms that have been described are now extinct. Other groups of algae are not abundant in fossil records (Falkowski and Knoll 2007).

5 Algal Systematics

The systematic arrangement of algae into Divisions or Phyla, and Classes is based on basic traits, such as the chemical composition of photosynthetic pigments, storage substances that accumulate, type of cell wall, and cytological characters like cellular ultrastructure, especially the characteristics of chloroplasts and their endosymbiotic origin (Fig. 2). Recently, classifications have been supported by molecular genetics and particularly the interpretation of the DNA base sequence of the chloroplast and the 5S, 18S and 28S ribosomal RNA sequences. Based on these characters, there is at least one phylum of prokaryotic algae, Cyanophyta, and ten phyla of eukaryotic algae: Glaucophyta, Chlorarachniophyta, Euglenophyta, Dinophyta, Cryptophyta, Haptophyta, Ochrophyta, Rhodophyta, Chlorophyta and Charophyta (Table 1).

5.1 Cyanophyta

The cyanophytes, or Cyanoabacteria, comprise a single class of Cyanophyceae. Cyanobacteria are related to gram-negative eubacteria due to their four-layered cell wall, constituted by the characteristic murein containing peptidoglycan. However, they share some characteristics with eukaryotic algae, such as the presence of chlorophyll and similar aerobic photosynthetic oxygen production. All blue-green algae are non-motile.

Their structural diversity comprises colonial and filamentous unicellular organisms (Figs. 5a, b). Unicellular species form cell colonies produced

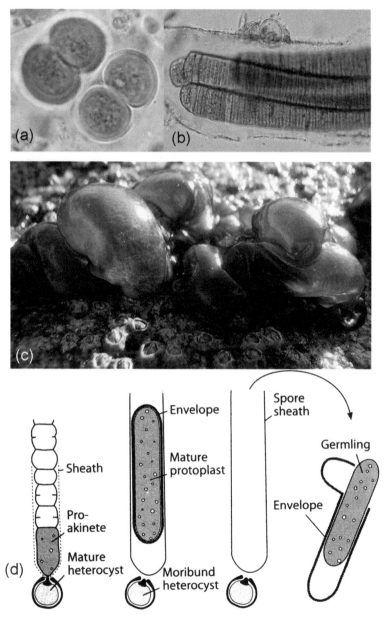

Figure 5. Cyanophyceae. (a) *Chroococcus minutus*. (b) *Blennothrix glutinosa*. (c) *Rivularia bullata*. (d) Development and germination of the akinete of *Gloeotrichia echinulata*; note the presence of a terminal heterocyst (a, courtesy S. Calvo; b, c, courtesy I. Bárbara; d, after Cmieck et al. 1984).

after several successive divisions, and they stay together by mucilage. The pigments needed for photosynthesis are chlorophyll *a* and, in some cases, also chlorophyll *b*, as well as accessory pigments called phycobilins found in the phycobilisomes and attached to the outer surface of the thylakoids (Table 2). The cytoplasm contains gas vacuoles, cyanophycin granules composed of a protein containing only arginine and aspartic acid, carboxysomes, polyhedral bodies related to the synthesis of RuBisCO, and cyanophycean starch or polyglucan granules. Cyanobacteria are able to fix atmospheric nitrogen, a capacity that has been linked with the presence of specialized cells called heterocytes (Fig. 5d). These specialized cells are found in filamentous cyanobacteria. When viewed under a light microscope, they appear hyaline, surrounded externally by a thick wall of three layers. Heterocytes can be terminal as in *Gloeotrichia*, or intercalary as in *Anabaena*. They are connected to adjacent cells by polar proteinaceous nodules provided with a channel through which cytoplasm passes. It has been shown that the lack of nitrogen in the medium increases the number of heterocysts, whereas nitrogen-rich media inhibit their development.

Reproduction is strictly asexual, by simple cell division or fragmentation of the colony and, in some cases, by special cells called endospores. There is no evidence of sexual reproduction or bacterial conjugation. Some species produce akinetes, resting cells found only in species capable of producing heterocysts. Akinetes have been related to cell reproduction, and are formed from vegetative cells which increase in size, produce thick walls and accumulate DNA and cyanophycin granules. Heterocytes can also act as spores. Some species develop small filamentous strands, hormogonia, which can produce new filaments. A few cyanobacteria contain chlorophyll *a* and *b* in thylakoids grouped by two, but other common structures in blue-green algae, such as phycobiliproteins, gas vesicles, or granules containing cyanophycin are not present. Few species are known with these features. Most of them are *Prochloron* species, living as symbionts of colonial tropical ascidians or free-living species, such as the marine planktonic species *Prochlorococcus marinus*.

5.2 Eukaryotic Algae

Glaucophyta

Glaucophyta are freshwater inhabitants, similar to cryptophytes in having chlorophyll *a* and phycobilins (Table 2). These algae, like the genus *Cyanophora*, have symbiotic cyanobacteria, known as cyanelles, instead of chloroplasts. Some authors consider that they represent the first step in the endosymbiotic theory of the origin of the chloroplast. The cyanelles are surrounded by a peptidoglycan wall like cyanobacteria.

Rhodophyta

Rhodophyta, or red algae, are characterized by phycobilins in phycobilisomes, free thylakoids and the absence of flagellate cells. Rhodophyta are a morphologically diverse group (Cole and Sheath 1990). About 10 genera are unicellular, while the rest are multicellular with simple or branched filaments, although most are pseudoparenchymatous uniaxial or multiaxial thalli (Fig. 6). Cell walls have mucilages which compact the adjacent branched filaments and give consistency to the thallus. The parenchymatous organization is present in some genera of the order Bangiales, like *Porphyra* (Fig. 8), and in some foliose thalli like *Delesseria*.

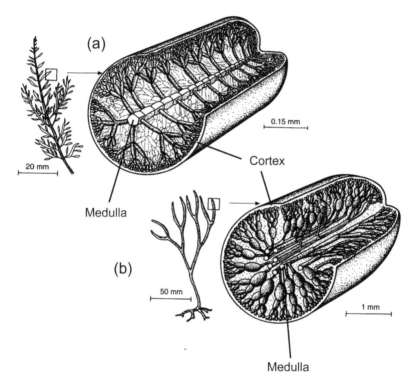

Figure 6. Diagrams of pseudoparenchymatous organization of red algae. (a) Uniaxial thallus. (b) Multiaxial thallus.

The cell wall of red algae is composed of cellulose and a high percentage of mucilage which is the source of products of commercial interest, such as agar and carrageenan. In multicellular red algae, the cell wall formed during cell division usually leaves a pit in the center. In mature cells, the pits are occluded with a protein plug that looks like a biconvex lens when observed with transmission electron microscopy (Fig. 7). There are several types of

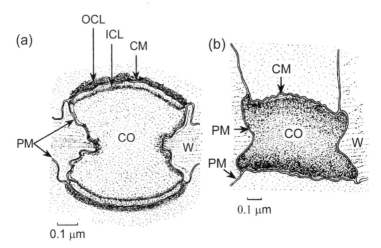

Figure 7. Diagrams of pit plug in the Rhodophyta. (a) The *Palmaria* type of pit plug, with cap membrane and two layered plug cap. (b) The *Rhodymenia* type without cap layers. CM: cap membrane. CO: core of pit plug. ICL: inner cap layer. OCL: outer cap layer. PM: plasma membrane. W: cell wall.

protein plugs which have been used in the classification of red algae due to their systematic value. One of the most common types is the plug covered by two plug caps, each composed of one internal and one external protein layer separated by the plasma membrane. At maturity, some rhodophytes produce secondary pits between cells of adjacent filaments. Corallinales, one of the most important groups of Rhodophyta, incorporate large amounts of calcium carbonate on the walls, as crystals of calcite or aragonite, which gives them an appearance of stone.

The chloroplasts have ungrouped thylakoids which contain chlorophyll *a* and, in some species, they also have chlorophyll *d*. They present several accessory pigments among which are phycobiliproteins (*r*-phycoerythrin, *r*-phycocyanin and allophycocyanin) and lutein (Table 2). Phycobiliproteins are contained in phycobilisomes, similar to those found in cyanobacteria. The main storage product is a specific polysaccharide, floridean starch, which accumulates in the cytoplasm. Rhodophyta are grouped into seven classes (Table 1). The first six classes include the simplest forms with haplontic or haplo-diplontic life cycles like *Porphyra* (Fig. 8), and most of them live in freshwater. The class Florideophyceae contains most of the Rhodophyta with a more complex morphology and is characterized by a life cycle of three generations: a haploid stage, the gametophyte, and two diploid stages. The diploid stages consist of a parasitic stage called carposporophyte which grows on the gametophyte and a second stage which is usually free, called tetrasporophyte, in which the tetrasporangia produce four meiotic spores or tetraspores (Fig. 9). Tetrasporophytes and

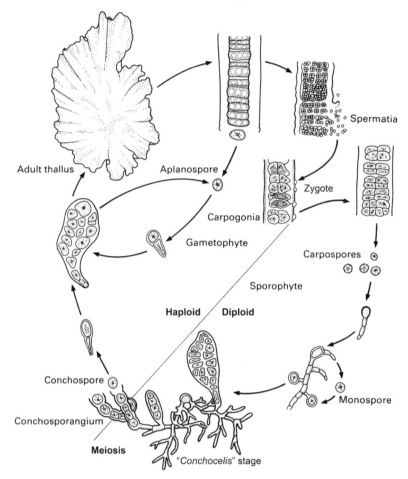

Figure 8. Life cycle in *Porphyra* sp. (After West and Hommersand 1981).

gametophytes can be isomorphic or heteromorphic. The carposporophyte is considered a parasite of the gametophyte, since feeder cells from the gametophyte are frequently involved in its development.

In sexual reproduction, the female gametangia, called carpogonia, are bottle-shaped with a narrow neck or trichogyne. The immobile male gametes, known as spermatia, are produced in spermatangia. The spermatia are passively transported by water currents until they contact the carpogonium. In Florideophyceae, the carpogonium comes directly from a vegetative cell or appears at the apex of a specialized filament two to five cells long, the carpogonial branch. Next to carpogonium, there

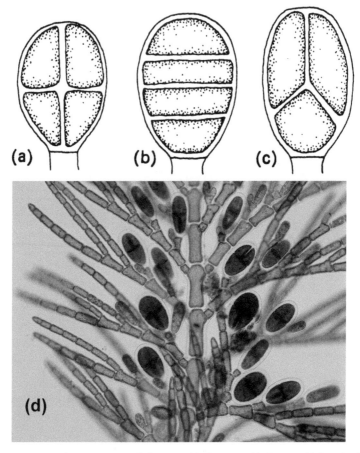

Figure 9. Patterns of tetrasporangial division. (a) Cruciate. (b) Zonate. (c) Tetrahedral. (d) Branches with several tetrahedral tetrasporangia *Antithamnion villosum* (d Courtesy I. Bárbara).

may be one or more auxiliary cells which the zygote is transferred to after fertilization. Asexual reproduction occurs in tetrasporophytes through diploid monospores.

Most of the more than 5,000 species of red algae are strictly marine species, while only a hundred are freshwater (Order Batrachospemales). Some genera contain marine and continental representatives, as in the case of *Audouinella*. Red algae look red due to phycobiliproteins that mask chlorophyll *a*. These pigments allow them to absorb the light spectrum band from violet to blue more easily. As this wavelength penetrates deeper into the water, rhodophytes can live at greater depths. Calcified red algae are the dominant vegetation in warm seas, where they are an important part of the coral reefs.

Rhodophyta have traditionally been associated with cyanobacteria, as their plastids derived from endosymbiotic cyanobacteria. Some phycologists think they may come from simple eukaryotic algae like cryptophytes or glaucophytes, as the lack of flagellate cells is a derived character. Rhodophyta constitute a homogeneous, well-circumscribed division. However, family relationships within the division are difficult to establish (Table 1). Two subphyla are recognized: the subphylum Cyanidiophytina with only one class, Cyanidiophyceae, the most primitive group of red algae with unicellular, freshwater representatives, and the subphylum Eurhodophytina, which includes the rest of the red algae, with 6 classes (Yoon et al. 2006). The classes Porphyridiophyceae and Rhodelophyceae contain unicellular species, most of which live in freshwater and have high concentrations of salt. The freshwater class Compsopogonophyceae presents simple or branched filaments like *Erythrotrichia* (Fig. 10A) or tubular thalli without pits. Stylonematophyceae are small branched filaments, without pits, which are common in marine and inland waters.

Bangiophyceae

Bangiophyceae are freshwater or marine. Their thalli can be formed by unbranched filaments, as in *Bangia*, branched filaments as in *Asterocystis* (Fig. 3f) or by a parenchymatous lamina as in *Porphyra*. The blade of *Porphyra* is the gametophyte generation, which alternates with a sporophyte generation formed by branched filaments, known as the *Conchocelis* phase (Fig. 8). *Porphyra* is morphologically the most complex genus of this class. Its reproductive cycle was discovered by Drew (1949) who recognized that the microscopic filamentous shell-boring alga *Conchocelis rosea* was the sporophyte phase of *Porphyra*. In the sporophyte, meiosis occurs in some fertile cells or conchosporangia, producing immobile spores called conchospores, whose development results in a macroscopic gametophyte with a laminar appearance. Most vegetative cells of *Porphyra* gametophytes have reproductive capacity. In the gametophyte, some cells originate 32, 64 or 128 spermatia. In the same or in another individual, some cells develop carpogonia with a small protrusion, and fertilization occurs when one of the spermatia fixes to the carpogonium protrusion. The resulting zygote divides by mitosis producing 4–16 carpospores which are released by the rupture of the carpogonium wall. Carpospores are initially naked and amoeboid, and they fix to substrate to develop the conchocelis phase. In some cases, the first zygote division is meiotic, resulting in four carpospores that directly give rise to macroscopic gametophytes, bypassing the conchocelis phase. Some sporophytes can produce diploid spores or monospores, reproducing the conchocelis phase. The alternation of generations in *Porphyra* is controlled by photoperiod. *Porphyra* is of great commercial importance. It is grown in large marine farms, and is known by the Japanese name Nori.

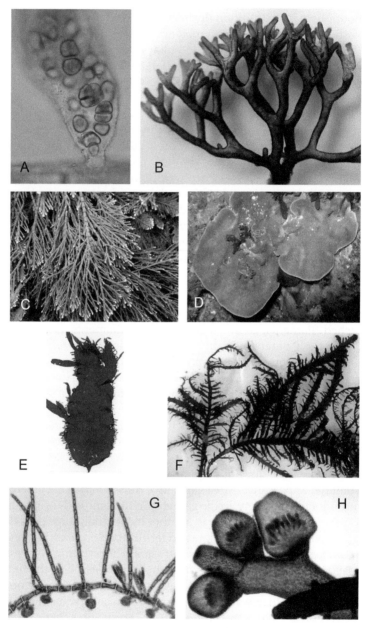

Figure 10. Rhodophytes diversity. A. *Erythrotrichia bertholdii*, thallus attached by a single cell. B. *Liagora viscida* partially calcified thallus. C. *Corallina officinalis*, articulate calcified thallus. D. *Lithophyllum stictaeforme*, crustose calcified thallus. E. *Grateloupia turuturu*. F. *Gelidium spinulosum*. G. *Callithamniella tingitana*, detail of heterotrichous filaments. H. *Chondria coerulescens*, detail of cistocarps with carpospores (Courtesy I. Bárbara).

Florideophyceae

The class Florideophyceae comprises basically heterotrychous filaments with apical growth that originate pseudoparenchymatous thalli constructed by densely branched filaments. The thallus can be unixial, derived from a single apical cell, or multiaxial, derived from a group of apical cells (Fig. 6). Thalli are terete or laminar (Fig. 10). In the most complex species, an outer cortex formed by small photosynthetic cells and an inner medulla can be observed. Their cells have pits and numerous parietal or axial plastids without pyrenoids. Life cycles are usually trigenetic with isomorphic or hetermorphic alternation of generations.

Polysiphonia is a good example of the life cycle of the class Florideophyceae. Three generations or phases alternate in the cycle: two diploid phases, the carposporophyte and tetrasporophyte, and one haploid phase, the gametophyte (Fig. 11). In the apical parts of the male gametophyte, haploid spermatangia are produced which release spermatia. Female individuals

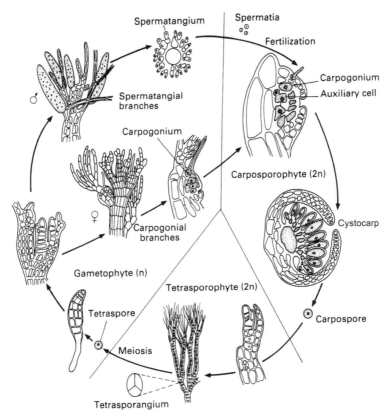

Figure 11. Life cycle in *Polysiphonia* sp. (After West and Hommersand 1981).

originate carpogonia provided with a long trichogyne which appears at the tip of carpogonial branches. Fertilization occurs when one of the spermatia reaches the trichogyne. Once karyogamy has occurred, the zygote migrates to a predetermined cell of the carpogonial branch called the auxiliary cell. Mitotic divisions of the zygote give rise to diploid branched filaments or gonimoblast, which at maturity is enveloped by a layer of cells from the gametophyte, forming a kind of urn or cystocarp with a pore at the end. Cystocarp formation involves a cluster of cells called a procarp that originates from the auxiliary cell located at the carpogonial branch, known as a mature carposporophyte or gonimocarp. Diploid spores or carpospores are formed here and released through the pore of the cystocarp. They germinate to produce diploid individuals, the tetrasporophytes, which are morphologically identical to gametophytes. Tetrasporophytes originate tetrasporangia, each with four haploid tetraspores resulting from meiotic division. Upon release, the tetraspores close the cycle by generating male and female gametophytes.

The classification of Florideophyceae in orders is based on the ultrastructure of pit plugs (Fig. 7), the presence or absence of a carpogonial branch and location of carpogonia, the presence of auxiliary and nutricial cells, and postfertilizational changes in carposporophyte development. Type of tetrasporangium is another taxonomic feature. Five subclasses and nearly 30 orders of florideophyceans have been proposed (Le Gall and Saunders 2006). Most phycologists agree on at less ten large orders: Acrochaetiales, Nemaliales, Batrachospermales, Gelidiales, Corallinales, Gigartinales, Halymeniales, Rhodymeniales and Ceramiales.

The order Acrochaetiales is the simplest group, consisting of branched filaments. The carpogonium originates from a vegetative cell, and the carposporophyte develops directly from the fertilized carpogonium without an auxiliary cell. The life cycle shows isomorphic alternation of generations, and tetrasporangia are cruciate. Protein bodies of pits plugs have a cap consisting of inner and outer layers which are similar in thickness. *Audouinella* is a cosmopolitan genus living in fresh and marine waters. The thalli are small, uniseriate branched filaments formed by a few cells surrounded by mucilage. Asexual reproduction occurs by monospore formation.

Nemaliales have thalli of multiaxial structure. Pit plugs have a two-layered cap. The life cycle is trigenetic with heteromorphic alternation of generations. The carpogonium is on a carpogonial branch without an auxiliary cell. Tetrasporangia are cruciate. *Nemalion*, *Liagora* and the tropical genus *Galaxaura* are abundant in the sea shore. *Nemalion* is a mucilaginous rarely branched thallus with blunt apices. *Liagora* is a terete branched thallus with a lubricus texture, impregated with carbonate calcium (Fig. 10B). *Galaxaura*, with subdichotomous thallus, is calcified and has similar gametophytes and tetrasporophytes.

Corallinales is a cosmopolitan group which is very important in tropical seas as their representatives are actively involved in the formation of coral reefs (Johansen 1981, Woelkerling 1988). Its members are multiaxial and have calcium carbonate deposits on their walls. Pit plugs are characterized by their dome-shaped outer caps. The tetrasporangia are zonate. The basal cell or supporting cell of the two-celled carpogonial branch is the auxiliary cell. Reproductive organs are in conceptacles, cavities opening to the exterior by one or more pores. It is a group with great morphological diversity with crustose species like *Lithophyllum*, *Lithothamnion* or *Melobesia* (Fig. 10D) and articulate species like *Corallina*, *Jania* or *Amphiroa* (Fig. 10C).

The order Gelidiales includes uniaxial cartilaginous species of great economic interest for agar production. Pit plugs only present the inner layer of the cap. One of the cells of the nutritional branch acts as the auxiliary cell. Tetrasporangia are cruciate. The most representative genus is *Gelidium* (Fig. 10F). In Gracilariales, the principal genus of the order is the agarophyte *Gracilaria* with uniaxial, terete or flattened thalli. In this group, the supporting cell originates a two-celled carpogonial branch and several sterile branches. The carposgorangia are surrounded by a pericarp forming a surface cistocarp.

The orders Gigartinales and Ceramiales are the largest orders of red algae. Thalli of Gigartinales are uniaxial or multiaxial. Pit plugs are devoid of caps. The auxiliary cell is an intercalary cell of the nutritional branch, and it is generated before fertilization. Tetrasporangia can be cruciate or zonate. In this group, some genera like *Chondrus*, *Mastocarpus* and *Gigartina* are of economic interest, since they are used for obtaining the phycocolloid carrageenan.

In Rhodymeniales, pit plugs are also devoid of caps. The auxiliary cell originates before carpogonium fertilization and is characterized by the formation of a cystocarp. Tetrasporangia are cruciate or tetrahedral. Many of its species, like *Rhodymenia*, have laminar shapes. The Halymeniales order is close to Rhodymeniales. It consists of multiaxial thalli with isomorphic alternation of generations. Carpogonial branches are of two or four cells, and auxiliary cells are intercalary. Cystocarps are composed of carposporangia buried within rudimentary pericarps. Tetrasporangia are cruciate. The most numerous genera are *Halymenia* and *Gratelopia* (Fig. 10E).

The order Ceramiales differs from Rhodymeniales because the auxiliary cell is generated after fertilization from the supporting cell of the four-celled carpogonial branch. After fertilization, a cystocarp or gonimocarp is developed. Tetrasporangia are cruciate or tetrahedral. The pit plugs have no cap layers. The order contains more than 2,000 species and nine families which are widely distributed. Ceramiaceae include widely distributed genera of delicate filaments, such as *Antithamnion* (Fig. 9d), *Ceramium* or *Callithamniella* (Fig. 10G), and Rhodomellaceae include genera as common as *Polysiphonia* or *Chondria* (Fig. 10H).

Cryptophyta

Cryptophytes are grouped in the class Cryptophyceae, consisting of 12 genera and about 200 species. They include asymmetric unicellular marine and freshwater algae, with flagellate and naked cells, provided with chlorophyll *a* and *c*, several complementary pigments (Table 2) and phycobiliproteins. These are not found in phycobilisomes but between the chloroplast thylakoids. Most have a single chloroplast, but some have lost them and are therefore heterotrophic.

The genus Cryptomonas combines the typical characteristics of cryptophytes (Fig. 12): a bilobate chloroplast with a pyrenoid located in the junction of the two lobes, and a reduced nucleus or nucleomorph, both surrounded by the rough endoplasmic reticulum. Hence, these chloroplasts

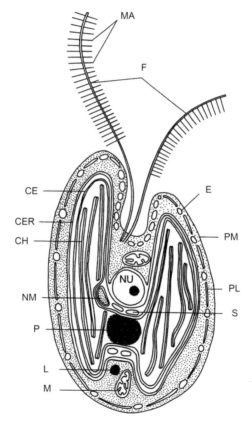

Figure 12. Diagram of *Cryptomonas*. CE: chloroplast envelope. CER: chloroplast endoplasmic reticulum. CH: chloroplast. E: ejectosomes. F: flagella. L: lipid. M: mitochondria. MA: mastigonemes. NM: nucleomorph. NU: nucleus. P: pyrenoid. PL: protein plate. PM: plasma membrane. S: starch grain (After Lee 2008).

have been considered endosymbionts. The cell has two subapical flagella, both with mastigonemes which are embedded in a depression called the vestibulum. The vestibulum extends into a slot in which the eyectosomes are found. Eyectosomes are defensive structures that can be projected when the cell is excited. The cell covering or periplast is formed by the plasma membrane that surrounds a coat of numerous rectangular or polygonal protein plates. The shape and number of plates differ from one species to another. Cryptophytes are often sensitive to light, and species such as *Thecomonas* have stigma. Fertilization in *Cryptomonas* is isogame, and the life cycle is haplontic with zygotic meiosis.

Dinophyta

The division Dinophyta or Pyrhophyta comprises a single class Dinophyceae with 150 genera and about 4,000 species. Dinoflagellates are unicellular, biflagellate organisms especially abundant in the sea and brackish water. Although some are naked, most have a cell cover or amphiesma constituted by the plasma membrane under which a layer of polygonal vesicles is located, where cellulosic plate formation takes place. The cell wall or theca has two grooves, a transverse groove or cingulum and a longitudinal groove called the sulcus (Fig. 13). Two pleuronematic flagella emerge from the area where the grooves intersect. The longitudinal flagellum protrudes from the longitudinal sulcus and is covered by two rows of fibrillar mastigonemes, while the transverse flagellum extends horizontally around the cell and is covered by only one row of hairs. In some cases, the cell wall is perforated by pores through which trichocysts are discharged.

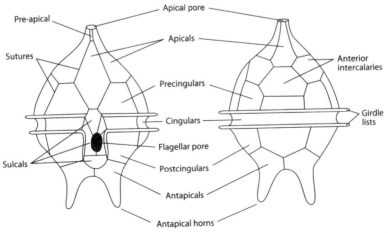

Figure 13. Thecal plate terminology of ventral (left) and dorsal (right) view of the dinoflagellate *Peridinium* sp. (After Taylor 1990).

Trichocysts are membrane-bound organelles containing proteinaceous threadlike structures. These structures can be projected from the cell under stimulation, propelling it in the opposite direction acting as a defensive mechanism. Chloroplasts contain chlorophyll *a*, although some species also contain *c*, and various accessory pigments of which the most important is peridinin (Table 2). Starch is the main storage product that accumulates outside the chloroplast. Plastids are surrounded by three membranes with the thylakoids grouped in stacks of three, as in euglenophytes. In the process of cell division, nuclear envelope remains intact during mitosis, as in euglenophytes and cryptophytes. Dinoflagellates and euglenophytes exhibit large chromosomes that remain condensed during the interphase of the mitotic cycle. Asexual reproduction occurs by longitudinal bipartition. Each of the two daughter cells carries a part of the mother theca and reconstructs the missing part. In sexual reproduction, fertilization is through isogamy and sometimes anisogamy. The general reproductive life cycle is haplontic with zygotic meiosis, but some species are known to have a diplontic cycle (Fig. 14). When environmental conditions are unfavorable, some dinoflagellates produce resting spores with thick walls known as cysts. Fossil cysts are called hystrichospheres, and they formed exploitable deposits similar to diatomaceous earth.

Dinoflagellates are an important part of the plankton of tropical marine waters, and they are essential in the formation of coral reefs. Some species are responsible for blooms known as red tides, caused by changes in environmental conditions and increased nutrients. They are linked to situations of calm in the sea that originate concentrations of dinoflagellates

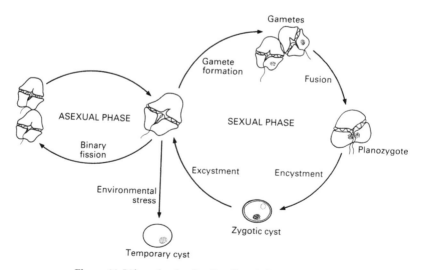

Figure 14. Life cycle of a dinoflagellate (After Walker 1984).

exceeding 100 million cells per liter, coloring the water red. Some red tides produce large economic losses, since certain dinoflagellates such as *Dinophysis acuta, Prorocentrum lima* and *Gymnodinium catenatum*, excrete toxins that kill other marine organisms such as fish or bivalve filter feeders, which are often the subject of intensive farming. Dinoflagellates excrete two major toxins: the paralyzing type or saxitoxin (PSP) and the diarrheal type like okadaic acid (DSP).

Haptophyta (=Prymnesiophyta)

This group comprises about 500 species, all of which are included in the class Haptophyceae. They are unicellular colonial and filamentous flagellates. Some haptophytes are naked, while in others the cell surface is covered by ellipsoidal calcareous scales formed from the Golgi apparatus. As in *Chrysochromulina* (Fig. 3a), flagellate forms typically have two apical, isokont flagella covered with small scales of polysaccharides like the rest of the plasma membrane. They also have another appendage called the haptonema, which contains only seven microtubules surrounded by three plasma membranes. In some species, the haptonema attaches the cell to the substrate. Sometimes the haptonema is very small. In *Isochrysis*, only traces of it remain inside the cell. In *Pleurochrysis carterae*, the life cycle is haplo-diplontic with a pseudofilamentous *"Apistonema"* stage (Fig. 15). Haptophytes and the class Chrysophyceae have many common features, such as chloroplasts, and they are biochemically and structurally similar. Although most are part of marine plankton, some are parasitic, like *Prymnesium parvulum* which attaches to the gills of fish causing death by asphyxiation. States often go from flagellate or amoeboid to sessile phases. Fossils from the Jurassic period are known by the name of coccolithophores, which formed important deposits that are exploited for applications similar to diatomaceous earth.

Ochrophyta

Ochrophytes are algae of diverse organization and include unicellular, colonial, filamentous and parenchematous thalli. They are characterized by the presence of chlorophylls *a* and *c* in their plastids as well as xanthophylls and carotenoids that mask the chlorophylls (Table 2). Due to the presence of these pigments, many ochrophytes have a yellowish-green, gold or brown appearance. As storage products, they accumulate oils and chrysolaminarin, but never starch. Cell walls contain cellulose, and in certain species they contain silica. Cells possess one or more plastids, each with an envelope formed by two membranes of chloroplast and two membranes of chloroplast

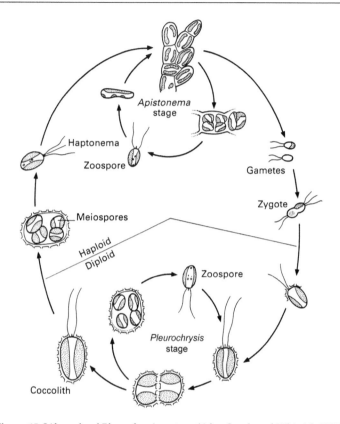

Figure 15. Life cycle of *Pleurochrysis carterae* (After South and Whittick 1987).

endoplasmic reticulum (Fig. 1a). Thylacoids, in stacks of three, in most ochrophytes are surrounded by a band of thylakoids, girdle lamella, just beneath the innermost plastid membrane (Fig. 1c). Ochrophytes have heterokont flagellated cells with two different flagella, a smooth flagellum and one with mastigonemes. In this group of algae, mastigonemes consist of three parts, a basal, tubular and apical part formed by fibrils.

Cells with heterokont flagella appear in groups of heterotrophic and autotrophic organisms, which are known as Stramenopila. The golden-brown and brown algae, together with the fungi Oomycetes with cellulosic walls, were assembled under the name Heterokontophyta. Heterokontophytes, dinoflagellates, haptophytes and certain nonphotosynthetic heterokont protozoans constitute the group known as the kingdom Cromobionta. Photosynthetic Stramenopila are a monophyletic group, which is referred to as Ochrophyta in this text (Riisberg et al. 2009, Guiry and Guiry 2013).

This division contains at least thirteen classes, some of which appear in the category of division in other classifications: Bacillariophyceae,

Bolidophyceae, Chrysophyceae, Synurophyceae, Eustigmatophyceae, Raphidophyceae, Dictyochophyceae, Pelagophyceae, Pinguiophyceae, Phaeothamniophyceae, Chrysomerophyceae, Xanthophyceae and Phaeophyceae (Table 3). Phylogenetic analysis using rDNA nucleotide sequences of the 16S subunit have shown that the classes Chrysophyceae, Synurophyceae, Eustigmatophyceae, Raphidophyceae, Pelagophyceae and Dictyochophyceae are evolutionarily close. Phaeophyceae, Xanthophyceae, Phaeothamniophyceae, Pinguiophyceae and Chrysomerophyceae are also related, and Bacillariophyceae and Bolidophyceae form an isolated group.

Bacillariophyceae

Bacillariophyceans, or diatoms, constitute the largest class of Ochrophyta. About 10,000 benthic and planktonic species are known, and they can be found in both freshwater and marine environments. This group is responsible for 25% of primary production of the sea. Diatoms are unicellular, and in some cases, live in colonies with a filamentous appearance, formed by numerous loosely-joined individuals (Fig. 16). The most distinctive feature of diatoms is their rigid translucent wall, or frustule, consisting of silicon dioxide and traces of other substances, such as sugars, amino acids and uronic acid. Cellulose is never present. The frustule is composed of two overlapping halves or valves (Fig. 17a). Each valve is attached to a circular band known as a cingulum or girdle. The larger valve and its cingulum constitute the epitheca and cover the lower valve and its cingulum, forming the hypotheca. The hypotheca and the epitheca fit inside each other like a box and its lid. When looking at the frustule from the top or bottom, the faces of the diatom can be seen in a valvar view, while from the side, a cingular view can be observed. The ornamentation of the frustule is very complex. Each species has a configuration of specific spines, pores or striae. Diatoms can be classified in two basic groups based on frustule symmetry: pennate diatoms which exhibit bilateral symmetry, and centric diatoms which are radially symmetric (Fig. 3b). The valves of pennate diatoms often present a longitudinal groove or raphe, which connects the two polar nodules and crosses the central nodule (Fig. 17a).

Diatoms with only one or two raphes are able to move due to the production of a mucilaginous material that flows out of the raphe and holds the cell to the substrate. The contraction of this material and the production of new mucilage cause a sliding movement of the cell. In other pennate diatoms, there is a groove, or pseudoraphe, that does not pass through the cell wall. The use of scanning microscopy has revealed the complexity of frustule structures (Fig. 17b). The tiny walls of centric diatoms are formed by chambers known as loculi. These have a thin wall which is finely perforated in the upper or lower face. In pinnate diatoms, light microscopy shows

Table 3. Ochrophyta (photosynthetic Stramenopiles). Classes, pigments and habitats.

Classe	Common name	Chlorophylls	Xanthophylls	Habitats
Bacillariophyceae	Diatoms	a, c_1, c_2	Fucoxanthin	Marine and freshwater
Bolidophyceae	Bolidophyceans	a, c_1, c_2, c_3	Fucoxanthin	Marine
Chrysophyceae	Chrysophyceans	a, c_1, c_2	Fucoxanthin Violaxanthin	Freshwater, rare marine
Synurophyceae	Synurophyceans	a, c_1	Fucoxanthin Violaxanthin	Freshwater
Eustigmatophyceae	Eustigmatophyceans	a	Violaxanthin Vaucheriaxanthin	Freshwater, rare marine
Raphidophyceae	Raphidophyceans/Chloromonads	a, c_1, c_2	Fucoxanthin Violaxanthin Vaucheriaxanthin Heteroxanthin Diatoxanthin	Marine and freshwater
Dictyochophyceae	Dictyochophyceans/Silicoflagellates	a, c_1, c_2	Fucoxanthin Diatoxanthin	Marine, rare freshwater
Pinguiophyceae	Pinguiophyceans	a, c_1, c_2	Fucoxanthin Violaxanthin	Marine
Phaeothamniophyceae	Phaeothamniophyceans	a, c_1, c_2	Fucoxanthin Diatoxanthin	Freshwater
Chrysomerophyceae	Chrysomerids	a, c_1, c_2	Fucoxanthin Violaxanthin	Brackish water
Pelagophyceae	Pelagophyceans	a, c_1, c_2	Fucoxanthin	Marine
Xanthophyceae	Xanthophyceans/Tribophyceans	a, c_1, c_2	Heteroxanthin Diatoxanthin Vaucheriaxanthin	Freshwater, rare marine
Phaeophyceae	Phaeophyceans	a, c_1, c_2	Fucoxanthin Violaxanthin	Marine, rare freshwater

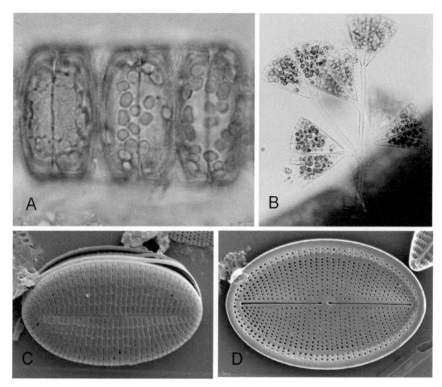

Figure 16. Diatoms. (A) lineal colony of *Melosira*. (B) Dendritic colony of *Lycmophora*. (C) *Cocconeis placentula* epivalve. (D) *Cocconeis placentula* hipovalve (A, B, courtesy I. Bárbara; C, D, courtesy of M. Hernández).

cell wall ornamentation with striae, formed by the arrangement of areoles in rows. The protoplasm is located inside the frustule; in their plastids thylakoids form packs of three surrounded by the girdle lamella (Fig. 1c). Photosynthetic pigments are chlorophylls *a* and *c*, as well as the carotenoids fucoxanthin, diatoxanthin and violaxanthin (Table 3). The most common storage products are chrysolaminarin and lipids.

Asexual reproduction occurs by bipartition. When mitosis occurs, the two valves are separated and each produces a new hypotheca, so that the hypotheca of the parental cell always acts as the epitheca of the daughter cells. After successive cell divisions over time, there is an effect on the average cell size of the diatom population. This phenomenon has been observed in natural populations. The mean diameter of the population progressively decreased until a minimum mean diameter was reached, at which point cell size suddenly increased due to the formation of spores called auxospores. Auxospores were generally the result of sexual reproduction processes, but in some cases they were produced asexually.

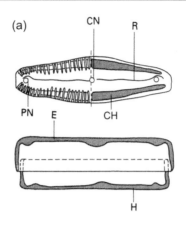

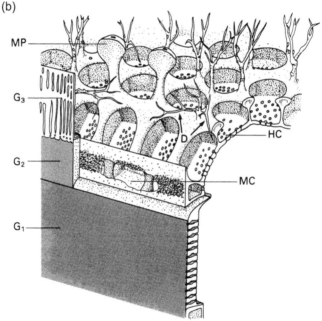

Figure 17. (a) Diagram of the frustule from a pennate diatom. (b) Diagram of the frustule structure of *Triceratium flavus*. CN: central nodule. CH: chloroplast. D: dendritic structures. E: epitheca. G_{1-3}: girdle bands. H: hypotheca. HC: hexagonal chamber. MC: marginal canal. MP: marginal process. PN: polar nodule. R: raphe (After South and Whittick 1987).

Sexual reproduction of diatoms is rare in nature and usually occurs as described. The vegetative diploid cell that reaches the minimum size undergoes meiotic division, and two or three of the four haploid nuclei become degenerate, thus producing one or two haploid cells that behave as gametes. Gametes are amoeboid in pennate diatoms, while in centric diatoms they are flagellate with a flagellum (Figs. 1a, b). Fertilization is accompanied by the loss of the valves, and the resulting zygote enlarges and is surrounded by a thick silica wall forming an auxospore. Auxospore germination results in a larger vegetative cell than the parental one. The life cycle is diplontic with isogamy in pennate diatoms and oogamy in centric diatoms (Fig. 18). In some cases, auxospore formation is a consequence of selfing: the fusion of two haploid nuclei produced by the same diploid cell. Auxospore formation can also be caused by changes in environmental conditions of temperature, light and available nutrients.

The systematic arrangement of diatoms has traditionally been based on morphology and consists of a single class called Bacillariophyceae. Round et al. (1990) proposed that this group should be divided into three classes: Coscinodiscophyceae, Bacillariophyceae, and Fragilariophyceae. However, this division is not supported phylogenetically, and they are treated as subclasses in this text. The published phylogenetic trees show that the primary radiation occurred among centric diatoms. Pennate diatoms evolved later from ancestors with a centric structure. The subclass Coscinodiscophycidae includes circular or irregular diatoms with radially symmetric ornamentation and without raphe. Vegetative cells have numerous plastids, they are always immobile, and male gametes are flagellate. They often form large colonies of cells joined by mucilage, as in *Chaetoceros*. They are almost all marine organisms. Bacillariophycidae

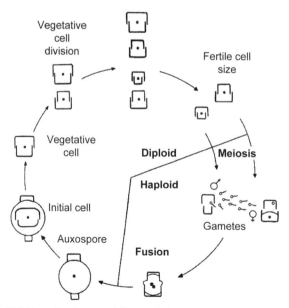

Figure 18. Life cycle of a central diatom (After Hasle and Syvertersen 1997).

include bilateral symmetric diatoms, like *Achnanthes*, which are cuneiform or navicular with raphe or pseudoraphe on one or both sides of the frustules. Most have one or two elongated plastids occupying a large part of the cell. They are more abundant in inland waters. Most of them are benthic, but some live in marine phytoplankton. They can form stalked colonies, like *Gomphonema*, living on plants or animals. Fragilariophycidae include bilateral symmetric diatoms like *Synedra* with cuneiform or navicular cells without raphe. Most genera have many small, discoid plastids.

The siliceous frustules of diatoms have been well preserved in the fossil record. Marine centric diatoms are known from the Cretaceous, and pennate diatoms appeared in the Paleocene. Diatomaceous fossil deposits, known as diatomite, are exploited in many parts of the world and have important industrial applications as abrasives, refractory ceramic and filters. Phylogenetically close to bacillariophyceans, the class Bolidophyceae is a small group of unicellular flagellates without a silica cover. Both classes constitute the Khakista group (Table 1).

Chrysophyceae and related groups

Chrysophyceans are mostly unicellular flagellate organisms. Some have a single functional flagellum like *Chromulina*, while others can form colonies like the freshwater genus *Uroglena* (Fig. 3d). Some are amoeboid, like *Chrysoamoeba*, or coccoid (Kristiansen 2005). Chloroplasts have chlorophylls *a* and *c*, and accessory pigments β-carotene and xanthophylls (Table 3), which give them a golden color. The main storage product is chrysolaminarin. The presence of cellulose in cell walls is common, and some species are covered with silica scales or protected by an organic sheath called the lorica. Very few chrysophyceans are naked. Silica is present in cysts which are common to several genera. Cysts, or statospores, are shaped like a small, externally ornamented bottle closed by a nonsilicified plug. These cysts form under adverse environmental conditions. Mitotic division is the most common mechanism of asexual reproduction. Sexual reproduction is rare, but when it occurs, fertilization is isogame. Vegetative cells are haploid, and meiosis is the first division of the zygote. In some cases, zygotes are statospores, representing a resting phase. Most of the 1,000 described species live in unpolluted freshwaters. Some are strictly marine algae and part of the nanoplankton. Some *Ochromonas* species are heterotrophs, mixotrophs or phagotrophs. Chrysophyceans have many affinities with synurophyceans and eustigmatophyceans.

Synurophyceae are flagellate algae covered with silica-scales which have been segregated by chrysophyceans, due to their lack of chlorophyll c_2, flagellar swelling on the basis of photosensitive flagella and the loss of the ability to carry out phagocytosis. Biflagellate unicellular forms can live in colonies. As they inhabit relatively unpolluted freshwater, they are good indicators of water quality. Of the 200 species that compose this group, most of them belong to the cosmopolitan genera *Mallomonas* and *Synura* and are often covered with very ornate silica-scales. Sexual reproduction occurs by isogamy, vegetative cells are haploid, and zygotes may remain as cysts when environmental conditions are unfavorable. The class Eustigmatophyceae comprises about twelve coccoid unicellular species, almost all of which are freshwater planktonic species. They produce naked zoospores. Most bear a single pleuronematic flagellum, but some have two flagella. They are characterized by the presence of an extraplastidial stigma, the lack of girdle lamella in the chloroplast and violoxanthin as the main xanthophyll. Raphidophyceae are unicellular flagellate algae, also known as Chloromonads. Their plastids contain pigments similar to those of chrysophyceans, but the cells have trichocysts like in dinoflagellates. Unlike other ochrophytes, the endoplasmic reticulum membrane that envelops the chloroplast is not connected to the membrane surrounding the nucleus. There are about 15 species, most of which are marine algae. Some species like *Heterosigma* produce blooms.

Dictyochophyceae

Dictyochophyceans are naked unicellular, marine or continental uniflagellate, characterized by a flagellum that has a widening at the base called the paraflagellar rod (Fig. 19a). This class includes algae known as silicoflagellates which have a siliceous skeleton with a star-shaped or basket form (Fig. 19b). The cytoplasm surrounds the skeleton and contains golden discoid plastids. Their siliceous skeletons have been preserved since the Cretaceous period. *Dictyocha* is the most representative genus, for which more than 100 species have been described, although there are only two living cosmopolitan species at present. This class includes unicellular ochrophytes called pedinelids. Their body is covered with scales, and chloroplasts are located in a ring surrounding the nucleus, which occupies a central position (Fig. 19a). Some species are capable of emitting pseudopodia and catch small organic particles.

Xanthophyceae and related groups

The organization of xanthophyceans, or Tribophyceae, varies from simple filaments to amoeboid cells to siphonous thalli. Plastids are green or yellow-green due to the presence of chlorophyll *a* and several xanthophylls, none of which are fucoxanthin (Table 3). As storage products, they accumulate chrysolaminarin, paramilo, sugars and oils. Their cell walls contain cellulose

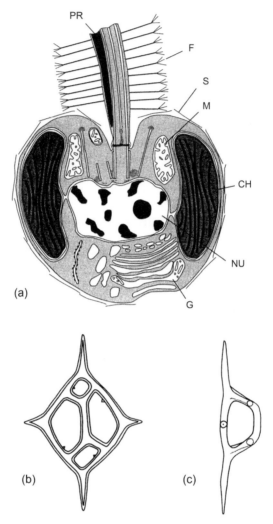

Figure 19. (a) Diagram of *Mesopedinella*. (b, c) Siliceous skeleton of *Dictyocha fibula*, frontal and lateral view. CH: chloroplast. F: flagellum with mastigonemes. G: golgi. M: mitochondria. NU: nucleus. PR: paraflagellar rod. S: scales (After Daugbjerg 1996).

and often silica scales. Flagellate cells have two heterokont flagella with lateral or apical insertion. Sexual reproduction is only known in three genera with a haplontic life cycle. Some species form resistant cysts with silica walls, closed by a cap. Xanthophyceans, despite their green color, can be distinguished from chlorophytes by their lack of chlorophyll *b* and by their heterokont flagella. Most of the 600 known xanthophyceans species live in freshwater and moist soil, and only a few are marine species. The freshwater and marine genus *Vaucheria* has a cylindrical body consisting of a branched coenocytic filament with many discoid plastids and numerous nuclei (Fig. 20). The class Phaeothamniophyceae includes freshwater filamentous forms, which can be simple or branched, without chrysolaminarin. They produce biflagellate zoospores as in *Phaeothamnion*. The heterotrichous genera *Giraudyopsis* and *Chrysomeris* belong to the class Chrysomerophyceae, characterized by a lack of alginates. Pinguiophyceae include marine unicellular flagellates as in *Phaeomonas*, or no flagellate as in *Pinguiochrysis*, with a high content in omega-3 fatty acids.

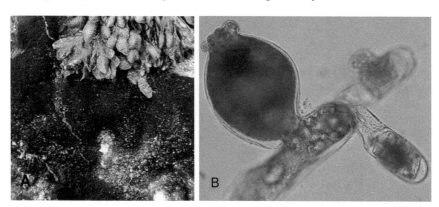

Figure 20. *Vaucheria coronata.* A Habit with *Fucus spiralis.* B Oogonium (Courtesy I. Bárbara).

Phaeophyceae

Phaeophyceans are known as brown algae, whose color varies from yellow to almost black due to the presence of the xanthophyll fucoxanthin, which conceals the rest of the pigments (Table 3). As characteristic storage products, they accumulate the polysaccharide laminarin, an insoluble polymer composed mainly of β-linked glucans. The cell wall consists of an inner layer of cellulose fibers and an outer external layer of mucilage, comprising colloidal substances called phycocolloids, such as alginates, which are salts of alginic acid, and the substance fucoidan, which is mainly composed of sulfated polysaccharides. Both alginates and fucoidan are of commercial interest. In the phaeophycean *Padina*, calcium carbonate deposits are present in the form of aragonite. The cells contain a single nucleus with small chromosomes and one or more chloroplasts, whose structure can be laminar, perforated, discoid or lenticular. Thylakoids are arranged in packs of three, surrounded by a girdle lamella. Plastids have their own membrane and two membranes from the endoplasmic reticulum. Pyrenoids, when present, accumulate reserve polysaccharides around them.

Many phaeophyceans such as *Ectocarpus* are branched filaments. When branched filaments of one or several axial filaments are joined by mucilages, they form pseudoparenchymatic thalli called haplostichous, as in *Castagnea* or *Leathesia*. The thalli called polystichous are parenchymatic, as in *Sphacelaria* (Fig. 3g). Polystichous thalli originate by the division of their cells in all directions, causing a thickening of the thallus and cell differentiation: an outer layer consisting of pigmented cortical cells and an inner medullar layer of unpigmented cortical cells. The greater morphological complexity of Phaeophyceae, and of algae as a whole, is found in the order Laminariales, in which thalli have three distinct parts: the holdfast, stipe and blade (Fig. 21a).

Reproductive cells are motile, generally pyriform and provided with two lateral or subapical flagella. One of them is acronematic, directed towards the apex, while the other is covered with tiny hairs or mastigonemes arranged in two rows and is directed posteriorly. Near the base of flagella there is a reddish eyespot located in the plastid. Asexual reproduction is by means of zoospores, fragmentation or specialized multicellular structures called propagules, which are able to produce adult individuals. Sexual reproduction occurs by isogamy, anisogamy or oogamy. Life cycles are isomorphic haplo-diplontic or heteromorphic with a dominance of gametophytes or sporophytes (Fig. 24). Fucales and Durvilleales have a diplontic cycle (Fig. 25). Two types of sexual reproductive structures can be observed in Phaeophyceae. One type has multilocular structures, in which each cavity produces a flagellate cell by mitosis. This type can be produced either in gametophytes or in sporophytes. Other reproductive structures are unilocular, formed by a single cell, in which 16 to 128 flagellate haploid spores are formed by meiosis. This type is mainly produced in sporophytes.

Phaeophyceans are found almost exclusively in marine habitats and are an important component of the benthic vegetation in the rocky shores of the northern and southern hemispheres. Some occupy the intertidal zone, as in certain fucoids, and can resist desiccation for hours or even for days (*Pelvetia canaliculata*). Laminariales form extensive subtidal "forests". Some brown algae live more than a hundred meters deep due to fucoxanthin, a pigment that allows them to use the blue part of the radiation spectrum. Only five genera live in freshwater, but many are found in brackish waters of estuaries. Some phaeophyceans have air bladders or vesicles in their thalli which increase their buoyancy, allowing them to live upright and rooted to the substrate, as in *Ascophyllum nodosum*, or floating, as in *Sargassum natans*. Large concentrations of the latter algae gave its name to a region of the Atlantic Ocean known as the Sargasso Sea. Brown algae provide habitat and a food source for many marine animals. Traditionally, they have been used by man as fertilizers for their high phosphate content. Some, like Laminariales, accumulate iodine in a concentration 10,000 times higher than that found in the sea.

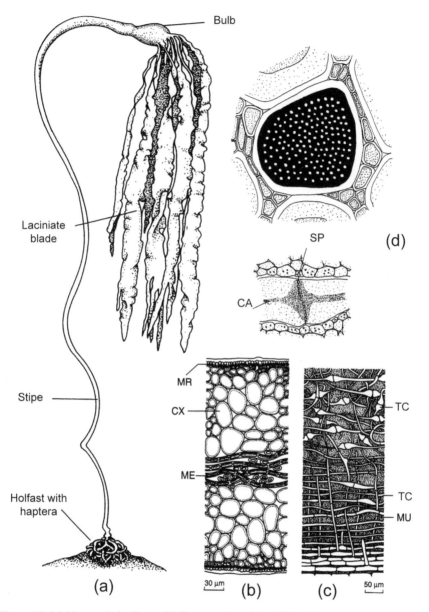

Figure 21. (a) *Nereocystis luetkeana*. (b) Cross section of the blade of *Laminaria*. (c) Detail of medulla from the stipe of *Laminaria*. (d) Longitudinal and transverse sections of the specialized cell known as trumpet cell, with sieve plate of pores and callose. CA: callose. CX: cortex. ME: medulla. MR: meristoderm. MU: mucilage. SP: sieve plate. TC: trumpet cell (b, c, d, after Lee 2008).

Brown algae consist of more than 250 genera and about 1,500 species (Fig. 22). Most phaeophyceans with simple thalli and isomorphic reproductive cycles belong to the order Ectocarpales. *Ectocarpus* is cosmopolitan; thalli are uniseriate branched filaments with heterotrichous

Figure 22. Diversity in brown algae. A. *Desmarestia dudresnayii*. B. *Undaria pinnatifida*. C. *Laminaria ochroleuca*. D. *Sargassum muticum* (A, B, D courtesy I. Bárbara and C courtesy J. Cremades).

organization and diffuse intercalary growth. The haploid gametophytes and diploid sporophytes are morphologically identical. The gametophytes produce multilocular gametangia that generate male and female gametes, which are morphologically identical but have a different behavior. Male gametes are chemotactically attracted towards female gametes by the substance, ectocarpene. After gamete fusion, the zygote, without a resting stage, originates a sporophyte, in which either unilocular meiosporangia or plurilocular sporangia are produced. The latter originates diploid asexual spores that can generate diploid sporophytes. Gametes can also produce new gametophytes by parthenogenesis, and even haploid zoospores can fertilize to produce sporophytes. In *Ectocarpus*, multilocular structures are formed on both haploid and diploid thalli.

The order Cutleriales is characterized by biflagellate gametes. The male gamete is larger than the female one, which produces the hormone multifidene that attracts male gametes. The order includes species with life cycles with heteromorphic alternation of generations, as in *Cutleria*, in which the gametophyte is predominantly composed of dichotomously branched flat thalli that produce micro and megagametangia grouped in sori. The sporophyte is flat and crustose and was described as a different genus named *Aglaozonia*. In *Zanadinia*, the life cycle shows isomorphic alternation of generations.

Desmarestiales contains a single family with two genera, *Desmarestia* (Fig. 22A) and *Himantothallus*, both with pseudoparenchymatous thalli and heteromorphic life cycles, microscopic gametophytes and oogamous sexual reproduction. *Desmarestia* is an important component of the Antarctic and Arctic flora in which there are vicarious species. This order is distributed worldwide in temperate and cold waters.

Sphacelariales consists of several genera distributed from temperate to tropical waters. The thallus is a small tuft of branches with a parenchymatous construction in which growth is by a prominent apical and pigmented cell. For asexual reproduction, the thallus develops specialized branches called propagules. Fertilization can be oogamous, as in *Cladostephus* and *Halopteris*, or anisogamous as in *Sphacelaria* (Fig. 3g).

In the order Dictyotales, thalli are flat, dichotomous or fan-shaped, growing by one or more apical cells. Sporangia produce four or eight spores. Sexual reproduction occurs by oogamy (Fig. 23). *Dictyota* exhibits flat dichotomous thalli with an external monostromatic layer of assimilating cells and one thick inner layer of unpigmented cells. The life cycle of *Dictyota* involves the isomorphic alternation of generations. Gametophytes are dioecious, and the oogonia are arranged in sori, each containing a single immobile cell. In multilocular gametangia, male thalli develop pyriform sperm with two laterally inserted flagella. Only the pleuronematic flagellum is externally visible, since the other flagellum is reduced to the basal body.

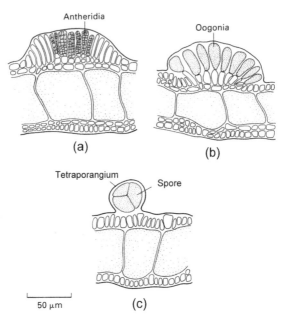

Figure 23. Diagrams of reproductive structures of *Dictyota* in cross section of the thallus. (a) Sorus with antheridia. (b) Sorus with oogonia. (c) Tetrasporangium (After South and Whittick 1987).

Fertilization takes place during spring tides in the warmer months, when eggs and sperm are discharged. Sporophytes are externally identical to gametophytes. They produce unilocular sporangia that contain four haploid meiotic spores. Spores are immobile. Sporangia and gamentangia are grouped in sori, surrounded by unicellular hairs. Many species of this order, such as *Dictyota dichotoma* or fan-shaped *Padina pavonia,* are common in temperate and warm coasts.

The order Laminariales is characterized by a life cycle involving the heteromorphic alternation of generations (Fig. 24), sexual reproduction by oogamy and plastids without pyrenoids. Sporophytes reach several meters in height and have a marked morphological differentiation. They are attached to the substrate by a system of rhizoids, called haptera, from which a cylindrical stipe rises, ending in a widening laciniate blade consisting of several layers of cells. Growth in length is due to an intercalary meristem located between the end of the stipe and the blade. *Laminaria hyperborea* can live for several years and each year it renews its blade. In winter a new meristem inserted in the base of the blade generates a new blade, and the old one is destroyed at the apex. The thalli increase in thickness by a layer of cells with meristematic activity known as meristoderm (Fig. 21b). The stipe of *Laminaria* is cylindrical. In cross section, it shows a meristoderm

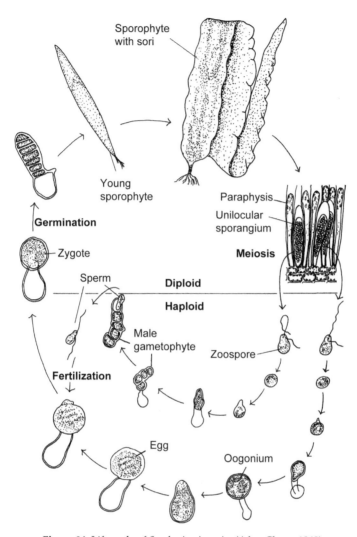

Figure 24. Life cycle of *Saccharina japonica* (After Cheng 1969).

that produces small pigmented cells outwards and cortical unpigmented cells inwards. Cortical cells increase in size from the outside inwards. Mucilaginous ducts are located in the outer cortical layers. These ducts are formed by a ring of secretory cells. Inside the stipe, a medulla is formed by elongated cells with a narrow lumen, known as hyphae or trumpet cells, arranged in a network (Fig. 21c). Trumpet cells are widened at their ends and consist of sieve plates (Fig. 21d), through which there is a transport of substances similar to vascular plants. Some species of *Laminaria* can live for several years and produce a growth ring per year in their old stipes. Gametophytes in *Laminaria*, as in all Laminariales, are microscopic. Female gametophytes consist of few cells. Oogonia produce a single egg, which when mature leaves the oogonium through a pore, but remains attached to the oogonium wall. Male gametophytes are branched filaments. Male unicellular gametangia produce a single biflagellate gamete. After fertilization, the diploid sporophytes grow on the female gametophyte (Fig. 24).

Laminariales are subtidal algae which form large populations in temperate and cold seas, comprising the biggest algae such as *Macrocystis pyrifera* which can reach 50 meters in length. Species of *Laminaria* (*L. hyperborea, L. digitata, L. ochroleuca* Fig. 22C) and *Saccharina* are common in Atlantic coasts. In Pacific coasts, other genera, such as *Macrocystis, Nereocystis* and *Eisenia* are dominant. Laminariales, collected in nature or cultivated, are the main source of alginates and other compounds such as mannitol and iodine. Some species are used as food, such as the kelp known by the Japanese name Kombu.

The order Fucales comprises phaeophyceans in which the life cycle is diplontic, and thalli are thought to grow by an apical cell that generates a promeristem. In these algae, adult plants are diploid so that meiosis is gametic (Fig. 25). In *Fucus,* the thalli are dichotomous, ribbon-shaped, with a central thickening or midrib, fixed to the substrate by means of a disc. The reproductive organs are arranged at the apex of the thallus, in

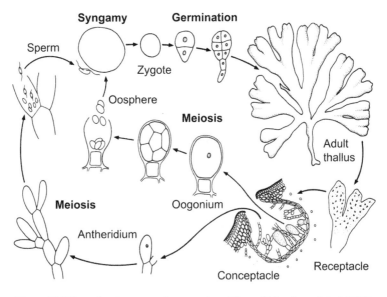

Figure 25. Life cycle of a monoecious species of *Fucus* (After Scagel et al. 1982).

widened parts known as receptacles, constituted by multiple cavities or conceptacles sunk in the thallus. Inside the conceptacles, male gametes are found in the antheridia and eggs in oogonia, together with sterile filaments or paraphyses that project outside the conceptacle through an ostiole. Antheridia are on branched filaments, producing 64 biflagellate sperms after meiosis. Antheridia have two walls. The breakage of the outer wall releases the sperm in a package, and then the second wall breaks and the sperm swim attracted by the sexual pheromone, fucoserratene, produced by the egg. The oogonia have three walls that surround eight eggs. Breakage of the outer envelope releases the entire assembly, and after the disruption of the remaining walls, the eggs float until they are fertilized. The zygote produces a cellulose wall, and then it attaches to the substrate and begins to divide to give a new sporophyte. Species of the genus *Fucus* (*F. vesiculosus*, *F. spiralis*, *F. serratus*) are common in cold and temperate coasts of the northern hemisphere, while other fucoids, as numerous *Sargassum* species live in tropical waters worldwide (Fig. 22D). *Cystoseira* is a genus that may have originated in the Mediterranean and is characteristic of well structured coastal communities. *Caulocystis* and *Horrmosira* are common in the waters of the South Pacific.

Phaeophyceae are thought to have evolved from ancestors that had reproductive life cycles with the isomorphic alternation of generations, fertilization by anisogamy or isogamy and simple morphological types, like *Ectocarpus*. They evolved towards forms with a progressive reduction of the gametophyte and increased morphological complexity of the sporophyte, as in *Laminaria*. Diplontic life cycles and fertilization by oogamia, as in *Fucus*, are considered the most advanced characters after an extreme reduction of the haploid generation.

Euglenophyta

In euglenophytes, a single class, Euglenophyceae, is recognized. It includes about 800 species, most of which are freshwater species, but some are estuarine and intertidal. They are free-living motile cells. Only a third of the species have chloroplasts, and the rest are heterotrophic. Autotrophic euglenophytes contain chlorophylls *a* and *b* (Table 2) and, thus, have been related to Chlorophyta, but this is the only shared character. Plastids are surrounded by two membranes and a third membrane from the endoplasmic reticulum. The energy storage product is paramylon, a polysaccharide found outside the chloroplast. Chrysolaminarin can also accumulate in a liquid form in vacuoles. The most representative genus is *Euglena*. The cell is fusiform with a flexible periplast formed by a number of protein bands, which enables the cell to change its shape while moving. Periplast bands are helically arranged and located beneath the plasma membrane. The cell shows an anterior bottle-shaped invagination called the reservoir. At the base of the reservoir, there is a contractile vesicle as well as two flagella, but only one of them emerges through the neck. Near the base of the reservoir, there is a stigma and a photoreceptor, which are considered to be related to the ability of these algae to move in response to light. The stigma consists of red-orange lipid droplets and is independent of the chloroplast. The stigma covers the photoreceptor, which looks like a crystalline body surrounded by the flagellum membrane. In the nucleus, chromosomes are permanently condensed. Sexual reproduction in euglenophytes has not yet been observed. In order to survive when external conditions are unfavorable, some species produce resistant cysts surrounded by a thick mucilaginous sheath that they produce.

Chlorarachniophyta

Chlorarachniophytes are marine unicellular coccoid, flagellate, or amoeboid cells with filose pseudopodia, some of which form plasmodial colonies. There are five known genera and six species, grouped in the class Chlorarachniophyceae. *Chlorarachnion reptans* contain multiple chloroplasts with chlorophyll *a* and *b*, each with a laterally-located pyrenoid surrounded by a vesicle. The chloroplast has four membranes, of which the external membrane has no ribosomes. There is a wide periplastidial compartment situated between the two membranes of the chloroplast and the two ones of the endoplasmic reticulum. This compartment contains a nucleomorph surrounded by its nuclear membrane in a depression of the pyrenoid. The periplastidial compartment could be interpreted as the remains of secondary endosymbiosis involving a green alga. Sexual reproduction is performed by the fusion of two amoeboid cells of different size that generate a coccoid zygote surrounded by a thick wall. After meiosis, the zygote produces amoeboid or flagellate cells provided with chloroplasts and a single pleuronematic flagellum which surrounds the cell.

Chlorophyta

Green algae have chlorophyll *b* (Table 2) and are distinguished from other algae because their main storage product is starch, which accumulates in the chloroplast and not in the cytoplasm. Starch formation is often associated with a pyrenoid. The chloroplast is surrounded by a double membrane. Green algae comprise two divisions, the freshwater group Charophyta (Streptophyta, pro parte), which is not addressed in this text, and Chlorophyta with six classes: Prasinophyceae, Chlorodendrophyceae, Chlorophyceae, Trebouxiophyceae, Ulvophyceae and Dasycladophyceae

(Table 1). The separation between the divisions and classes of Chlorophyta is based on the type of cell division, the ultrastructure of flagellar roots and, recently, on molecular analysis (Leliaert et al. 2012).

Only 10% of Chlorophyta are marine algae. Some groups are almost exclusively freshwater and others are marine. Their species include unicellular flagellates like *Tetraselmis* (Fig. 29), immobile like *Chlorocytis*, multicellular colonies like *Dictyosphaerium*, simple or branched filaments like *Cladophora* (Fig. 30B), pseudoparenchymatous thalli like *Codium* (Fig. 30A) and parenchymatous laminae like *Ulva* (Fig. 32). Most are microscopic, but some species like *Caulerpa* can reach over 10 meters long. In general, polysaccharide cellulose is present in their walls. The most common auxiliary pigment found in the chloroplast is lutein. Some groups contain specific pigments, such as siphonaxanthin (Table 2). Thylakoids are grouped in stacks of 3 to 5. The chloroplasts of flagellate cells show an eyespot. Contractile vacuoles are common in unicellular and colonial flagellate chlorophytes.

Vegetative or reproductive flagellate cells are isokont with smooth flagella. Asexual and sexual reproduction has been described in green algae. Sexual reproduction can occur by isogamy, anisogamy or oogamy. There are two basic types of cell division in green algae. In most chorophytes, the daughter nuclei remain close together during cytokinesis and the spindle collapses, then producing a new set of microtubules that are arranged parallel to the plane of cell division, the phycoplast, whose function is to keep the nuclei separate. The new cell wall originates from an invagination or furrow, which progresses from the side walls of the cell toward the center (Fig. 26a). In other cases, the new wall cell is generated by accumulation of materials from the Golgi vesicles leading to a cell plate that grows from the center to the periphery, in which case the wall is traversed by plamodesmata; this type of cell division occurs in most Charophyta. During cytokinesis, the daughter nuclei are separated and the mitotic spindle does not collapse. The new cell wall is generated by means of an invagination that progresses from the side walls or, as in some charophytes and embryo plants, the new wall is generated by the accumulation of material from the vesicles of the Golgi apparatus. The accumulated material moves through microtubules, called phragmoplasts, resulting in a growing cell plate from the center to the periphery, also traversed by plamodesmata (Fig. 26b). The persistence or not of the nuclear membrane during mitosis is also a major character in the systematics of green algae (Figs. 26c, d). In most cases, the nuclear membrane remains during cell division (closed mitosis), but in other cases, like in some charophytes, it disappears (open mitosis). In Trebouxiophyceae, centrioles are in the metaphase plane instead of at the poles (Fig. 26d).

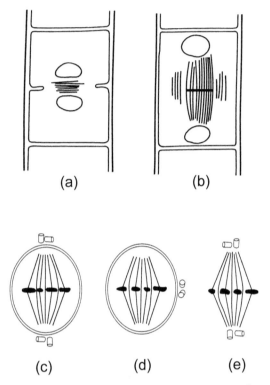

Figure 26. Diagrams of the two basic types of cytokinesis in green algae. (a) Phycoplast, microtubules lying parallel to the cleavage furrow. (b) Phragmoplast microtubules perpendicular to the cell plate which develop from the center toward periphery; and the mitosis in different lineages of green algae. (c) Closed mitosis. (d) Closed mitosis with centrioles located in the plane of metaphase. (e) Open division.

The characteristics of flagellar roots vary depending on whether the position of flagella is apical or lateral (Fig. 27). If there are two flagella located at the apex, the microtubules that "anchor" the flagella form four groups of two or more microtubules that are arranged in a cross shape. Flagellar roots often present rhizoplasts, a fiber system with a striate appearance when observed by electron microscopy, composed of contractile proteins (Fig. 27a). In some Prasinophyceae and charophytes, flagella are inserted laterally and microtubular roots are formed by a wide band and a narrow one that cross a great part of the cell. There is also a rectangular multilayered structure at the base of the wide band of microtubules (Fig. 27b).

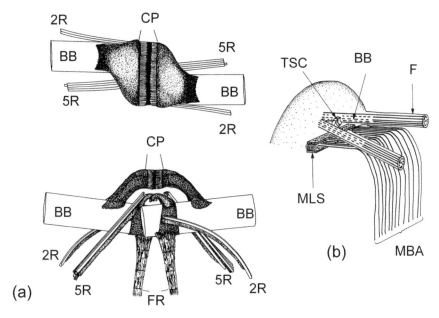

Figure 27. Ultrastructure of flagellar apparatus in green algae. (a) *Batophora* (Ulvopphyceae), anterior and lateral view, 11o'clock/5 o'clock configuration of root system. (b) *Chaetosphaeridium* (Coleochaetophyceae) zoospore, side view of unilateral type of flagellar apparatus. CP: caping plate connecting the basal bodies. BB: basal body. MBA: microtubular band. MLS: multilayered structure. F: flagella. FR: fibrose roots or rhizoplasts. 2R and 5R: two and five stranded mictotubular roots. TSC: transversely striate fiber connecting the basal bodies (a, After Roberts et al. 1984; b, after Hoek et al. 1995).

Prasinophyceae

Prasinophyceans contain a small group of unicellular green algae covered by one or several layers of organic scales. Most are flagellate, provided with one to eight flagella that often emerge from a depression. The flagellar root system can be cross-shaped or show a multilayer structure (Fig. 27b). Species of this group live both in marine and freshwater. A common genus is *Pyramimonas*, a unicellular flagellate with the body of the cell covered by three layers of different types of organic scales (Fig. 28). Prasinophytes are considered an ancestral group of green algae.

Chlorophyceae

Chlorophyceans show great morphological diversity and live almost exclusively in inland waters. They are characterized by cell division forming phycoplast, motile cells with two or four apical flagella, as in *Platymonas*,

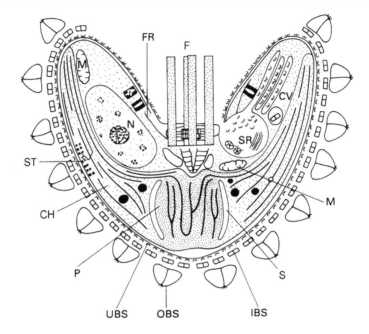

Figure 28. Diagram of *Pyramimonas lunata*, note the three bands of scales. CH: chloroplast. CV: cylindrical vesicle. F: flagella. FR: flagellar microtubular root. IBS: intermediate body scales. M: mitochondria. N: nucleus. OBS: outermost body scales. UBS: undermost body scales. P: pyrenoid. S: starch grain. SR: scale reservoir. ST: stigma (After South and Whittick 1987).

whose microtubular root system is crossed and without rhizoplasts. Mitosis is closed (Fig. 26c). The cell wall is composed of cellulose and an external layer of crystalline glycoproteins.

Chlorodendrophyceae are closely related and include flagellate unicellular algae enclosed by an organic wall or theca, as in the genus *Tetraselmis* (Fig. 29) or the sessile stalked *Prasynocladus*.

Ulvophyceae

Ulvophyceans are cosmopolitan and predominantly marine. Most of their over 1,200 species are macroscopic benthic algae which have a life cycle with alternation of generations. Their mobile cells have apical flagella with a crossed microtubular flagellar root system and rhizoplasts (Fig. 27a). During cell division, the spindle does not collapse in telophase, and the new wall is produced by furrow. Ulvophyceans are grouped in at least six orders according to their morphology, number of nuclei per cell, cellular and reproductive features and molecular sequence analysis. The terrestrial order Trentepohliales is not discussed here.

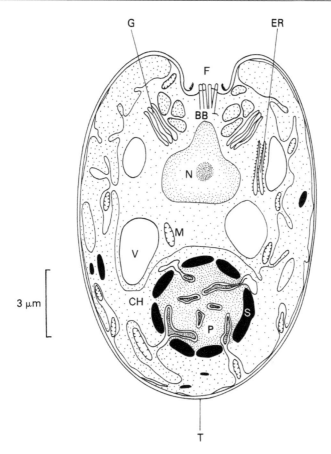

Figure 29. Diagram of *Tetraselmis astigmatica*. BB: basal body. CH: chloroplast. ER: endoplasmic reticulum. F: flagella. G: golgi. M: mitochondria. N: nucleus. P: pyrenoid. S: starch grain. T: theca. V: vacuole (After South and Whittick 1987).

The order Ulotrichales comprises unbranched filaments like *Ulothrix* or *Urospora* (Fig. 30C), two marine and freshwater genera, and branched filaments like *Gomontia* or monostromatic *Monostroma*. Cells are uninucleate and contain a parietal, band-like chloroplast with one or several pyrenoids. The filament is attached to substrate by a special basal cell called the holdfast. All cells are able to divide except the holdfast. In sexual reproduction, biflagellate isogametes produced by a cell of the filament are released through a pore. Gametes fuse in pairs to form quadriflagellate zygotes, which develop a thick wall originating a resting cell. After meiosis, new thalli are produced. Asexual reproduction can occur by fragmentation of filaments or by the production of quadriflagellate zoospores.

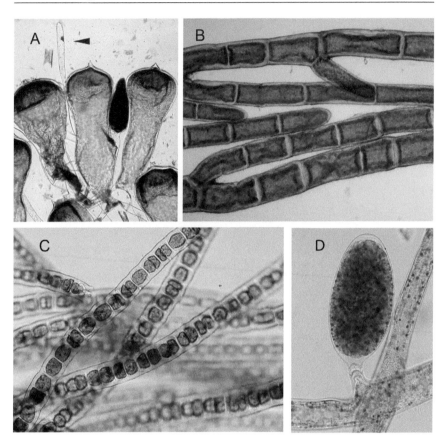

Figure 30. Diversity in Ulvophyceae. A. *Codium fragile,* utricles, gametangium and a hair (arrow). B. *Cladophora hutchinsiae,* details of branching formation. C. *Urospora penicilliformis,* unbranched filaments. D. *Pedobesia solieri,* coenocytic thallus and sporangium (Courtesy I. Bárbara).

The order Ulvales comprises parenchymatic thalli, which can be tubular as in *Blidingia,* laminar with a single cell layer as in *Ulvaria,* or two layers thick as in *Ulva.* These genera have diffuse growth, and all cells except the basal ones have reproductive capacity. Life cycle of *Ulva* presents isomorphic alternation of generations. Every gametophyte cell can produce several gametes. Female gametes are bigger than male gametes. After fertilization, a quadriflagellate zygote is produced. This zygote fixes to the substrate to form a cylindrical thallus that becomes laminar at maturity (Fig. 31). Diploid sporophytes produce quadriflagellate zoospores by meiosis. Asexual reproduction can occur by parthenogenetic gametes.

Cladophorales includes about 400 species with filamentous and pseudoparenchimatous thalli. The body is constituted by septate filaments

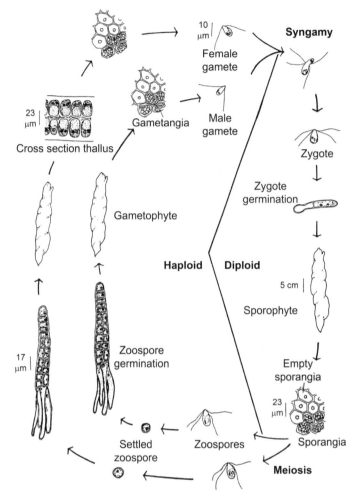

Figure 31. Life cycle in *Ulva arasaki* (After Chihara 1969).

formed by large cylindrical cells. Each cellular segment contains multiple nuclei and a reticulate chloroplast with many pyrenoids. The family Cladophoraceae comprises single filaments as in *Chaetomorpha* or branched filaments as in *Cladophora* (Fig. 30B). The life cycles are haplo-diplontic with the isomorphic alternation of generations as in *Cladophora* or pseudo-heteromorphic as in *Spongomorpha* in which a unicellular zygote (*Codiolum*) lives for several months after germination. Fertilization is by isogamy or anisogamy with biflagellate gametes. Quadriflagellate zoospores arise after meiotic division. Cladophoraceae live in marine, brackish and inland waters. The family Siphonocladaceae includes multicellular Cladophorales that presents a characteristic type of cell division known as segregative cell division. In *Siphonocladus,* such division takes place from unicellular young thalli constituted by a coenocytic vesicle. In the process of segregative cell division, cell content is divided to form spherical vesicles that secrete an enveloping membrane. The vesicles increase in size until they contact each other, finally producing lateral protrusions which grow to form a thallus with a branched appearance. Their chloroplasts contain a particular xanthophyll, called siphonaxanthin. Sexual reproduction is by isogamy. *Siphonocladus, Anadyomene* and *Valonia* are the most common genera of this order and are typical of tropical waters.

The order Bryopsidales includes about 500 species of multinucleate, siphonous ulvophyceans of varied morphology which were previously treated as orders and are now treated as families. Thalli lack cross-walls except at sites of reproductive cell formation. Chloroplasts are discoid, numerous and may or not have pyrenoids. Some genera, like *Udotea* and *Caulerpa,* contain two types of plastids: pigmented plastids and unpigmented plastids or leucoplasts in which starch accumulates. Some taxa contain polysaccharides like xylans or mannans in their cell walls instead of cellulose. Members of Bryopsidales, like *Halimeda,* show flattened articulated thalli, whose walls are calcified in the form of araganonite crystals. If the thallus is injured, rupture can result in the loss of large amounts of cytoplasm, but repair of the wound occurs rapidly, by contraction of protoplasm and plug formation. This plug formation is related to the terpenoid caulerpenyne which also has the function of defense against herbivores. The majority of Bryopsidales are seaweeds which occur in warm tropical to temperate waters and live on the littoral or in deeper waters.

The large genus *Caulerpa* is a uniaxial siphonous form with a rhizoidal horizontal axis from which erect photosynthetic axes emerge. *Caulerpa* is characterized by the occurrence of cylindrical wall ingrowths or trabeculae traversing the central lumen of siphon and providing mechanical support. Chloroplasts and leucoplast are present. Sexual reproduction by anisogametes is known, and zygote development produces a new diploid thallus. *Caulerpa* grows in sandy beds from tropical to template sea waters. In *Derbesia* and *Pedobesia,* the thallus is constituted by a uniaxial system with pinnate branches (Fig. 30D). The chloroplasts are pigmented and have a pyrenoid that accumulates starch. This genus is characterized by a heteromorphic life cycle involving a branching siphonous sporophyte and a vesicular coenocytic gametophyte, known as the *Halicystis* phase. Gametophytes produce small pale gametes and larger dark green gametes in different individuals. Sporophytes produce multinucleate zoospores provided with a crown of subapical flagella. *Codium,* a widespread marine genus, is common in temperate waters (Fig. 30A). Its thalli exhibit varied forms from hollow spheres to prostrate crusts and branched cylindrical axes, which can be up to 1 m in length. They are multiaxial, composed of interwoven siphons arranged in a colorless medullary region and a pigmented cortical one. Chloroplasts are discoid without pyrenoids. The tips of coenocytic siphons are dilated into enlarged utricles that form a compact palisade-like surface layer. Gametangia are produced laterally on the utricles. Sexual reproduction is by biflagellate anisogametes. Other pseuparenchymatous thalli are the fan-shaped of *Udotea* and the terete stalk of *Penicillus* (Fig. 32).

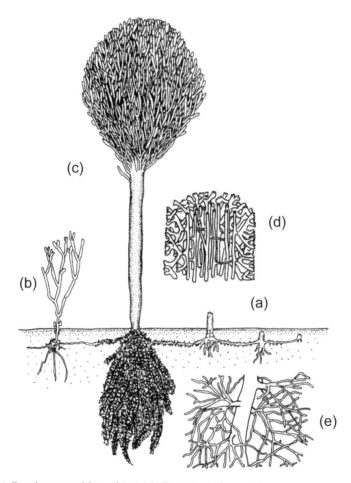

Figure 32. Development of the multiaxial thallus in *Penicillus capitatus*. (a) Subsurface siphonous filaments develop new thalli. (b) The young thallus forms dichotomous branches. (c) Adult thallus. (d) Interweave of filaments form a pseudoparenchimatous thallus and (e) an extensive holhfast (After Friedmann and Roth 1977).

Dasycladophyceae

Dasycladales is the only order of this class. It comprises species with thalli formed by a nonseptate primary axis with branches in whorls, exhibiting radial symmetry. *Acetabularia* has uninucleate diploid thalli that become coenocytic and haploid at maturity. The thallus is formed by a cylindrical axis attached to substrate by a holdfast of rhizoidal branches. At the apex, whorls of adherent branches form the so-called cap rays. At maturity, the nucleus contained in the axis is thought to divide by meiosis, and the resulting haploid nuclei then undergo mitosis and move towards the cap rays. A septum is formed in each cap ray, and it then becomes a gametophore and later a cyst (Fig. 33). Mature cysts release isogamous biflagellate gametes by abscision of an operculum. The zygote generates vegetative thalli. The genus *Cymopolia* is multinucleate. The nuclei have a heterogeneous

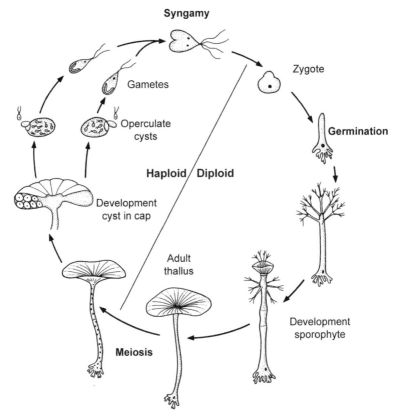

Figure 33. Life cycle in *Acetabularia mediterranea* (After South and Whittick 197).

appearance in the different parts of the thallus. Dasycladales occurs in shallow waters of tropical to temperate seas. They are often calcified, for which there are numerous known fossil genera.

6 Ecology, Distribution and Uses of Marine Algae

Algae are able to colonize any media on Earth, but most are strictly aquatic. They all require an aquatic environment for reproduction. Many of them are marine organisms, while others live in freshwater. Unicellular and colonial algae usually live suspended in the water as part of plankton. The size of planktonic algae varies between 0.2 and 200 μm, and several categories have been recognized: macroplankton (greater than 200 μm), microplankton (between 200 and 20 μm), nanoplankton (between 20 and 10 μm), ultraplankton (between 10 and 2 μm) and picoplankton (between 2 and 0.5 μm). Other algae grow attached to a substrate as part of the benthos, more specifically, its plant fraction or phytobenthos. They can live on rocks (epilithic), slimes (epipelic), plants (epiphytes), animals (epizootic), inside rocks (endolithic) or inside plants or animals as symbionts or parasites. Most algae are photoautotrophs but some, mostly unicellular algae, are both heterotrophic and optional photoautotrophic.

Algae always live in the photic zone at a depth where light can reach. Sea benthic algae occupy much of the continental shelf. In clear waters, they can reach great depths, like in the Caribbean Sea where red algae have been found at depths of over 260 m (Littler et al. 1986). The ability of algae to live at different depths is closely related to the composition of their photosynthetic pigments (Table 2). Light intensity of aquatic ecosystems corresponds to wavelengths of the blue-green region of the spectrum. As accessory pigments phycoerythrin, β-carotene, fucoxanthin and siphonoxanthin are able to absorb blue-green light, the benthic algal communities in deep water are composed of red, brown and green algae. Algae tend to increase the total amount of pigments and the accessory pigments: chlorophyll ratio with increasing depth. For instance, the red alga *Chondrus crispus* increases its amount of phycoerythrin with depth (Sagert et al. 1997). As the adaptation of algae to depth is related to the amount and quality of light they receive, the depth each alga can reach would be related to the minimum light intensity needed for photosynthesis.

The different benthic algae can adapt to the amount of radiation and the spectrum of light, causing the known bathymetric zonation, which also involves other factors such as wave strength, tidal range and type of substrate. The difference in sea level between high and low tide varies from one place to another on the globe, and in each case, depends on latitude and the extension of the basin of the site. Several zones can be distinguished on the coast: the intertidal or eulittoral zone which is the region of the shore between the highest and the lowest tides, subject to tidal influence; the subtidal or infralittoral zone where algae are not subject to emersion; and the supralittoral zone where some algae and lichens are able to live above the maximum tide level (Lüning 1990). Only some algae can survive the extreme conditions of drying, increased light intensity and extreme temperatures during low tide. As environmental conditions vary gradually from the upper to the lower zones of the littoral, algae are distributed in bands or belts.

Water temperature and salinity, together with light and nutrient availability, are the most important factors in the geographical distribution of algae (Lüning 1990). Temperature is the most important factor in the distribution of marine algae, since they are limited by maximum and minimum temperatures that allow them to survive or complete their reproductive life cycle. Cold and temperate seas show a predominance of brown algae and higher biomass than warm seas, where diversity is greater and red and green algae groups are predominant. The local distribution of benthic algae is determined by the degree of wave exposure and the substrate. Phytoplankton composition and biomass vary considerably depending on physico-chemical changes in the water body where it lives. Light and the amount and availability of nutrients are the most important factors (Barsanti and Gualtieri 2005). When nutrients are plentiful (eutrophic waters), algal biomass increases substantially. However, in nutrient-poor waters (i.e., oligotrophic), a low density of phytoplankton is produced, and hence the water has a crystalline appearance. Other critical factors for phytoplankton development are turbulence, bacteria and herbivorous zooplankton. Water turbulence and cell structures (e.g., cell vacuoles, arms, bristles, spines, flagella) allow phytoplankton, which is denser than water, not to sediment, so it can receive light. Bacteria transform organic matter into nutrients for phytoplankton, and zooplankton herbivores feed on many algae partly controlling their populations.

Most of the algae that establish symbiotic relationships are species of dinoflagellates, ochrophytes, chlorophytes and rhodophytes. They interact with diverse organisms such as protozoa, invertebrates, fungi and other algae. Interactions are mainly nutritional. The algae bring oxygen and reduced compounds like maltose, glucose, amino acids or glycerol to the partner which, in turn, brings inorganic matter and a stable environment to the algae. For instance, marine chlorophyceans are found living inside platyhelminthes and nudibranchs. Some dinoflagellates, called zooxanthellae, live in symbiosis with sponges, jellyfish and marine coelenterates. The dinoflagellate *Symbiodinium* is essential in the formation of coral reefs, because it stimulates calcification of the colony. Other dinoflagellates, like *Noctiluca*, can store thousands of small green algae inside. Although the specificity between the two partners is not absolute, some organisms are dependent on the other, but not the contrary. For instance, the larvae of the marine platyhelminth *Convoluta roscoffensis* cannot grow if they are not infected by the chlorophycean *Platymonas convolutae*, but the latter can live freely. In the symbiosis between algae and fungi to form lichens, the fungus gives the highest biomass, and cyanophyceans and chlorophyceans only constitute a photosynthetic layer. In other cases, algal biomass exceeds that of the fungus, as occurs between algae living in the supralittoral zone, like the phaeophycean *Pelvetia* or the chlorophycean *Prasiola* with species of ascomycetes. Symbiosis has been interpreted as one of the most important evolutionary mechanisms (Margulis 1998). One of the clearest examples is *Cyanophora*, a glaucophyte that can be considered a recent endosymbiosis between a protozoa and a cyanophycean which acts as a chloroplast.

The direct use of algae by man as food, animal feed or fertilizer has a long tradition on the Pacific coast both in Asia and South America. Today its use is rising considerably in other regions, such as Europe. Some species of calcified red algae form extensive fields in the seabed. They are collected, crushed and then used to decrease the acidity of agricultural soils. Fossil deposits of diatoms and dinoflagellates are used as abrasives, industrial filters, and in the construction of refractory or insulating materials. Algae for human consumption are used in the preparation of soups and to accompany fish and meat dishes in many Eastern countries. Of the approximately 300 algae used by man, the best known are marketed under the Japanese names kombu (*Laminaria* spp. or *Saccharina* spp.), wakame (*Undaria pinnatifida*) and nori (*Porphyra* spp.). Many names are popular in other countries, such as sea lettuce (*Ulva* spp.) in Europe, dulse (*Palmaria palmata*) in Ireland or cochayuyo (*Durvillaea antarctica*) in Chile.

The greatest economic interest of seaweeds is their content in polysaccharides known as phycocolloids. These polysaccharides form colloids in water that are stable at room temperature (Guiry and Blunden 1991). The most used phycocolloids are agar, carrageenan and alginic acid, which have many applications such as stabilizers and emulsifiers in food, in the pharmaceutical and textile industries, as well as in the manufacture of paints, paper, and other industrial uses (Akatsuka 1990, Prud'homme van Reine and Trono 2001). The main compounds used by industry are alginates, alginic acid compounds, obtained from brown algae in which phycocolloids often represent 70% of the weight of the algae. The phycocolloid carrageenan is a sulfated galactan polymer extracted from red algae of genera such as *Chondrus*, *Iridaea*, *Gigartina* and *Eucheuma*. It is used as a thickener and stabilizer in many products like soups, creams, drinks and cosmetics. Agar, a sulfated polysaccharide polymer, is obtained from *Gelidium* (Fig. 10F), *Gracilaria* and *Pterocladia* among others. Agar contains less sulphate than carrageenan and produces more consistent gels. It is used as culture media for microorganisms in laboratory and in pharmaceutical activities. The high demand for some algae and their high market price has led to their cultivation even in industrialized countries. The cultivation of brown algae is currently widespread. Laminariales is cultivated in several countries, especially in China and Japan, and such cultures are spreading to Western countries as in the case of *Undaria pinnatifida* (Fig. 22B). The cultivation of red algae, especially *Gracilaria* and *Eucheuma,* is common since the current demand exceeds natural production. Colonial or unicellular algae are cultivated in open ponds to obtain vitamins, proteins and pigments, such as carotenes. Algae are essential in marine aquaculture, constituting the first phase of the food chain and the base for optimal production of these industries. Seaweed farming to obtain biofuel is one of the most promising prospects in the use of algae.

Many algae are good indicators of coastal water quality. For instance, diatoms and some macroscopic algae are one of the criteria to obtain the Blue Flag qualification on beaches. Some green algae like species of *Ulva* are abundant in eutrophic waters where they can produce green tides. On the contrary, the phaeophycean *Cystoseira* is characteristic of unpolluted water (Torres et al. 2008). In the last 100 years, many unicellular and macroscopic algae have been dispersed from one region of the globe to another, due to human activity. In European shores, the appearance of the Australian red algae *Asparagopsis armata* was detected 80 years ago and is now very abundant. Following the development of bivalve cultivation in industrialized countries, seaweeds from the Pacific have invaded the European Atlantic coast as is the case of the Japanese kelp *Sargassum muticum* and the red algae *Lomentaria hakodatensis* or *Grateloupia turuturu* (Fig. 10E). Sometimes, the introduction of an alien species can have a negative impact on the native flora, as occurred when the tropical chlorophyte *Caulerpa taxifolia* was introduced in Mediterranean marine ecosystems (Meinesz 1999). In other cases, the accidental introduction of a species has led to the cultivation of algae of economic interest like *Undaria pinnatifida* (Lembi and Waaland 1988).

Unicellular algae can be harmful when they appear in high concentrations, sometimes reaching several million cells per liter, since dense populations can produce anoxic conditions in deeper water arising from algal decomposition. This situation is known to be caused by several algal species belonging to the haptophycean genera *Phaeocystis*, *Chrysochromulina,* and several cyanophytes. Also, many algae produce substances that are toxic for other organisms, and some of them can even cause death in higher animals and man (Hallegraeff et al. 1995).

7 Field and Laboratory Methods for the Study of Marine Algae

Some algae grow very easily in the laboratory using simple culture media. The techniques to study benthic marine algae and phytoplankton are very different due to both the size of individuals and the particular characteristics of their habitat. There is a great amount of literature on methods to study marine algae. There are texts related to benthic algae (Lewis 1964, Hellebust and Craigie 1978, Round 1981, Litter and Litter 1985, Lobban et al. 1988, Alveal et al. 1995, Daves 1998, Lobban and Harrison 2000, Knox 2000, Kim 2011), phytoplankton (Sournia 1978, Round 1981, Innamorati et al. 1990, Alveal et al. 1995), algae cultivation (Perez et al. 1992, Alveal et al. 1995, Barsanti and Gualtieri 2005, Richmond 2008) and measuring the chemical and physical characteristics of seawater (Strickland and Parsons 1968, Parsons et al. 1989).

7.1 Phytobenthos

Macroscopic algae grow on a substrate in the photic zone of the coast, with rare exceptions like free-living *Sargassum natans* (Phaeophyceae) or *Ulva* spp. in estuaries. They are collected from the littoral, especially in areas subject to wide tidal variation, which can be more than two meters in spring tides. Knowledge of the date, time and amplitude of tides is needed, because these localities must be accessed a few hours before low tide to be able to reach the specimens. Benthic algae can be collected using waterproof boots, gloves and tools like small knives, spatulas or scrapers. Snorkel and scuba equipment are also needed.

Sampling data should include information on geographic and climatic data, such as GPS localization, wind speed and direction, rainfall, water movement (wave height, dominant direction), characteristics of the substrate or sediments (substrate type, particle size) and physical data of the water (temperature, salinity, pH, transparency). Water samples should also be collected to obtain measurements of dissolved oxygen and nutrients (Carbon, Nitrogen, and Phosphorus). Algal samples should be kept in plastic bags or bottles, labeled with a waterproof and light resistant marker. If we want to keep material alive until we reach the laboratory, it must be kept oxygenated and in cold and dark conditions during storage and transportation. Long-term storage may be performed by fixing the material in formaldehyde (4–10%) or in a mixture of glycerin-formaldehyde by pressing and drying or by freezing. Samples dried with silica gel are used for molecular analysis.

In order to determine species zonation, biomass, composition and structure of algae communities, sampling is performed following transects from the upper to the lower zone of the coast. Samples can be taken along the transect collecting all individuals in a continuous band of several cm wide, or samples can be taken discontinuously using quadrats. Quadrat size can range from 100 cm² to 1 m², and they are generally subdivided into smaller squares. The use of quadrats simplifies the observation of the composition and structure of algae communities, as it allows the use of statistical analyses and obtains data on the frequency, abundance, biomass, sociability and morphological variations of species present in the transect. In order to avoid over-dimensioning the study effort, the minimum necessary number of samples and size of quadrats should be analyzed. A performance curve is useful for determining sample number. The concept of sample minimum area (Margalef 2005) is very useful for estimating algal communities. Non-invasive studies of vegetation can follow the same method using the cover of individuals without requiring their collection. The elevation of the sampling point relative to zero tide level is important to know the percentage of emersion, hours a day, a month or a year the specimens are subject to.

7.2 Phytoplankton

Techniques for collecting and studying marine phytoplankton depend on their biological and geographical conditions. Boats or ships are generally needed to collect phytoplankton. In qualitative and quantitative studies, sampling is performed using networks or bottles. Nets are used to filter large volumes of water. There are many different nets whose use depends on the size of the species, their relative abundance or type of trawl. Oceanographic bottles are used to obtain samples of a given volume to a specific depth. The advantage of the bottle compared to nets is that more accurate data on the water column are obtained. The disadvantage is the low amount of algae material collected, and the need for the subsequent concentration of the sample. There are several models of sampling bottles. Nansen, Niskin and Van Dorn are the most used designs, all of which can close to the desired depth. The capacity of the bottle varies from one liter to over 100 liters, and several bottles can be grouped forming a rosette allowing collections at different depths in a single set. These mechanisms allow other parameters such as salinity, temperature, pH, dissolved oxygen, nitrate, and methane or chlorophyll fluorescence to be measured at the same time. Another possibility involves pumping water from a desired depth, and then filtering it. This procedure allows the continuous sampling of phytoplankton. Sediment traps, located at the bottom or attached to buoys, enable the continuous recording of information covering periods of months or years, and the amount of material that descends to the sea bottom is directly related to organic production. This method is only applicable to phytoplankton with hard walls or skeletons, such as diatoms, dinoflagellates and silicoflagellates (Lange and Boltovskoy 1995).

The techniques for the conservation and fixation of phytoplankton samples often depend on the taxonomic group under study. In general, samples should be stored in dark bottles. One of the most common fixatives is 4% aqueous formaldehyde whose solution is neutral, but tends to acidify with time. Thus, it is useful for the conservation of diatoms, but not for other groups like coccolithophores, whose calcareous scales are destroyed in an acid medium. The use of lugol-acetic fixative is the most common preservative in samples of phytoplankton and maintains the structure of colonial algae. The disadvantage is the need to add lugol periodically. The microscopic study of phytoplankton is usually done by an inverted microscope and with sedimentation chambers (Utermöhl method; Lund et al. 1957) to analyze species diversity and abundance in a given volume of sample.

7.3 Physico-chemical and Physiological Methods

Salinity can be measured in the field using several methods. The simplest, but imprecise method is the use of an Abbe refractometer, which compares the refractive index of light in saltwater to that of distilled water. Other possibilities are the use of a conductivity meter or salinometer, which measures the electrical resistance of the ion concentration, or the hydrometer, which determines the density of seawater at a known temperature and pressure. A simple laboratory method is the Winkler titration method, in which silver chloride is used to provide the chlorinity of the sample which can then be related to salinity, since the proportion of ions in seawater is stable. The concentration of dissolved oxygen in seawater is about 0.9%, depending on biological activity, photosynthesis and respiration. Together with the amount of carbon dioxide produced during respiration, this is a good parameter for measuring the productivity of algal populations. Dissolved oxygen can be measured by laboratory equipment like a respirometer (a Clark electrode), an oxygen meter or by titration. There are also kits that allow measurements to be taken in the field. Samples are placed in calibrated glass bottles, transparent and opaque, to measure photosynthesis and respiration. The difference between these two values is a measure of phytoplankton productivity commonly used in field studies. Photosynthesis can also be measured, incubating samples with ^{13}C, which is a stable isotope, or with ^{14}C, which is a radioactive one.

In macroalgae, the percentage of dry weight is a simple method to obtain information on the biomass of algal communities, on dominant species or for industrial purposes. After the cleaning of epiphytes and debris, the collected algae can be weighed to obtain their wet weight. To obtain dry weight, the algae should be dried at a temperature not over 60°C. Another useful parameter is the daily growth rate of algae. This parameter can be estimated using biomass increase, obtained by relating initial and final weight after a given number of days. Levels of chlorophylls and other pigments are measured for diverse purposes. They can be measured in the laboratory by means of spectrophotometric and chromatographic techniques from a sample of known weight. Spectrophotometric measures the concentration of pigments since each of them has peak absorption at a specific wavelength. Chlorophylls are extracted with several organic solvents (e.g., acetone, methanol, ethanol, dimethyl sulfoxide, etc.) and homogenized by centrifugation. The concentration of other compounds such as lipids, proteins, carbohydrates and cellulose can be easily obtained in the laboratory. Nowadays, the extraction of DNA and RNA for molecular analysis is also common in phycological research.

Acknowledgments

To Ignacio Bárbara, Silvia Calvo Javier Cremades and Mariona Hernández by the photographs included in this chapter, also to Carmen Prada, Ignacio Bárbara and Miguel Álvarez for their critical review of the manuscript.

Keywords: marine algae, anatomy, morphology, reproduction, systematics

References Cited

Akatsuka, I. (ed.). 1990. Introduction to Applied Phycology. SPB Academic Publ., The Hague.

Alveal, K., M.E. Ferrario, E.C. Oliveira and E. Sar (eds.). 1995. Manual de métodos ficológicos. Universidad de Concepción, Concepción, Chile.

Barsanti, L. and P. Gualtieri. 2006. Algae, Anatomy, Biochemistry and Biotechnology. CRC, Taylor & Francis, Boca Raton, Florida.

Bold, H.C. and M.J. Wynne. 1978. Introduction to the Algae. Prentice Hall, Englewood Cliffs, New Jersey.

Cheng, T.H. 1969. Production of kelp. A major source of China's exploitation of the sea. Economic Botany 23: 215–236.

Chihara, M. 1969. *Ulva arasaki*, a new species of green algae: Its life history and taxonomy. Bulletin of the National Science Museum, Tokyo, Ser. B (Botany) 12: 849–862.

Cmieck, H.A., G.F. Leedale and C.S. Reynolds. 1984. Morphological and ultrastructural variability of planktonic cyanophyceae in relation to seasonal periodicity. I. *Gloeotrichia echinulata*: Vegetative cells, polarity, heterocysts, akinetes. British Phycological Journal 19: 259–275.

Cole, K.M. and R.G. Sheath (eds.). 1990. The Biology of the Red Algae. Cambridge University Press, Cambridge.

Daugbjerg, N. 1996. *Mesopedinella arctica* gen. et sp. nov. (Pedinellales, Dictyochophyceae) I: Fine structure of a new marine phytoflagellate form Artic Canada. Phycologia 35: 435–445.

Dawes, C.J. 1998. Marine Botany. Hohn Willey & Sons, New York.

Drew, K.M. 1949. *Conchocelis*-phase in the life history of *Porphyra umbilicalis* (L.) Kuetz. Nature 164: 748–749.

Falkowski, D.G. and A.H. Knoll. 2007. Evolution of Primary Producers in the Sea. Elsevier Academic Press, New York.

Friedmann, E.I. and W.C. Roth. 1977. Development of the siphonous green alga *Penicillus* and the *Espera* state. Botanical Journal of the Linnean Society 74: 189–214.

Graham, L.E., J.M. Graham and L.W. Wilcox. 2009. Algae, Second edition. Benjamim Cumming, Pearson, San Francisco.

Guiry, M.D. 2012. How many species of Algae are there? Journal of Phycology 48: 1057–1063.

Guiry, M.D. and G. Blunden. 1991. Seaweed Resources in Europe. Uses and Potential. John Wiley & Sons Ltd., Chichester.

Guiry, M.D. and G.M. Guiry. 2013. AlgaeBase. World-wide electronic publication, National University of Ireland, Galway. http://www.algaebase.org; searched on 15 April 2013.

Hallegraeff, G.M., D.M. Anderson and A.D. Cembella (eds.). 1995. Manual of Harmful Marine Microalgae. IOC Manuals and Guides No. 33. UNESCO, Paris.

Hasle, G.R. and E.E. Syvertersen. 1997. Marine diatoms. pp. 5–385. *In:* Tomas, C.R. (ed.). Identifying Marine Phytoplankton. Academic Press, San Diego.

Hellebust, J.A. and J.S. Craigie (eds.). 1978. Handbook of Phycological Methods. Physiological and Biochemical Methods: Macroalgae. Cambridge University Press, Cambridge.

Hoek, van den C., D.G. Mann and H.M. Jahns. 1995. Algae, An Introduction to the Phycology. Cambridge University Press, London.

Innamorati, M., I. Ferrari, D. Marino and M. Ribera d'Alcala. 1990. Metodi nell'Ecologia del Plancton Marino. Nova Thalassia 11: xxvii+372.

Johansen, H.W. 1981. Coralline algae, a first synthesis. CRC Press, Boca Raton, Florida.

Johri, R.M., S. Lata and S. Sharma. 2004. A Textbook of Algae. Dominant Cop., New Delhi, India.

Kim, S.-K. 2011. Handbook of Marine Macroalgae: Biotechnology and Applied Phycology. Wiley & Sons, New York.

Knox, G. 2000. The Ecology of Seashores. CRC Press, Boca Raton, Florida.

Kristiansen, J. 2005. Golden Algae. A Biology of Chrysophytes. ARG Gantner Verlag K.G. Dehra Dun, India.

Lange, C. and D. Boltovskoy. 1995. Trampas de sedimento. pp. 93–118. *In:* Alveal, K., M.E. Ferrario, E.C. Oliveira and E. Sar (eds.). Manual de métodos ficológicos. Universidad de Concepción, Concepción, Chile.

Le Gall, L. and G.W. Saunders. 2006. A nuclear phylogeny of the Floridophyceae (Rhodophyta) inferred from combined EF2, small subunit and large subunit ribosomal DNA: Establishing the new red algal subclass Corallinophycidae. Mol. Phylog. Evol. 43: 1118–1130.

Lee, R.E. 2008. Phycology. Cambridge University Press, Cambridge.

Leliaert, F., D.R. Smith, H. Moreau, M.D. Herron, H. Verbruggen, C.F. Delwiche and O. De Clerck. 2012. Phylogeny and molecular evolution of the green algae. Critical Reviews in Plant Sciences 31: 1–46.

Lembi, C.A. and J.R. Waaland. 1988. Algae and Human Affairs. Cambridge University Press, Cambridge.

Lewis, J.R. 1964. The Ecology of Rocky Shores. Hodder and Stoughton, London.

Litter, M.M. and D.S. Litter (eds.). 1985. Handbook of Phycological Methods. Ecological Field Methods: Macroalgae. Cambridge University Press, Cambridge.

Litter, M.M., D.S. Litter, S.F. Blair and J.N. Norris. 1986. Deep-water plant communities from an uncharted seamount off San Salvador Island, Bahamas: distribution, abundance, and primary productivity. Deep-Sea Research Part A Oceanographic Research Papers 33: 881–892.

Lobban, C.S. and P.J. Harrison. 2000. Seaweed Ecology and Physiology. Cambridge University Press, Cambridge.

Lobban, C.S., D.J. Chapman and B.P. Kremer (eds.). 1988. Experimental Phycology. A Laboratory Manual. Cambridge University Press, Cambridge.

Lund, J.W.G., C. Kipling and E.D. Le Cren. 1957. The inverted microscope method of estimating algal numbers, and the statistical basis of estimations by counting. Hydrobiologia 11: 143–170.

Lüning, K. 1990. Seaweeds, Their Environment, Biogeography and Ecophysiology. Wiley & Sons, New York.

Margalef, R. 2005. Ecologia. Omega, Barcelona, Spain.

Margulis, L. 1998. Symbiotic Planet: A New Look at Evolution. Basic Books, New York.

Margulis, L., J.O. Corliss, M. Melkonian and D.J. Chapman (eds.). 1989. Handbook of Protoctista. Jones and Bartlett Publ., Boston.

Meinesz, A. 1999. Killer Algae. University of Chicago Press, Chicago.

Parsons, T.R., Y. Maita and C.M. Lalli. 1989. A Manual of Chemical and Biological Methods for Seawater Analysis. Pergamon Press, Oxford.

Perez, R., R. Kaas, F. Campello, S. Arbault and O. Barbaroux. 1992. La culture des algues marines dans le monde. IFREMER, Plouzane, France.

Prud'homme van Reine, W.F. and G.C. Trono (eds.). 2001. Plant Resources of South–East Asia 15(1). Cryptogams: Algae. Backhuys Publishers, Leiden, The Netherlands.

Richmond, A. (ed.). 2008. Handbook of Microalgal Culture: Biotechnology and Applied Phycology. Blackwell Publishing, Oxford.

Riisberg, I., Russell, J.S. Orr, R. Kluge, K. Shalchian-Tabrizid, H.A. Bowerse, V. Patilb, B. Edvardsena and K.S. Jakobsen. 2009. Seven gene phylogeny of heterokonts. Protist 160: 191–204.

Roberts, K.R., K.D. Stewart and K.R. Mattox. 1984. Structure and absolute configuration of the flagellar apparatus in the isogametes of *Batophora* (Dasycladales, Charophyta). Journal of Phycology 20: 183–191.

Round, F.E. 1981. The Ecology of Algae. Cambridge University Press, Cambridge.

Round, F.E., R.M. Crawford and D.G. Mann. 1990. The Diatoms. Biology and Morphology of the Genera. Cambridge University Press, Cambridge.

Roy, S., C.A. Llewellyn, E. Skarstad Egeland and G. Johnsen. 2011. Phytoplankton Pigments: Characterization, Chemotaxonomy and Applications in Oceanography. Cambridge University Press, Cambridge.

Sagert, S., R.M. Forster, P. Feuerpfiel and H. Schubert. 1997. Daily course of photosynthesis and photoinhibition in *Chondrus crispus* (Rhodophyta) from different shore levels. Europ. Journal of Phycology 32: 244–255.

Scagel, R.F., R.J. Bandoni, J.R. Maze, G.E. Rouse, W.G. Schofield and J.R. Stein. 1982. Nonvascular Plants. An Evolutionary Survey. Wadsworth, Belmont.

Sournia, A. (ed.). 1978. Phytoplankton Manual. UNESCO, Paris.

South, G.R. and A. Whittick. 1987. Introduction to Phycology. Blackwell Scientific Publ., Oxford.

Strickland, J.D.H. and T.R. Parsons. 1968. A Practical Handbook of Seawater Analysis. Bulletin 167. Fisheries Research Board of Canada, Ottawa.

Taylor, F.J.R. 1990. Dinoflagellata (Dinomastigomicota). pp. 419–437. *In*: Margulis, L., J.O. Corliss, M. Melkonian and D.J. Chapman (eds.). Handbook of Protoctista. Jones and Bartlett Publishers, Boston.

Torres, M.A., M.P. Barros, S.C.G. Campos, E. Pinto, S. Rajamani, R.T. Sayre and P. Colepicolo. 2008. Biochemical biomarkers in algae and marine pollution: A review. Ecotoxicology and Environmental Safety 71: 1–15.

Walker, L.M. 1984. Life histories, dispersal and survival in marine planktonic dinoflagellates. pp. 19–34. *In*: Steidinger, K.A. and L.M. Walker (eds.). Marine Planckton Life Cycle Strategies. CRC Press, Boca Raton, Florida.

West, J.A. and M.H. Hommersand. 1981. Rhodophyta life histories. pp. 133–193. *In*: Lobban, C.S. and M.J. Wynne (eds.). The Biology of Seaweeds. Blackwell Scientific Publications, Oxford.

Woelkerling, W.J. 1988. The Coralline Red Algae: An Analysis of the Genera and Subfamilies of Nongeniculate Corallinaceae. Oxford University Press, London.

Yoon, H.S., K.M. Müller, R.G. Sheath, F.D. Ott and D. Bhattacharya. 2006. Defining the major lineages of red algae (Rhodophyta). Journal of Phycology 42: 482–492.

Yoon, H.S., R.A. Andersen, S.M. Boo and D. Bhattacharya. 2009. Stramenopiles. pp. 721–731. *In*: Schaechter, M. (ed.). Encyclopedic of Microbiology, Vol. 5. Elsevier, Oxford.

CHAPTER 2

Searching for Ecological Reference Conditions of Marine Macroalgae

Rui Gaspar, * *João M. Neto* and *Leonel Pereira*

1 Introduction

Worldwide, increase of human pressures and consequent degradation of the ecological quality of aquatic systems have been contributing to the awareness that the integrity of marine ecosystems is under threat (e.g., Crain et al. 2009) and hence, in decline from earlier natural or pristine conditions. In this context, the idea of a 'natural' or 'pristine' ecological condition that can be linked to a historic past without significant human pressures, leads to the idea of ecological reference conditions, against which the current degraded ecological conditions can be seen in perspective and compared.

The ecological degradation of marine environments can be caused by multiple stressors, such as the water pollution resultant from urban, industrial and agricultural waste, causing problems like toxic chemical pollution, suspended sediments and excessive nutrients in the water (e.g., UNEP/GPA 2006). The anthropogenic impacts induce changes on the coastal assemblages, namely on the spatial and temporal patterns of the organisms' distribution, decreasing the species diversity and affecting the physical and biological structure of natural habitats (Benedetti-Cecchi et al. 2001 and references therein). It affects as well, the ecosystem goods and services, such as food provision, disturbance prevention, nutrients recycling or leisure and recreation (e.g., Beaumont et al. 2007).

IMAR-CMA (Institute of Marine Research), Department of Life Sciences, FCTUC, University of Coimbra, 3001-455 Coimbra, Portugal.
* Corresponding author: rui.miguel.m@gmail.com

Macroalgae species, being sessile and aquatic organisms, can integrate and reflect along time the water related environmental characteristics of the shores they occupy (Ballesteros et al. 2007). In this sense, contemporary anthropogenic pressures may shift the macroalgae communities from pristine conditions to degraded quality states. These communities show similar patterns when exposed to disturbance: decrease in the species richness, reduction in the complexity of the community structure and changes in the patterns of variability; the decrease or disappearance of the most sensitive species, like large canopy-forming and slow-growing perennial species, tend to be replaced by opportunistic, short-lived and fast-growing species, with lower structural complexity, such turf-forming, filamentous, leaf-like or other annual macroalgae (e.g., Murray and Littler 1978, Diez et al. 1999, Tewari and Joshi 1988, Benedetti-Cecchi et al. 2001, Thibaut et al. 2005, Mangialajo et al. 2008, Gorman and Connell 2009, Perkol-Finkel and Airoldi 2010).

As a matter of fact, macroalgae have been widely used to evaluate the effects of anthropogenic disturbances, such as climate change (e.g., Lima et al. 2007, Hawkins et al. 2008, Fernandez 2011, Harley et al. 2012), habitat loss (e.g., Airoldi 2003, Airoldi and Beck 2007) or, more commonly, changes resultant from human-polluted waters (e.g., Tewari and Joshi 1988, Gorostiaga and Diez 1996, Rodriguez-Prieto and Polo 1996, Roberts et al. 1998, Soltan et al. 2001, Panayotidis et al. 2004, Melville and Pulkownik 2006, Yuksek et al. 2006, Arévalo et al. 2007, Scanlan et al. 2007, Krause-Jensen et al. 2008, Juanes et al. 2008, Orfanidis et al. 2011, Neto et al. 2012).

For example, human activities have been closely linked with the increase of nutrients (e.g., nitrogen based nutrients) in marine waters towards eutrophication, which represents an important issue for ecology and environmental management (e.g., Vitousek et al. 1997, de Jong 2006, Kelly 2008). Eutrophication leads to the proliferation of opportunist and tolerant macroalgae (e.g., Morand and Merceron 2005) featured by fast-growing and high nutrient uptake rates (e.g., Wallentinus 1984), at the expense of seagrasses and perennial macroalgae (e.g., Duarte 1995, Schramm 1996). The opportunistic turf-forming species (genera examples include green macroalgae *Ulva, Chaetomorpha, Cladophora* and *Monostroma*, and red macroalgae *Ceramium, Gracilaria* and *Porphyra* or brown macroalgae *Ectocarpus* and *Pilayella*) can bloom into nuisance proportions (Morand and Merceron 2005, Scanlan et al. 2007). Contrarily, canopy macroalgae, such as the perennial genus *Cystoseira*, are highly sensitive to anthropogenic disturbances and tend to disappear from sites nearby urban areas and with higher levels of nutrient concentration (e.g., Benedetti-Cecchi et al. 2001, Sales and Ballesteros 2009).

Therefore, as macroalgae communities can reflect ecological degradation, their study can be useful to quantify the level of marine ecosystem integrity or to what extent the conditions have changed from their former pristine environment. It is for such reasons that for water policies such as the European Water Framework Directive (WFD 2000/60/EC), these organisms are key biological elements when assessing the ecological quality condition of coastal waters.

Yet difficulties may arise when trying to effectively describe ecological reference conditions, given the limited knowledge about coastal environments, of the dynamism of their natural processes, and the linkage between different levels of anthropogenic pressure and the varying conditions of ecological degradation. This work aims to briefly review some investigative approaches concerning the challenge of describing marine macroalgae under ecological reference conditions.

2 The Spatial and Temporal Uncertainty of Macroalgae Natural Patterns

Despite their benthic sessile feature, the macroalgae species exhibit populations whose distribution along coastal rocky shores is not uniform, either in space or time, but rather, in a non-random and dynamic way (e.g., Lobban and Harrison 1994). This results from complex ecological processes, such as succession patterns, where different species have different recruitment, growth and mortality rates (e.g., Foster et al. 2003, Cervin et al. 2005). Naturally, a variety of different 'mosaics', 'patches' or 'assemblages' of macroalgae species can be observed. These assemblages tend to differ in shores, both spatially and temporally, according to the presence or absence (composition) of different species and accordingly to the relative abundance of each species. For example, different macroalgae species tend to exhibit vertical patterns of distribution, from the uppermost to the lowermost tide levels, giving the idea of different zones of species or zonation patterns. This is because different species have different adaptive responses to several physical (e.g., emersion or exposure to the atmosphere), chemical (e.g., salinity) and biotic (e.g., competition, grazing) factors (e.g., Lobban and Harrison 1994, Dawes 1998), which can influence unevenly the different locations on the shore. Being aquatic species, the complexity of the macroalgae communities' structure (composition and abundance) tends to increase from land to sea, as the aquatic environments' characteristics become more effective at lower levels on the shore; the environmental factors variability tends to be higher at upper shore levels (e.g., higher emersion times tend to create more desiccation problems), causing to the species, several vertical stress gradients (Raffaelli and Hawkins 1999).

Ultimately, the distributional area of each species can be seen as the complex expression of its ecology and evolutionary history, determined by diverse factors operating with different intensities and at different scales (Soberón and Peterson 2005). Many environmental factors can influence macroalgae but, depending on the spatial and time scales considered, some factors may, more than others, influence the particular distribution of each species. As species differ in their adaptations to particular environments, the outcome of interactions depends on the species identity (Viejo 2008) and, at the end, on its population distribution patterns, either locally or in a broader global geographical scale. In this connection, there are considerable gaps in the understanding about rocky shore communities, including the need of cross country comparisons, the role of key species, the macroalgae vs. filter-feeders interactions, the early life history of the species, the effect of the nearshore upwelling and downwelling water mass or the effect of human impacts (Schiel 2004).

Undoubtedly, much scientific knowledge has been produced regarding marine macroalgae, including environmental, biogeographical, ecological or ecophysiological issues. Yet complete knowledge of macroalgae distribution patterns (from single species to populations to entire communities) and the prevailing environmental factors driving macroalgae species assemblages to arrange in a particular way on a given shore or in a broader geographical scales, may continue on the marine biologists' agendas as research challenges. Moreover, knowledge of these dynamic patterns grows more uncertain when considering that environmental factors do change, due to complex and interrelated processes that occur both naturally and, not less important, as a result of human activities.

3 The Concept of Reference Conditions

Several terms with equivalent meanings have been applied to describe the reference conditions present in a site (e.g., unimpaired, unperturbed, undegraded, unaltered, undisturbed, nearly undisturbed or least affected). The concept of ecological reference conditions have been applied in several contexts, including environmental considerations (from "pristine" to the "best available" state of a water body, in a historical or a spatial context, etc.) and also including a wide range of economic, social and political considerations (Economou 2002). Economic and social aspects of marine management may have equal or perhaps even greater weight than the ecological aspects, since we have to protect and maintain the natural ecological characteristics while concurrently delivering the services and benefits required by society (Elliott 2011).

For water ecosystems such as coastal environments, the ecological reference conditions are associated to a set of biological, physical, chemical and hydrologic quality elements. They should reflect the status of those quality elements under pristine, or not impacted conditions, i.e., conditions to be found in the absence of, or under minimal anthropogenic disturbance. Biological reference conditions (in the sense of the WFD) can be derived from biological communities that inhabit sites without (or with minor) anthropogenic disturbances and should reflect an ecological state that corresponds to very few pressures, i.e., without the effects of major industrialisation, urbanisation or intensification of agriculture and with only very minor modifications of the physical, chemical, hydromorphological and biological characteristics (e.g., WFD CIS 2009).

This can be interpreted as sites having an absence of pressures or a presence of high ecological quality. However, even large areas such as Europe may not have many pristine places to be used as reference sites. Furthermore, if the human impacts are considered on a global climate change, it can be acknowledged that the pristine marine habitats are currently scarce and may be impossible to achieve (Borja et al. 2013).

Consequently, different approaches can be used in order to define ecological reference conditions. The WFD Common Implementation Strategy (WFD CIS 2003) suggested an hierarchical criteria for defining reference conditions using various methods in the following order: i) an existing undisturbed (pristine) site or a site with only very minor disturbance; ii) historical data and information; iii) models; and (iv) expert judgment. Also, legislation such as the Marine Strategy Framework Directive (MSFD) is seeking: (v) environmental targets, rather than (undisturbed) reference conditions; this is because it is recognized that humans are part of the marine ecosystem (as users) and, then, their activities can create a certain impact, making it impossible to find pristine areas (Borja et al. 2012).

Important advantages and disadvantages of those methodologies have been recently discussed under the relationships of the DPSIR paradigm (between Drivers-Pressure-State of change-Impact-Responses variables), in assessing the environmental quality status of the marine waters, and having an 'ecosystem approach' or 'holistic approach' methodological point of view (see Borja et al. 2012). Within that framework, each methodological approach (pristine sites, historical data, modelling, best professional judgment, baselines state set at a date in the past or at a current state) aims to relatively different directions (or target goals) for ecological reference and recovery (Fig. 1).

Hence, it is difficult to define how far back the baseline has to be set, or how pristine a condition a benchmark should characterize (Hawkins et al. 2010). It depends both on the level of ecological conditions that are aimed by human society and on the availability of historical data and knowledge of the system. In fact, 'pristine state' or 'naturalness' can be best viewed as

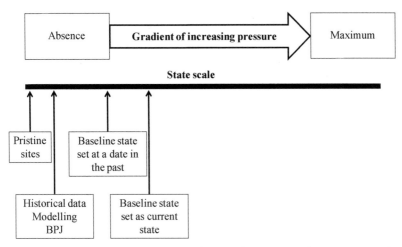

Figure 1. Environmental status can be regarded as a gradual transition from pristine conditions (high status in the absence of human pressures) to an irrecoverable status (bad status, in a maximum of human pressure). Assessment systems need to set reference conditions (pristine sites, historical data, modelling, best professional judgment—BPJ) or baseline targets (set at a date in the past or as current state) along the pressure (and subsequent state) gradient to assist in status assessments (Adapted from Borja et al. 2012).

having dynamic attributes, which may need to be periodically redefined in response to new and better scientific knowledge (van Hoey et al. 2010 and references therein).

Still the need to describe the natural condition of a system—whose features may establish a standard, target goal, or reference condition —becomes as important as science needs controls to compare results (Economou 2002). Ultimately, to describe a reference condition is to point out directions for ecological restoration and protection (Borja et al. 2012). Reference conditions are just that, a reference, which can be linked to ecological conditions that remain pristine or unchanged, but also as defined environmental goals that can be considered (somehow) reasonable to be achieved.

Regardless of the degree of 'naturalness' or 'pristineness' implied by a specific definition of a benchmark, the accuracy and precision when assessing ecological quality conditions are dependent on the degree to which those benchmarks can be quantified and predicted (Hawkins et al. 2010). Bearing this in mind, several investigation approaches to describe ecological reference conditions regarding the marine macroalgae are briefly presented below.

4 Investigation Approaches to Describe the Ecological Reference Conditions of Marine Macroalgae

4.1 Establishing Biotypes Ranges for Ecological Homogeneity: The Type-Specific Reference Condition Approach

Marine ecosystems are subject to many different changes resulting from both natural processes and human activities (e.g., Aubry and Elliott 2006). In fact, one of the problems associated with using a reference condition approach is that high natural variability may be present, making it difficult to distinguish between anthropogenic and undisturbed environmental effects on biotic communities. The combination of abiotic and biotic attributes, such as different climates and biotypes, contribute to define different eco-regions, but these eco-regions themselves embrace high biological variability due to environmental heterogeneity or historical/phylogenetic factors (Economou 2002). For example, the temperature plays an important role on the survival, recruitment, growth and reproduction of macroalgae (Breeman 1988). This is also reflected by the global geographical distributions of macroalgae species, which are typically delimited by certain seawater isotherms (Lüning et al. 1990). The sea surface temperatures (SST) outline boundaries among different bio-geographical regions, preventing the uninterrupted spread of macroalgae species beyond their present distributions to rocky shores all over the world (Eggert 2012). In this regard, considering for example, the North East Atlantic geographical region by itself, it has heterogeneous coastal waters, with diverse species composition. It includes zones as diverse as Norway, in the North, and the Canary Islands in the South, which makes difficult to establish a common reference condition for the region as a whole (Ramos et al. 2012). Actually, on the Portuguese coastline, several coldwater species (i.e., species abundant in Northern Europe) and several warm-water species (i.e., species that are commonly found in the Mediterranean Sea and in Northwestern Africa) reach their southern or northern distributional limits, respectively (Lima et al. 2007).

To minimize the variability associated with geographical differences, a common approach is to organize the environmental information on a narrower spatial scale. For an effective use of the reference data, the variability within the monitoring network over which the same reference conditions apply must be small enough, in order to enable the effects of anthropogenic activities to dominate (Economou 2002). Hence, different type-specific reference conditions should be established. The biological reference conditions must summarize a range of possibilities and values over periods of time and across the geographical extent of each water body type; the reference conditions represent part of nature's continuum and must reflect the natural variability (WFD CIS 2003).

In fact, in order to predict and better distinguish the anthropogenic effects from the natural environmental effects on biological communities, a more appropriate geographical area or type scale allows to define biological communities with higher homogeneity even though type-specific areas encompass communities presenting high variability features, such as the differences resultant from seasonal species, the effects of phenotypic plasticity, or the intra-specific genetic differences. On the other hand, to describe different reference conditions at higher discrimination levels (small areas like site level) would imply important disadvantages (costs, attainability, data availability, time availability, etc.) (Economou 2002).

4.2 From Species Richness to Grouping Species into Representative Taxa Listings

It can be agreed that the analysis of species composition and abundance is an unavoidable methodological approach to describe the structure of their community assemblages. In this respect, the WFD outlined the following composition and abundance criteria that should be related to type-specific reference conditions: (1) taxonomic composition corresponds completely to, or almost completely to undisturbed conditions (where all sensitive *taxa* should be present); and (2) there are no detectable changes in macroalgae abundances due to anthropogenic disturbances (WFD CIS 2003).

Nevertheless, the absence of reference values, or the awareness of the ecological state of a certain system against which comparisons can be made, makes these assessment requirements complicated to address (Borja et al. 2004). The sensitive species are not easy to define and the species composition changes for natural reasons even under undisturbed conditions (Wells et al. 2007). Moreover, to identify all macroalgae to a species level and describe their relative abundance (such as species coverage or biomass) over time and space is a task that involves several problems such as the time consumption involved, the need for taxonomic expertise or the costs of monitoring designs. Representativeness and time consumption are central issues in monitoring designs. Sorting, identifying and quantifying the macroalgae species samples are time and labour demanding and need good taxonomical expertise (Ballesteros et al. 2007) to accurately record their species richness (Wells et al. 2007).

Taxonomical identifications with a species-level resolution are a time-consuming task and therefore, the need to find simpler ways to assess species richness may be inevitable (Wilkinson and Wood 2003). Macroalgae species richness decreases along increasing disturbance gradients, shifting the composition of their communities (e.g., Díez et al. 2012) while, on the other hand, the numerical macroalgae species richness—not the list of actual species present—remains approximately constant in the absence of disturbance, which provides an excellent rationale for using it as a measure of ecological quality (Wells et al. 2007).

In effect, most of the macroalgae species are easily identifiable in the field with easy-to-acquire expertise and, therefore, allow the monitoring of large areas with relatively little effort (Ballesteros et al. 2007). The use of both non-destructive data collection and easy-to-apply methods has an effective cost-benefit relationship (e.g., Wells et al. 2007, Juanes et al. 2008), being scientifically rigorous and at the same time, allowing the execution of wider monitoring plans (Guinda et al. 2008). An alternative way to record qualitative species data has been achieved by selecting comprehensive and representative species listings that can be more commonly found within certain geographical areas or typologies and simultaneously associated with expected undisturbed conditions (Table 1) (e.g., Wells et al. 2007, Guinda et al. 2008, Juanes et al. 2008, Gaspar et al. 2012, Bermejo et al. 2012).

When developing a representative and comprehensive species list for a certain area, the species' natural variety must be considered. Also, the list of species must reflect the water conditions over time, and should be sensitive enough to detect those changes in the composition of macroalgae communities. Representativeness can be included by merging the available historical data with the contemporary monitoring data, allowing for the selection of more common *taxa*. Selected *taxa* can be grouped considering taxonomical, morphological or functional similarities (see Section 4.4 below), while maintaining the natural proportions of the main macroalgae taxonomical groups (Chlorophyta, i.e., green macroalgae, Heterokontophyta-Phaeophyceae, i.e., brown macroalgae and Rhodophyta, i.e., red macroalgae) (see Section 4.3 below) that are normally associated with a given water body type (Gaspar et al. 2012). Departing from previous specific works done for certain geographical areas, these approaches may define *a priori* which are the main characteristic species of macroalgae that constitute well-defined, conspicuous populations, along with opportunistic species related to anthropogenic disturbances, as well as invasive species (e.g., *Sargassum muticum* in European coastal waters) (Juanes et al. 2008). Furthermore, taking into account the seasonal variability of macroalgae communities, composition data for those selected *taxa* can be studied from undisturbed areas during their local seasonal period of maximum development (spring/summer). Last but not least, during that period of time and at the same undisturbed areas, data concerning macroalgae abundance can be estimated, such as the coverage of characteristic species and the coverage of opportunistic species (Juanes et al. 2008, Gaspar et al. 2012).

Table 1. Examples of representative species listings established for different geographical areas within European Atlantic coasts, namely for the coasts of Southern England, Republic of Ireland and Wales, Northern Ireland, Scotland and Northern England (Wells et al. 2007), Northern Spain (Cantabrian) (Guinda et al. 2008, Juanes et al. 2008), Northern Portugal (Gaspar et al. 2012), and Southern Spain (Andalusia) (Bermejo et al. 2012). The species' opportunistic character is indicated (Op). Species examples are given for particular species groupings at the table end.

Southern England, Republic of Ireland & Wales	Northern Ireland	Scotland and Northern England
Chlorophyta	**Chlorophyta**	**Chlorophyta**
Blidingia sp. (Op)	*Blidingia* sp. (Op)	*Blidingia* sp. (Op)
Bryopsis plumosa	*Chaetomorpha linum* (Op)	*Chaetomorpha linum* (Op)
Chaetomorpha linum (Op)	*Chaetomorpha mediterranea* (Op)	*Chaetomorpha melagonium*
Chaetomorpha mediterranea (Op)	*Cladophora albida*	*Cladophora rupestris*
Chaetomorpha melagonium	*Cladophora rupestris*	*Cladophora sericea*
Cladophora rupestris	*Cladophora sericea*	*Enteromorpha* sp. (Op)
Cladophora sericea	*Enteromorpha* sp. (Op)	*Sykidion moorei*
Enteromorpha sp. (Op)	*Monostroma grevillei*	*Ulva lactuca* (Op)
Ulva lactuca (Op)	*Rhizoclonium tortuosum*	
	Spongomorpha arcta	
	Ulothrix sp.	
	Ulva lactuca (Op)	
Phaeophyceae (Heterokontophyta)	**Phaeophyceae (Heterokontophyta)**	**Phaeophyceae (Heterokontophyta)**
Ascophyllum nodosum	*Alaria esculenta*	*Alaria esculenta*
Chorda filum	*Ascophyllum nodosum*	*Ascophyllum nodosum*
Cladostephus spongious	*Asperococcus fistulosus*	*Asperococcus fistulosus*
Dictyota dichotoma	*Cladostephus spongious*	*Chorda filum*
Ectocarpus sp. (Op)	*Dictyota dichotoma*	*Chordaria flagelliformis*
Elachista fucicola	*Ectocarpus* sp. (Op)	*Cladostephus spongious*
Fucus serratus	*Elachista fucicola*	*Desmarestia aculeata*
Fucus spiralis	*Fucus serratus*	*Dictyosiphon foeniculaceus*
Fucus vesiculosus	*Fucus spiralis*	*Dictyota dichotoma*
Halidrys siliquosa	*Fucus vesiculosus*	*Ectocarpus* sp. (Op)
Himanthalia elongata	*Halidrys siliquosa*	*Elachista fucicola*

Table 1. contd....

Table 1. contd.

Southern England, Republic of Ireland & Wales	Northern Ireland	Scotland and Northern England
Phaeophyceae (Heterokontophyta)	**Phaeophyceae (Heterokontophyta)**	**Phaeophyceae (Heterokontophyta)**
Laminaria digitata	*Himanthalia elongata*	*Fucus serratus*
Laminaria hyperborea	*Laminaria digitata*	*Fucus spiralis*
Laminaria saccharina	*Laminaria saccharina*	*Fucus vesiculosus*
Leathesia difformis	*Leathesia difformis*	*Halidrys siliquosa*
Pelvetia canaliculata	*Pelvetia canaliculata*	*Himanthalia elongata*
Pilayella littoralis (Op)	*Petalonia fascia*	*Laminaria digitata*
Ralfsia sp.	*Pilayella littoralis* (Op)	*Laminaria hyperborea*
Saccorhiza polyschides	*Ralfsia* sp.	*Laminaria saccharina*
Scytosiphon lomentaria	*Scytosiphon lomentaria*	*Leathesia difformis*
	Sphacelaria sp.	*Litosiphon laminariae*
	Spongonema tomentosum	*Pelvetia canaliculata*
		Pilayella littoralis (Op)
		Ralfsia sp.
		Scytosiphon lomentaria
		Spongonema tomentosum
Rhodophyta	**Rhodophyta**	**Rhodophyta**
Aglaothamnion/Callithamnion	*Aglaothamnion/Callithamnion*	*Aglaothamnion/Callithamnion*
Ahnfeltia plicata	*Ahnfeltia plicata*	*Ahnfeltia plicata*
Calcareous encrusters	*Audouinella purpurea*	Calcareous encrusters
Catenella caespitosa	*Audouinella* sp.	*Callophyllis laciniata*
Ceramium nodulosum	Calcareous encrusters	*Ceramium nodulosum*
Ceramium shuttlewoorthanium	*Catenella caespitosa*	*Ceramium shuttlewoorthanium*
Ceramium sp.	*Ceramium nodulosum*	*Chondrus crispus*
Chondrus crispus	*Ceramium shuttlewoorthanium*	*Corallina officinalis*
Corallina officinalis	*Chondrus crispus*	*Cryptopleura ramosa*

Cryptopleura ramosa	*Coralina officinalis*	*Cystoclonium purpureum*
Cystoclonium purpureum	*Cryptopleura ramosa*	*Delesseria sanguinea*
Dilsea carnosa	*Cystoclonium purpureum*	*Dilsea carnosa*
Dumontia contorta	*Dilsea carnosa*	*Dumontia contorta*
Erythrotrichia carnea	*Dumontia contorta*	*Erythrotrichia carnea*
Furcellaria lumbricalis	*Furcellaria lumbricalis*	*Furcellaria lumbricalis*
Gastroclonium ovatum	*Gelidium* sp.	*Lomentaria articulata*
Gelidium sp.	*Hildenbrandia rubra*	*Lomentaria clavellosa*
Gracilaria gracilis	*Lomentaria articulata*	*Mastocarpus stellatus*
Halurus equisetifolius	*Mastocarpus stellatus*	*Membranoptera alata*
Halurus flosculosus	*Melobesia membranacea*	*Odonthalia dentata*
Heterosiphonia plumosa	*Membranoptera alata*	*Osmundea hybrida*
Hildenbrandia rubra	*Odonthalia dentata*	*Osmundea pinnatifida*
Hypoglossum hypoglossoides	*Osmundea hybrida*	*Palmaria palmata*
Lomentaria articulata	*Osmundea pinnatifida*	*Phycodrys rubens*
Mastocarpus stellatus	*Palmaria palmata*	*Phyllophora* sp.
Membranoptera alata	*Phyllophora* sp.	*Plocamium cartilagineum*
Nemalion helminthoides	*Plocamium cartilagineum*	*Plumaria plumosa*
Osmundea hybrida	*Plumaria plumosa*	*Polyides rotundus*
Osmundea pinnatifida	*Polysiphonia fucoides*	*Polysiphonia fucoides*
Palmaria palmata	*Polysiphonia lanosa*	*Polysiphonia lanosa*
Phyllophora sp.	*Polysiphonia* sp.	*Polysiphonia* sp.
Plocamium cartilagineum	*Porphyra umbilicalis* (Op)	*Porphyra leucosticta* (Op)
Plumaria plumosa	*Rhodomela confervoides*	*Porphyra umbilicalis* (Op)
Polyides rotundus	*Rhodothamniella floridula*	*Ptilota gunneri*
Polysiphonia fucoides		*Rhodomela confervoides*
Polysiphonia lanosa		*Rhodothamniella floridula*
Polysiphonia sp.		
Porphyra umbilicalis (Op)		
Rhodomela confervoides		
Rhodothamniella floridula		

Table 1. contd....

Table 1. contd.

Northern Spain (Cantabria)	Northern Portugal	Southern Spain (Andalusia)
Chlorophyta	**Chlorophyta**	**Chlorophyta**
Blidingia/Derbesia (Op)	Bryopsis spp. (Op)	Bryopsis spp. (Op)
Bryopsis plumosa (Op)	Other Filamentous Chlorophyta (1) (Op)	Chaetomorpha spp. (Op)
Chaetomorpha spp. (Op)	Cladophora spp. (Op)	Cladophora spp. (Op)
Cladophora spp. (Op)	Codium spp.	Codium spp. erect (i)
Codium adhaerens	Ulva spp. ('Sheet-type')/Ulvaria obscura/	Codium spp. Encrusting (ii)
Codium tomentosum/C. fragile	Prasiola stipitata (2) (Op)	Codium bursa
Enteromorpha spp. (now Ulva spp.) (Op)	Ulva spp. ('Tubular-type'/Blidingia spp. (3)	Derbesia spp. (Op)
Ulva spp. (Op)	(Op)	Flabellia petiolata
		Pedobepsia simplex
		Enteromorpha spp. (Op)
		Ulva spp. (Op)
		Valonia utricularis
Phaeophyceae (Heterokontophyta)	**Phaeophyceae (Heterokontophyta)**	**Phaeophyceae (Heterokontophyta)**
Bifurcaria bifurcata	Bifurcaria bifurcata	Cladostephus spongiosus
Cladostephus spongiosus–verticillatus	Cladostephus spongiosus	Colpomenia sinuosa
Colpomenia spp./Leathesia spp.	Colpomenia spp./Leathesia marina	Cystoseira compressa
Cystoseira baccata	Cystoseira spp.	Cystoseira spp.
Cystoseira tamariscifolia	Desmarestia ligulata	Cystoseira usneoides
Dictyota dichotoma	Dictyopteris polypodioides	Dictyota dichotoma
Ectocarpaceae/Sphacelaria spp. (Op)	Dictyota spp.	Dictyopteris polypodioides
Fucus spiralis	Filamentous Phaeophyceae (4) (Op)	Fucus spiralis
Fucus vesiculosus	Fucus spp.	Halopteris spp.
Laminaria spp.	Halopteris filicina/H. scoparia	Saccorhiza polyschides
Nemalion helminthoides	Himanthalia elongata	Padina pavonica
Pelvetia canaliculata	Laminaria spp.	Laminaria ochroleuca
Ralfsia verrucosa	Pelvetia canaliculata	Ectocarpus and Sphacelaria (Op)

Saccorhiza spp. Sargassum muticum Scytosiphon spp. Stypocaulon (Halopteris) scoparia	Ralfsia verrucosa Saccorhiza polyschides	
Rhodophyta	**Rhodophyta**	**Rhodophyta**
Epiphytic filamentous (a) (Op)	Acrosorium ciliolatum/Callophyllis laciniata/	Delesseriaceae (iii)
Small folioses (b)	Cryptopleura ramosa	Asparagopsis armata
Champiaceae (c)	Ahnfeltia plicata	Botryocladia botryoides
Calcareous encrusters (d)	Ahnfeltiopsis spp./Gymnogongrus spp.	Caulacanthus ustulatus
Asparagopsis armata	Apoglossum ruscifolium/Hypoglossum	Ceramium spp. (Op)
	hypoglossoides	Chondracanthus acicularis
Catenella caespitosa (Op)	Asparagopsis armata/Falkenbergia rufolanosa	
Caulacanthus ustulatus	Bornetia spp./Griffithsia spp.	Corallina sp.
Chondracanthus (Gigartina) acicularis	Calliblepharis spp.	Gelidium microdon
Chondria coerulescens	Catenella caespitosa/Caulacanthus ustulatus	Gelidium spinosum
Chondrus crispus	Champiaceae (5)	Gelidium corneum
Corallina elongata/C. officinalis/Jania	Chondracanthus acicularis	Gelidium pusillum
Falkenbergia/Trailliella	Chondracanthus teedei	Gymnogongrus and Ahnfetiopsis
Gelidium latifolium	Chondria spp.	Halopithys incurva
Gelidium pusillum	Chondrus crispus	Halurus equisetifolius
Gelidium corneum (sesquipedale)	Calcareous encrusters (6)	Hildenbradia rubra
Gigartina spp.	Calcareous erect (7)	Jania rubens
Gymnogongrus spp.	Dilsea carnosa/Schizymenia dubyi	Laurencia obtusa
Halurus equisetifolius	Gelidiales (8)	Lithophyllum byssoides
Hildenbrandia spp.	Gigartina pistillata	Lithophyllum dentatum
Lithophyllum byssoides	Gracilaria spp.	Lithophyllum incrustans
Mastocarpus stellatus	Grateloupia filicina	Nemalion helminthoides
Osmundea (Laurencia) spp.	Halurus equisetifolius	Lomentaria articulata

Table 1. contd....

Table 1. contd.

Northern Spain (Cantabria)	Northern Portugal	Southern Spain (Andalusia)
Rhodophyta	**Rhodophyta**	**Rhodophyta**
Peyssonnelia spp. *Plocamium/Sphaerococcus* *Porphyra* spp. (Op) *Pterosiphonia complanata*	*Hildenbrandia* spp. *Laurencia* spp./*Osmundea* spp. *Mastocarpus stellatus/Petrocelis cruenta* *Nitophyllum punctatum* Other Filamentous Rhodophyta (9) (Op) *Phyllophora* spp./*Rhodymenia pseudopalmata* *Palmaria palmata* *Peyssonnelia* spp. *Plocamium cartilagineum/Sphaerococcus coronopifolius* *Porphyra* spp. (Op) *Pterosiphonia complanata* *Scinaia furcellata*	*Osmundea pinnatifida* *Osmundea hybrida* *Peyssonnelia* spp. *Plocamium cartilagineum* *Pterocladiella capillacea* *Pterosiphonia complanata* *Rhodymenia* and *Schottera* *Sphaerococcus coronopifolius*
(a) *Ceramium, Pleonosporium, Aglaothamnion, Callithamnion, Antithamnion, Antithamnionella, Polysiphonia, Dasya, Pterosiphonia.* (b) *Apoglossum, Hypoglossum, Acrosorium, Nytophyllum, Cryptopleura, Rhodophyllis, Stenogramme, Callophyllis, Kallymenia, Rhodymenia.* (c) *Champia, Lomentaria, Gastroclonium, Chylocladia.* (d) *Lithophyllum, Mesophyllum, Lithothamnion*	(1) *Chaetomorpha, Pseudendoclonium, Rhizoclonium, Ulothricales.* (2) *Ulva* spp. 'Sheet-type' in opposition to (3) 'Tubular-type' in the sense of 'ex-*Enteromorpha* spp.' (4) *Ectocarpales/Sphacelaria* spp. (5) *Champia, Chylocladia, Gastroclonium, Lomentaria.* (6) *Lithophyllum, Melobesia, Mesophyllum, Phymatolithon.* (7) *Amphiroa, Corallina, Jania.* (8) *Gelidium, Pterocladiella.* (9) *Acrochaetium,*	(i) *Codium tomentosum, C. fragile, C. vermilara* and *C. decorticatum.* (ii) *Codium adhaerens* and *C. effusum.* (iii) *Acrosorium uncinatum, Cryptopleura ramulosa* or *Haraldiophyllum bonnemaisonnii*

Aglaothamnion, Antithamnion, Bangia, Boergeseniella, Brongniartella, Colaconema, Callithamnion, Ceramium, Compsothamnion, Dasya, Erythrotrichiaceae, Herposiphonia, Heterosiphonia, Janczewskia, Leptosiphonia, Lophosiphonia, Ophidocladus, Pleonosporium, Plumaria, Polysiphonia, Pterosiphonia (except P. complanata), Pterothamnion, Ptilothamnion, Rhodothamniella, Streblocladia, Vertebrata	

4.3 Green, Brown and Red Species Patterns

The use of macroalgae species patterns based on their main taxonomical groups—the Chlorophyta (green macroalgae), the Heterokontophyta-Phaeophyceae (brown macroalgae) or the Rhodophyta (red macroalgae)—can be also an important approach beyond the use of species richness alone when considering the macroalgae composition of a given area.

Green, brown and red macroalgae species have distinct patterns of distribution around the globe. Contrary to most kinds of macroscopic organisms in terrestrial and marine habitats, macroalgae species richness does not always increase towards the tropics. Brown macroalgae species richness increases and green macroalgae species richness decreases towards higher latitudes, while red macroalgae species richness increases from the Arctic to the Tropics and from the Tropics to the Subantarctic (Santelices et al. 2009).

Ratios such as the Feldmann's (1937) R:P ratio (the number of red macroalgae to the number of brown macroalgae), the P:C ratio (the number of brown macroalgae to the number of green macroalgae), the R:C ratio (the number of red macroalgae to the number of green macroalgae) or others possibilities such Cheney's (1977) R+C/P ratio (the number of red and green macroalgae to the number of brown macroalgae) can display changes between geographical areas (e.g., Santelices et al. 2009). These aspects can be very useful, since for sites within a range of a particular area, it is possible that the same ratios might have a narrow range of values over a range of separate shores and so that a departure from that range might indicate adverse influences and therefore, less ecological quality (Wilkinson and Wood 2003).

In order to fully apply the above ratios, the total species taxonomic identification is needed. Nevertheless, through the use of representative species listings (see previous Section 4.2) similar range of values for species ratios can be calculated and associated with the particular area to which the species are listed. Not least important, the descriptors 'number' and 'proportion' of red, brown and green macroalgae species (e.g., number of green macroalgae, proportion of green macroalgae, etc.) may also be established individually as local reference values, e.g., data outputs resulting from the use of regional species listings in monitoring undisturbed sites (e.g., Gaspar et al. 2012). The species listings representativeness can be evaluated by comparing the relative proportions of red, brown and green macroalgae species between regional phycological checklists or floras against the *taxa* list selected for the same areas (Fig. 2).

In northern cold waters, the brown algae are naturally dominant, but at southern temperate waters, the red algae predominate (e.g., Lüning 1990, Boaventura et al. 2002). Depending on the latitudinal location and

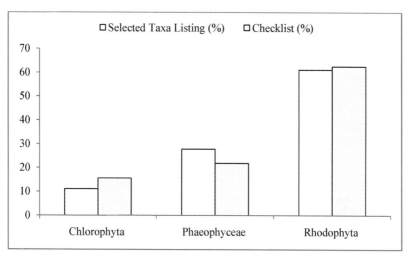

Figure 2. Comparison among relative proportions (%) of main macroalgal taxonomic groups (Chlorophyta, Phaeophyceae and Rhodophyta) and between the *taxa* list selected for Portuguese Northern Shores (see Table 1) and the *taxa* recorded by Araújo et al. (2009) regarding the northern Portugal checklist (data from Gaspar et al. 2012).

on the main macroalgae taxonomical group (red, brown or green species) considered, the use of the descriptor 'number', as the 'number' of red species, may correlate higher with environmental change than the use of the descriptor 'proportion', as the 'proportion' of red species. Considering the green species alone, they may depend less on bio-geographic factors, water quality or intertidal structure than the red and brown species: the proportion of green species decreases in parallel to the improvement of environmental conditions, where late-successional brown and red species thrive in large number; this can be very useful when results from different areas in the same eco-region are compared (Bermejo et al. 2012).

The number and the proportion of green species are likely to remain constant under stable environmental conditions; although the green species include many opportunistic species, they are also naturally present along the rocky shores. Under good environmental conditions, their presence is not normally dominant, and their biomass can be relatively low. These aspects change when the environmental conditions degrade. The opportunistic green species are able to respond more readily to changes, and dominate in coverage under worst environmental conditions and therefore may ultimately dictate the disappearance of other species (decreasing species richness and increasing the proportion of green species) (Gaspar et al. 2012).

The brown species are more likely to remain constant within certain ranges of environmental quality change (Wells et al. 2007). In the northern hemisphere, they show changes with latitude, decreasing from the north to the south in the presence of warmer waters but they decline under deviations from excellent quality standards (Gaspar et al. 2012).

The red species also show changes with latitude, increasing from northern cooler to southern warmer waters and in systems subject to less anthropogenic pressure, where they naturally dominate in numbers over the brown and the green species. Because of this, with increasing disturbances, the number of red species declines, although their proportions may not decline as clearly as in numbers, but ultimately that affects the diversity and the species richness (Gaspar et al. 2012).

4.4 The Use of Biodiversity Surrogates and Functional Traits

The use of biological surrogates as proxies for biodiversity patterns is gaining popularity, namely in marine systems where field surveys can be expensive and species richness high (Mellin et al. 2011). According to Smale (2010), a range of biodiversity (in the sense of species richness or diversity) surrogates have been proposed, which are usually derived by (1) selecting a subset of species (or higher taxon) from the whole assemblage that is thought to represent the richness of the whole assemblage (or another taxon); (2) analysing the entire assemblage to a coarser taxonomic resolution than species, thereby reducing the number of variables and level of expertise required; (3) employing the diversity of morphological or functional groups as proxies for species richness or evolutionary diversity; and (4) using environmental or habitat derived variable(s) to predict biological diversity.

Species identity (in the sense of species-level identifications) can be important for some studies (such as ones concerning biodiversity, conservation or introduced species) and then identification to the species-level will be needed. However, for communitywide impacts, where species identity may not be important (and actually may introduce more variation, making it difficult to find general similarities within regions), higher identification levels may be sufficient to reliably discern the similarities and differences among assemblages. Identification to a higher *taxa* level is cheaper, more efficient, requires less taxonomic expertise, and removes possible errors in identification (Konar and Iken 2009).

In order to adopt biodiversity surrogate measures, it must be assumed that the relationship between the surrogate(s) and the total richness or diversity of the assemblages is consistent in both space and time (Colwell and Coddington 1994). The use of taxonomical and functional surrogates for macroalgae species-level identifications has been studied (e.g., Goldberg et al. 2006, Konar and Iken 2009, Smale 2010, Rubal et al. 2011, Balata et al. 2011, Veiga et al. 2013) with the view to test the reliability of those surrogates in discerning similarities and differences among macroalgae assemblages along different spatial and temporal scales. Within this context, the spatial extent and design of any biodiversity monitoring program should be considered when choosing cost-effective alternatives to species-level data collection as the spatial and temporal scales will influence the efficacy of the biodiversity surrogates (Smale 2010).

In a recent work, Balata et al. (2011) departed from features such as the thallus structure, the growth form, the branching pattern and the taxonomic affinities, to further subdivide the seven traditional morphological groups proposed by Steneck and Dethier (1994) (Crustose, Articulated calcareous, Leathery, Corticated terete, Corticated foliose and Filamentous macroalgae and also a Microalgae group) in thirty-five newly defined groups (e.g., Siphonus Chlorophyta with thin compact filaments, Filamentous uniseriate Chlorophyta, Kelp-like Phaeophyceae, Prostrate Phaeophyceae not strictly adherent to substrate, Blade-like Rhodophyta with one or few layers of cells, Filamentous uniseriate and pluriseriate Rhodophyta with extensive prostrate filaments, Flattened Rhodophyta with cortication, Filamentous uniseriate and pluriseriate Rhodophyta with erect thallus, Smaller-sized articulated Rhodophyta, etc.). The newly defined groups are expected to show more uniform responses to environmental alterations than the traditional morphological groups, to be easily recognizable, and to represent a compromise between the respective advantages of species and traditional morphological groups (Balata et al. 2011).

Under different levels of environmental disturbance, macroalgae species display characteristic growth patterns, suggesting a link between species morphology and ecological function (e.g., Norton et al. 1982, Steneck and Watling 1982, Littler and Littler 1984, Dethier 1994, Steneck and Dethier 1994). For example, anthropogenic disturbances all over the world have been linked with the decline of large perennial species such the canopy-forming species of the Genus *Cystoseira*, considered to be sensitive to pollution; the calcareous red algae are considered to be tolerant to pollution; the simple forms such as filamentous and sheet-like algae proliferate in degraded environments (Díez et al. 2012 and references therein). The declining of kelps, fucoids and other complex canopy-forming species affect their role in coastal primary production, nutrient cycling, animal habitat creation or disturbance regulation (e.g., Steneck et al. 2002, Bertocci et al. 2010, Tait and Schiel 2011, Cheminée et al. 2013). One of the symptoms of the eutrophication in coastal waters and estuaries is the proliferation of fast-growing opportunistic macroalgae, resulting in blooms that change the community structure and function (e.g., Nelson et al. 2008, Teichberg et al. 2010).

If external morphology integrates and reflects the macroalgae function properties (e.g., primary productivity and growth rate, competitive ability, resistance to herbivores, resistance to physical disturbance, tolerance to physiological stress, successional stage, etc.), all of which should be interrelated with each other (Padilla 1985), then, groups of taxonomically distant species having similar morphological and functional characteristics, may be distinguished. However, the relationship between species and functional diversity remains poorly understood; but to comprehend this relationship is critically important, both for the mechanistic understanding of the community assemblages and for the appropriate expectations and approaches to protect and restore biological communities (Micheli et al. 2005).

An important functional-form model hypothesis became paradigmatic since Littler and Littler (1980) and Littler et al. (1983), having macroalgae species assigned into seven functional-form groups, which respond differently to photosynthesis and productivity (Littler 1980). Nevertheless, other explanations of the functional-form model in relation to other environmental factors, such as herbivory, succession stages of the community and desiccation stress, do not occur except for some species-specific interactions, or are explained by other factors regardless of the macroalgae morphology (Santelices 2009 and references therein).

Certainly, grouping *taxa* by a particular function can be very useful and often necessary for many ecosystem-level questions and modeling, and when there are too many species in a system to consider them all individually; however, to make functional group models more useful, groupings should be based on specific functions (e.g., nutrient uptake rates, photosynthesis rates, herbivore resistance, disturbance resistance, etc.) rather than gross morphology (Padilla and Allen 2000). In fact, two species belonging to exactly the same functional-form group in the sense of Littler et al. (1983) or situated in the same taxonomical group (Genus, Family) can display a completely different response to pollution. For example, *Cystoseira mediterranea* and *Corallina elongata* are perennial species but respond to disturbance in a completely different manner or *C. elongata* and *Jania rubens*, both in the Jointed Calcareous group and members of the Family Corallinaceae, exhibit different patterns of distribution along pollution gradients (Arévalo et al. 2007). As a matter of fact, to take into account species identity (species level identification) might be very important in ecological studies, just like distinguishing among different species of *Cystoseira*, not only because they can respond differently to human impacts, but also because they seem to have different 'engineering' effects on under-storey assemblages (Mangialajo et al. 2008).

Critical aspects to develop predictive measures of functional diversity should include (1) the choice of the functional traits with which organisms are distinguished; (2) how the diversity of that trait information is summarized into a measure of functional diversity; and (3) the measures of functional diversity should be validated through quantitative analyses and experimental tests (Petchey and Gaston 2006). Last but not least, the level of species diversity necessary for the functional redundancy, i.e., the capacity of one species to functionally compensate for the loss of another (and therefore preventing losses in ecosystem functioning if diversity declines due to disturbance), remains a critical question (Bernhardt and Leslie 2013).

In a recent approach, Orfanidis et al. (2011) assigned different macroalgae species into five different categories or Ecological Status Groups (ESGs) (Table 2) by using different traits—morphological (external morphology, internal anatomy, texture), physiological (surface area/volume ratio, photosynthetic/non-photosynthetic ratio, photosynthetic performance, growth, light adaptation) and life history (longevity, succession), all of which are important to nutrient and light responses—that were selected and respond accordingly along distributional data (including from reference sites) and across eutrophication gradients.

5 Conclusion

Marine ecosystems are being affected by multiple human impacts, facing ecological degradation and the concomitant decline of natural or pristine environments. In this connection, the concept of ecological conditions that can be linked to a historic past without significant human pressures—and implied in the idea of a 'natural' or 'pristine' ecosystem—give rise to the idea of ecological reference conditions, against which current degraded ecological conditions can be compared.

Reference conditions can be linked to ecological conditions that remain pristine or unchanged, but also as defined environmental goals that can be considered (somehow) reasonable to be achieved. It depends both on the level of ecological conditions that are aimed by human society and on the availability of historical data and knowledge of the system. Actually, 'pristine state' or 'natural state' can be best regarded as having dynamic attributes, which may need to be periodically re-defined in response to new and better scientific knowledge (van Hoey et al. 2010 and references therein).

Regardless of the degree of 'naturalness' or 'pristineness' implied by a specific definition of a benchmark, the accuracy and precision when assessing ecological quality conditions are dependent on the degree to which those benchmarks can be quantified and predicted (Hawkins et al. 2010). Macroalgae communities can reflect the ecological degradation and their study can be useful to quantify the level of integrity of marine ecosystems. Several lines of investigation can be taken into account, and combined in an integrative way, in order to search for marine macroalgae definitions

Table 2. Key functional traits used to assign macroalgae into different five Ecological Status Groups (ESGs). Some *taxa* examples are given at the table end (Adapted from Orfanidis et al. 2011).

Functional traits	ESG IA	ESG IB	ESG IC	ESG IIA	ESG IIB
Thallus morphology	thick	thick	calcareous upright and calcareous and non-calcareous crusts	fleshy	filamentous and leaf-like
Growth	slow	slow	slow	fast	fast
Light adaptation	sun-adapted	sun-adapted	shade-adapted	sun-adapted	sun-adapted
Phenotypic plasticity	no	yes	yes	yes	yes
Thallus longevity	perennial	perennial thallus basis or stipe	perennial thallus basis	annual	annual
Succession	late-successional	late-successional	late-successional	opportunistic	opportunistic
Taxa examples	*Chondrus, Cystoseira*	*Aglaozonia, Asperococcus, Cystoseira (C. barbata and C. compressa), Culteria, Erythroglossum, Halopitys, Plocamium, Rhodophyllis, Sargassum, Taonia*	*Amphiroa, Corallina, Dermatolithon, Halimeda, Hydrolithon, Jania, Lithophyllum, Melobesia, Mesophyllum, Peyssonnelia, Ralfsia, Titanoderma*	*Acrosorium, Ahnfeltiopsis, Asparagopsis, Boergeseniella, Caulacanthus, Champia, Chondracanthus, Chondria, Cladostephus, Colpomenia, Dictyopteris, Dictyota, Gastroclonium, Gelidium, Gigartina, Gracilaria, Grateloupia, Halopteris, Hypnea, Hypoglossum, Kallymenia, Laurencia, Lomentaria, Nitophyllum, Osmundea, Phyllophora, Pterocladiella, Schizymenia, Stypocaulon*	*Aglaothamnion, Anotrichium, Antithamnion, Blidingia, Bryopsis, Callithamnion, Ceramium, Chaetomorpha, Cladophora, Codium, Dasya, Derbesia, Ectocarpus, Griffithsia, Halurus, Herposiphonia, Lophosiphonia, Monostroma, Petalonia, Pleonosporium, Polysiphonia, Porphyra, Pterosiphonia, Rhizoclonium, Rhodothamnionella, Scytosiphon, Stylonema, Ulotrix, Ulva, Valonia*

under reference conditions. To search for it can be highly important to indicate directions for the conservation, restoration and management of the marine ecosystems.

Acknowledgments

This work was done under the financial support of a Ph.D. scholarship attributed to Rui Gaspar by the Portuguese Foundation of Science and Technology—FCT, and supported by João M. Neto through the programs POPH and QREN (FSE and national funds of MEC).

Keywords: Marine macroalgae, Reference conditions, Ecological quality, Coastal waters, Intertidal rocky shores, Environmental targets, Undisturbed conditions

References Cited

Airoldi, L. 2003. The effects of sedimentation on rocky coast assemblages. Oceanography and Marine Biology—An Annual Review 41: 161–236.

Airoldi, L. and M.W. Beck. 2007. Loss, status and trends for coastal marine habitats of Europe. Oceanography and Marine Biology—An Annual Review 45: 345–405.

Araújo, R., I. Bárbara, M. Tibaldo, E. Berecibar, P.D. Tapia, R. Pereira, R. Santos and I.S. Pinto. 2009. Checklist of benthic marine algae and cyanobacteria of northern Portugal. Botanica Marina 52: 24–46.

Arévalo, R., S. Pinedo and E. Ballesteros. 2007. Changes in the composition and structure of Mediterranean rocky-shore communities following a gradient of nutrient enrichment: descriptive study and test of proposed methods to assess water quality regarding macroalgae. Marine Pollution Bulletin 55: 104–113.

Aubry, A. and M. Elliott. 2006. The use of environmental integrative indicators to assess seabed disturbance in estuaries and coasts: Application to the Humber Estuary, UK. Marine Pollution Bulletin 53: 175–185.

Balata, D., L. Piazzi and F. Rindi. 2011. Testing a new classification of morphological functional groups of marine macroalgae for the detection of responses to stress. Marine Biology 158: 2459–2469.

Ballesteros, E., X. Torras, S. Pinedo, M. García, L. Mangialajo and M. De Torres. 2007. A new methodology based on littoral community cartography dominated by macroalgae for the implementation of the European Water Framework Directive. Marine Pollution Bulletin 55: 172–180.

Beaumont, N.J., M.C. Austen, J.P. Atkins, D. Burdon, S. Degraer, T.P. Dentinho, S. Derous, P. Holm, T. Horton, E. van Ierland, A.H. Marboe, D.J. Starkey, M. Townsend and T. Zarzycki. 2007. Identification, definition and quantification of goods and services provided by marine biodiversity: Implications for the ecosystem approach. Marine Pollution Bulletin 54: 253–265.

Benedetti-Cecchi, L., F. Pannacciulli, F. Bulleri, P.S. Morchella, L. Airoldi, G. Relini and F. Cinelli. 2001. Predicting the consequences of anthropogenic disturbance: large-scale effects of loss of canopy algae on rocky shores. Marine Ecology Progress Series 214: 137–150.

Bermejo, R., J.J. Vergara and I. Hernandez. 2012. Application and reassessment of the reduced species list index for macroalgae to assess the ecological status under the Water Framework Directive in the Atlantic coast of Southern Spain. Ecological Indicators 12: 46–57.

Bernhardt, J.R. and H.M. Leslie. 2013. Resilience to climate change in coastal marine ecosystems. Annual Review of Marine Science 5: 8.1–8.22.

Bertocci, I., F. Arenas, M. Matias, S. Vaselli, R. Araújo, H. Abreu, R. Pereira, R. Vieira and I. Sousa-Pinto. 2010. Canopy-forming species mediate the effects of disturbance on macroalgal assemblages on Portuguese rocky shores. Marine Ecology Progress Series 414: 107–116.

Boaventura, D.P., L.C. da Fonseca and S.J. Hawkins. 2002. Intertidal rocky shore communities of the continental Portuguese coast: analysis of distribution patterns. Marine Ecology 23: 69–90.

Borja, A., J. Franco, V. Valencia, J. Bald, I. Muxika, M.J. Belzunce and O. Solaun. 2004. Implementation of the European Water Framework Directive from the Basque Country (northern Spain): a methodological approach. Marine Pollution Bulletin 48: 209–218.

Borja, A., D.M. Dauerb and A. Grémare. 2012. The importance of setting targets and reference conditions in assessing marine ecosystem quality. Ecological Indicators 12: 1–7.

Borja, A., M. Elliott, P. Henriksen and N. Marbà. 2013. Transitional and coastal waters ecological status assessment: advances and challenges resulting from implementing the European Water Framework Directive. Hydrobiologia 704: 213–229.

Breeman, A.M. 1988. Relative importance of temperature and other factors in determining geographic boundaries of seaweeds: experimental and phenological evidence. Helgoland Marine Research 42: 199–241.

Cervin, G., P. Aberg and S.R. Jenkins. 2005. Small-scale disturbance in a stable canopy dominated community: implications for macroalgal recruitment and growth. Marine Ecology Progress Series 305: 31–40.

Cheminée, A., E. Sala, J. Pastor, P. Bodilis, P. Thiriet, L. Mangialajo, J.-M. Cottalorda and P. Francour. 2013. Nursery value of Cystoseira forests for Mediterranean rocky reef fishes. Journal of Experimental Marine Biology and Ecology 442: 70–79.

Cheney, D.F. 1977. R and C/P, a new and improved ratio for comparing seaweed floras. Journal of Phycology (Supplement): 12.

Colwell, R.K. and J.A. Coddington. 1994. Estimating terrestrial biodiversity through extrapolation. Philosophical Transactions of the Royal Society of London B 345: 101–118.

Crain, C.M., B.S. Halpern, M.W. Beck and C.V. Kappel. 2009. Understanding and managing human threats to the coastal marine environment. Annals of the New York Academy of Sciences 1162: 39–62.

Dawes, C.J. 1998. Marine Botany, 2nd Ed. John Wiley & Sons, New York.

de Jong, F. 2006. Marine Eutrophication in Perspective—On the Relevance of Ecology and Environmental Policy. Springer, Berlin.

Dethier, M.N. 1994. The ecology of intertidal algal crusts—variation within a functional-group. Journal of Experimental Marine Biology and Ecology 177: 37–71.

Diez, I., A. Secilla, A. Santolaria and J.M. Gorostiaga. 1999. Phytobenthic intertidal community structure along an environmental pollution gradient. Marine Pollution Bulletin 38: 463–472.

Díez, I., M. Bustamante, A. Santolaria, J. Tajadura, N. Muguerza, A. Borja, I. Muxika, J.I. Saiz-Salinas and J.M. Gorostiaga. 2012. Development of a tool for assessing the ecological quality status of intertidal coastal rocky assemblages, within Atlantic Iberian coasts. Ecological Indicators 12: 58–71.

Duarte, C.M. 1995. Submerged aquatic vegetation in relation to different nutrient regimes. Ophelia 41: 87–112.

Economou, A.N. 2002. Development, evaluation and implementation of a standardised fish-based assessment method for the ecological status of European rivers: a contribution to the Water Framework Directive (FAME), Defining Reference Conditions (D3), Final Report. National Centre for Marine Research, EL. fame.boku.ac.at. (http://fame.boku.ac.at/downloads/D3_reference_conditions.pdf).

Eggert, A. 2012. Seaweed responses to temperature. pp. 47–66. *In*: Wiencke, C. and K. Bischof (eds.). Seaweed Biology: Novel Insights into Ecophysiology, Ecology and Utilization. Ecological Studies. Springer-Verlag, Berlin, Heidelberg 219.

Elliott, M. 2011. Marine science and management means tackling exogenic unmanaged pressures and endogenic managed pressures—A numbered guide. Marine Pollution Bulletin 62: 651–655.

Feldmann, J. 1937. Recherches sur la végétation marine de la Méditerranée. La côte des Albères. Revue Algologique 10: 1–339.

Fernández, C. 2011. The retreat of large brown seaweeds on the north coast of Spain: the case of Saccorhiza polyschides. European Journal of Phycology 46(4): 352–360.

Foster, M.S., E.W. Nigg, L.M. Kiguchi, D.D. Hardin and J.S. Pearse. 2003. Temporal variation and succession in an algal-dominated high intertidal assemblage. Journal of Experimental Marine Biology and Ecology 289: 15–39.

Gaspar, R., L. Pereira and J.M. Neto. 2012. Ecological reference conditions and quality states of marine macroalgae sensu Water Framework Directive: An example from the intertidal rocky shores of the Portuguese coastal waters. Ecological Indicators 19: 24–38.

Goldberg, N.A., G.A. Kendrick and D.I. Walker. 2006. Do surrogates describe patterns in marine macroalgal diversity in the Recherche Archipelago, temperate Australia? Aquatic Conservation: Marine and Freshwater Ecosystems 16: 313–327.

Gorman, D. and S.D. Connell. 2009. Recovering subtidal forests in human-dominated landscapes. Journal of Applied Ecology 46: 1258–1265.

Gorostiaga, J.M. and I. Diez. 1996. Changes in the sublittoral benthic marine macroalgae in the polluted area of Abra de Bilbao and proximal coast (Northern Spain). Marine Ecology Progress Series 130: 157–167.

Guinda, X., J.A. Juanes, A. Puente and J.A. Revilla. 2008. Comparison of two methods for quality assessment of macroalgae assemblages, under different pollution types. Ecological Indicators 8: 743–753.

Harley, C.D.G., K.M. Anderson, K.W. Demes, J.P. Jorve, R.L. Kordas and T.A. Coyle. 2012. Effects of climate change on global seaweed communities. Journal of Phycology 48: 1064–1078.

Hawkins, C.P., J.R. Olson and R.A. Hill. 2010. The reference condition: predicting benchmarks for ecological and water-quality assessments. Journal of the North American Benthological Society 29: 312–343.

Hawkins, S.J., P.J. Moore, M.T. Burrows, E. Poloczanska, N. Mieszkowska, R.J.H. Herbert, S.R. Jenkins, R.C. Thompson, M.J. Genner and A.J. Southward. 2008. Complex interactions in a rapidly changing world: responses of rocky shore communities to recent climate change. Climate Research 37: 123–133.

Juanes, J.A., X. Guinda, A. Puente and J.A. Revilla. 2008. Macroalgae, a suitable indicator of the ecological status of coastal rocky communities in the NE Atlantic. Ecological Indicators 8: 351–359.

Kelly, J.R. 2008. Nitrogen in the Environment: Chapter 10: 271–332. *In*: Hatfield, J.L. and R.F. Follett (eds.). Nitrogen in the Environment: Sources, Problems, and Management, Second edition. Academic Press/Elsevier, Amsterdam, Boston.

Konar, B. and K. Iken. 2009. Influence of taxonomic resolution and morphological functional groups in multivariate analyses of macroalgal assemblages. Phycologia 48(1): 24–31.

Krause-Jensen, D., S. Sagert, H. Schubert and C. Bostro. 2008. Empirical relationships linking distribution and abundance of marine vegetation to eutrophication. Ecological Indicators 8(5): 515–529.

Lima, F.P., P.A. Ribeiro, N. Queiroz, S.J. Hawkins and A.M. Santos. 2007. Do distributional shifts of northern and southern species of algae match the warming pattern? Global Change Biology 13: 2592–2604.

Littler, M.M. 1980. Morphological form and photosynthetic performances of marine macroalgae: tests of a functional/form hypothesis. Botanica Marina XXII: 161–165.

Littler, M.M. and D.S. Littler. 1980. The evolution of thallus form and survival strategies in benthic marine macroalgae: field and laboratory tests of a functional form model. The American Naturalist 116: 25–44.

Littler, M.M. and D.S. Littler. 1984. Relationship between macroalgal functional form groups and substrate stability in a subtropical rocky intertidal system. Journal of Experimental Marine Biology and Ecology 74: 13–34.

Littler, M.M., D.S. Littler and P.R. Taylor. 1983. Evolutionary strategies in a tropical barrier reef system: functional-form groups of macroalgae. Journal of Phycology 19: 223–231.

Lobban, C.S. and P.J. Harrison. 1994. Seaweed Ecology and Physiology. Cambridge UniversityPress, New York, USA.

Lüning, K., C. Yarish and H. Kirkman. 1990. Seaweeds: Their Environment, Biogeography, and Ecophysiology. John-Wiley, New York.

Mangialajo, L., M. Chiantore and R. Cattaneo-Vietti. 2008. Loss of fucoid algae along a gradient of urbanisation, and structure of benthic assemblages. Marine Ecology Progress Series 358: 63–74.

Mellin, C., S. Delean, J. Caley, G. Edgar, M. Meekan, R. Pitcher, R. Przeslawski, A. Williams and C. Bradshaw. 2011. Effectiveness of biological surrogates for predicting patterns of marine biodiversity: a global meta-analysis. PLoS ONE 6(6): e20141.

Melville, F. and A. Pulkownik. 2006. Investigation of mangrove macroalgae as bioindicators of estuarine contamination. Marine Pollution Bulletin 52: 1260–1269.

Micheli, F. and B.S. Halpern. 2005. Low functional redundancy in coastal marine assemblages. Ecology Letters 8: 391–400.

Morand, P. and M. Merceron. 2005. Macroalgal population and sustainability. Journal of Coastal Research 21(5): 1009–1020.

Murray, S.N. and M.M. Littler. 1978. Patterns of algal succession in a perturbated marine intertidal community. Journal of Phycology 14: 506–512.

Nelson, T.A., K. Haberlin, A.V. Nelson, H. Ribarich, R. Hotchkiss, K.L. Van Alstyne, L. Buckingham, D.J. Simunds and K. Fredrickson. 2008. Ecological and physiological controls of species composition in green macroalgal blooms. Ecology 89(5): 1287–1298.

Neto, J.M., R. Gaspar, L. Pereira and J.C. Marques. 2012. Marine Macroalgae Assessment Tool (MarMAT) for intertidal rocky shores. Quality assessment under the scope of the European Water Framework Directive. Ecological Indicators 19: 39–47.

Norton, T.A., A.C. Mathieson and M. Neushul. 1982. A review of some aspects of form and function in seaweeds. Botanica Marina 25: 501–510.

Orfanidis, S., P. Panayotidis and K.I. Ugland. 2011. Ecological Evaluation Index continuous formula (EEI-c) application: a step forward for functional groups, the formula and reference condition values. Mediterranean Marine Science 12(1): 199–231.

Padilla, D.K. 1985. Structural resistance of algae to herbivores. A biomechanical approach. Marine Biology 90: 103–109.

Padilla, D.K. and B.J. Allen. 2000. Paradigm lost: reconsidering functional form and group hypotheses in marine ecology. Journal of Experimental Marine Biology and Ecology 250: 207–221.

Panayotidis, P., B. Montesanto and S. Orfanidis. 2004. Use of low-budget monitoring of macroalgae to implement the European Water Framework Directive. Journal of Applied Phycology 16: 49–59.

Perkol-Finkel, S. and L. Airoldi. 2010. Loss and recovery potential of marine habitats: an experimental study of factors maintaining resilience in subtidal algal forests at the adriatic sea. PLoS ONE 5(5): e10791.

Petchey, O.L. and K.J. Gaston. 2006. Functional diversity: back to basics and looking forward. Ecology Letters 9: 741–758.

Phillips, J.C., G.A. Kendrick and P.S. Lavery. 1997. A test of a functional group approach to detecting shift of macroalgal communities along a disturbance gradient. Marine Ecology Progress Series 153: 125–138.

Raffaelli, D. and S. Hawkins. 1999. Intertidal Ecology, 2nd Ed. Kluwer Academic Publishers, Dordrecht.

Ramos, E., J.A. Juanes, C. Galván, J.M. Neto, R. Melo, A. Pedersen, C. Scanlan, R. Wilkes, E. van den Bergh, M. Blomqvist, H.P. Karup, W. Heiber, J.M. Reitsma, M.C. Ximenes, A. Silió, F. Méndez and B. González. 2012. Coastal waters classification based on physical attributes along the NE Atlantic region. An approach for rocky macroalgae potential distribution. Estuarine, Coastal and Shelf Science 112: 105–114.

Roberts, D.E., A. Smith, P. Ajani and A.R. Davis. 1998. Rapid changes in encrusting marine assemblages exposed to anthropogenic point-source pollution: a beyond BACI approach. Marine Ecology Progress Series 163: 213–224.

Rodriguez-Prieto, C. and L. Polo. 1996. Effects of sewage pollution in the structure and dynamics of the community of Cystoseira mediterranea (Fucales, Phaeophyceae). Scientia Marina 60: 253–263.

Rubal, M., P. Veiga, R. Vieira and I. Sousa-Pinto. 2011. Seasonal patterns of tidepool macroalgal assemblages in the North of Portugal. Consistence between species and functional group approaches. Journal of Sea Research 66: 187–194.

Sales, M. and E. Ballesteros. 2009. Shallow Cystoseira (Fucales: Ochrophyta) assemblages thriving in sheltered areas from Menorca (NW Mediterranean): Relationships with environmental factors and anthropogenic pressures. Estuarine, Coastal and Shelf Science 84: 476–482.

Santelices, B., J.J. Bolton and I. Meneses. 2009. 6. Marine Algal Communities. pp. 153–192. *In*: Whitman, J.D. and K. Roy (eds.). Marine Macroecology. University Press, Chicago.

Scanlan, C.M., J. Foden, E. Wells and M.A. Best. 2007. The monitoring of opportunistic macroalgal blooms for the Water Framework Directive. Marine Pollution Bulletin 55: 162–171.

Schiel, D.R. 2004. The structure and replenishment of rocky shore intertidal communities and biogeographic comparisons. Journal of Experimental Marine Biology and Ecology 300: 309–342.

Schramm, W. 1996. The Baltic Sea and its transition zones. pp. 131–163. *In*: Schramm, W. and P.H. Nienhuis (eds.). Marine Benthic Vegetation: Recent Changes and the Effects of Eutrophication. Springer, Heidelberg.

Smale, D.A. 2010. Monitoring marine macroalgae: the influence of spatial scale on the usefulness of biodiversity surrogates. Diversity and Distributions 16: 985–995.

Soberón, J. and A.T. Peterson. 2005. Interpretation of models of fundamental ecological niches and species' distributional areas. Biodiversity Informatics 2: 1–10.

Soltan, D., M. Verlaque, C.F. Boudouresque and P. Francour. 2001. Changes in macroalgal communities in the vicinity of a Mediterranean sewage outfall after the setting up of a treatment plant. Marine Pollution Bulletin 42: 59–70.

Steneck, R.L. and L. Watling. 1982. Feeding capabilities and limitation of herbivorous mollusks: a functional group approach. Marine Biology 68: 299–319.

Steneck, R.L. and M.N. Dethier. 1994. A functional group approach to the structure of algal-dominated communities. Oikos 69: 476–498.

Steneck, R.S., M.H. Graham, B.J. Bourque, D. Corbett and J.M. Erlandson. 2002. Kelp forest ecosystems: biodiversity, stability, resilience and future. Environment Conservation 29: 436–459.

Tait, L.W. and D.R. Schiel. 2011. Legacy effects of canopy disturbance on ecosystem functioning in macroalgal assemblages. PLoS ONE 6(10): e26986.

Teichberg, M., S.E. Fox, Y.S. Olsen, I. Valiela, P. Martinetto, O. Iribarne, E.Y. Muto, M.A.V. Petti, T.N. Corbisier, M. Soto-Jiménez, F. Páez-Osuna, P. Castro, H. Freitas, A. Zitelli, M. Cardinaletti and D. Tagliapietra. 2010. Eutrophication and macroalgal blooms in temperate and tropical coastal waters: nutrient enrichment experiments with *Ulva* spp. Global Change Biology 16: 2624–2637.

Tewari, A. and H.V. Joshi. 1988. Effect of domestic sewage and industrial effluents on biomass and species-diversity of seaweeds. Botanica Marina 31: 389–397.

Thibaut, T., S. Pinedo, X. Torras and E. Ballesteros. 2005. Long-term decline of the populations of Fucales (*Cystoseira* spp. and *Sargassum* spp.) in the Alberes coast (France, North-western Mediterranean). Marine Pollution Bulletin 50: 1472–1489.

UNEP/GPA 2006. The State of the Marine Environment: Trends and Processes. United Nations Environment Programme, Global Programme of Action, The Hague.

van Hoey, G.V., A. Borja, S. Birchenough, L. Buhl-Mortensen, S. Degraer, D. Fleischer, F. Kerckhof, P. Magni, I. Muxika, H. Reiss, A. Schröder and M.L. Zettler. 2010. The use of benthic indicators in Europe: From the Water Framework Directive to the Marine Strategy Framework Directive. Marine Pollution Bulletin 60: 2187–2196.

Veiga, P., M. Rubal, R. Vieira, F. Arenas and I. Sousa-Pinto. 2013. Spatial variability in intertidal macroalgal assemblages on the North Portuguese coast: consistence between species and functional group approaches. Helgoland Marine Research 67(1): 191–201.

Viejo, R., F. Arenas, C. Fernández and M. Gómez. 2008. Mechanisms of succession along the emersion gradient in intertidal rocky shore assemblages. Oikos 117: 376–389.

Vitousek, P.M., J.D. Aber, R.W. Howarth, G.E. Likens, P.A. Matson, D.W. Schindler, W.H. Schlesinger and D.G. Tilman. 1997. Human alteration of the global nitrogen cycle: sources and consequences. Ecological Applications 7: 737–750.

Wallentinus, I. 1984. Comparisons of nutrient uptake rates for Baltic macroalgae with different thallus morphologies. Marine Biology 80: 215–225.

Wells, E., M. Wilkinson, P. Wood and C. Scanlan. 2007. The use of macroalgal species richness and composition in intertidal rocky seashores in the assessment of ecological quality under the European Water Framework Directive. Marine Pollution Bulletin 55: 151–161.

WFD CIS. 2003. Guidance Document No. 5: 'Transitional and coastal waters—Typology, reference conditions, and classification systems'. Common Implementation Strategy of the Water Framework Directive (http://www.waterframeworkdirective.wdd.moa.gov.cy/docs/GuidanceDocuments/Guidancedoc5COAST.pdf).

WFD CIS. 2009. Guidance Document Number 14. Guidance on the Inter-calibration Process 2008–2011. Implementation Strategy for the Water Framework Directive (2000/60/EC) (http://ec.europa.eu).

Wilkinson, M. and P. Wood. 2003. Type-specific reference conditions for macroalgae and angiosperms in Scottish transitional and coastal waters: Report to Scottish Environment Protection Agency. SEPA Project Reference 230/4136. Heriot-Watt University, Edinburgh, 105 pp.

Yuksek, A., E. Okus, I.N. Yilmaz, A. Aslan-Yilmaz and S. Tas. 2006. Changes in biodiversity of the extremely polluted Golden Horn Estuary following the improvements in water quality. Marine Pollution Bulletin 52: 1209–1218.

CHAPTER 3

Marine Macroalgae and the Assessment of Ecological Conditions

João M. Neto,[1,] José A. Juanes,[2] Are Pedersen[3]
and Clare Scanlan[4]*

1 General Characteristics of Macroalgae

Marine macroalgae belong to a polyphyletic group of living organisms divided into three main groups, the greens (Phylum: Chlorophyta), browns (Phylum: Ochrophyta) and the reds (Phylum: Rhodophyta) (see Chapter 1 for more information). While these are generally multi-cellular, some have microscopic phases, as part of an alternation of generations. These important marine primary producers, able to colonize a large latitudinal spectrum on worldwide seas, are morphologically highly varied. Some are large and robust, constituted mostly by a distinct stipe and lamina (or blade); others range from branched or unbranched monofilaments to dense parenchymatous thalli and calcareous and non-calcareous encrusting algae. All need to be submerged (at least temporarily) in saline water, to have a substratum to attach to and to have some nutrient supply to feed on. While most attach to hard substrata, some are epiphytic, and others may

---

[1] IMAR—CMA (Institute of Marine Research), University of Coimbra, Largo Marques de Pombal, 3004-517 Coimbra, Portugal.
  Email: jneto@ci.uc.pt
[2] Instituto de Hidráulica Ambiental (IH Cantabria), Universidad de Cantabria, c/Isabel Torres n° 15, Parque Científico y Tecnológico de Cantabria, 39011 Santander, Spain.
  Email: juanesj@unican.es
[3] NIVA—Norwegian Institute for Water Research, Gaustadalléen 21, 0349 OSLO, Norway.
  Email: are@niva.no
[4] SEPA (Scottish Environment Protection Agency), Inverdee House, Baxter Street, Aberdeen AB11 9QA, UK.
  Email: clare.scanlan@sepa.org.uk
* Corresponding author

occasionally grow unattached under specific conditions. Owing to their sessile nature, marine macroalgae are organisms influenced by the specific environmental conditions of the water which bathes them; water quality and local physicochemical conditions will determine species occurrence and community composition. Macroalgae are at the base of productive food webs (Harley et al. 2012), including some with recognized economic value (Graham 2004, Norderhaug et al. 2005). Marine macroalgae are linked to human culture and economic systems, providing ecosystem goods and services; they can be directly used as food or medicine (see Chapters 7–10), and may confer physical protection to coastal areas (Rönnbäck et al. 2007). Large brown macroalgae such as *Laminaria hyperborea* may form dense kelp forests on subtidal reefs which absorb wave energy, and may therefore be locally important in coastal protection (e.g., Angus 2014, Angus and Rennie 2014).

The distribution and abundance of marine macroalgae are ruled at a global scale by their sensitivity and ability to withstand different natural environmental conditions (van den Hoek 1982, Gorostiaga and Diez 1996, Arévalo et al. 2007), and locally by the ability they have to exploit the most abundant resources in their surrounding environment. The different members of the algal community may react differently to external stimuli (such as disturbance, nutrient enrichment), and as sessile organisms, they integrate water quality effects over time. As a result they are seen as good indicators of water quality. For this main reason, the most recent European environmental Directives have included macroalgae as Biological Quality Elements (BQE) that should be considered in the quality assessment of aquatic ecosystems.

2 Environmental Factors Influencing Macroalgae

Macroalgae respond to various climatic and physicochemical factors. Survival, growth, and reproduction of macroalgae depend on and vary with numerous key environmental variables such as temperature, desiccation, salinity, hydrodynamics and wave exposure, nutrients, carbon dioxide and pH (reviewed in Harley et al. 2012). These factors are largely responsible for shaping latitudinal patterns of algal distribution (van den Hoek 1982, Bartsch et al. 2012 and references therein; Ramos et al. 2012, 2014). The interactions of these parameters influence both the presence and abundance of individual taxa. Light (photosynthetically active radiation or PAR) and temperature are the most important natural parameters ruling the development of macroalgal communities. Without light, photosynthesis is not possible and temperature, as for all other organisms, determines the performance of macroalgae at the fundamental levels of enzymatic processes and metabolic function (Bell 1993, Harley et al. 2012 and references therein). Although macroalgae are generally well adapted to their thermal environment, any deviation from the optimum temperature range (daily and seasonal), particularly in situations of environmental stress, contributes to variation (Chung et al. 2007, Müller et al. 2009, Bartsch et al. 2012). It may also have a significant effect on macroalgal survival, slowing growth, delaying development, and leading to an increase in mortality (Bell 1993, Harley et al. 2012 and references therein), which may culminate in species' loss (Bartsch et al. 2012). The same is true for salinity, where a deviation from the optimum range may lead to a differential development of some species relative to others (Martins et al. 1999, Chung et al. 2007). Carbon dioxide (CO_2) is another essential element for photosynthesis. In terrestrial ecosystems an excess of CO_2 may be considered as a type of fertilizer, stimulating photosynthesis, but in marine systems, where bicarbonate ions (HCO_3^-) rather than CO_2 are used by most macroalgae as a photosynthetic substrate, other problems may arise due to excessive concentrations in seawater (Gattuso and Buddemeier 2000). Although the increase of carbon dioxide concentration could drive changes in seawater carbonate chemistry and lead to a reduction in pH (Gattuso and Buddemeier 2000, Kroeker et al. 2010), the response of different marine organisms to this change will vary. They show a variety of essential functional mechanisms affected by carbonate chemistry (e.g., dissolution and calcification rates, growth rates, development and survival) that allow their responses to vary so broadly. This means it is difficult and challenging to predict how marine ecosystems will respond to ocean acidification (Kroeker et al. 2010). Since the responses of marine autotrophs are dependent on light intensity and nutrient concentrations, some negative impacts of ocean acidification could potentially be buffered by the increase of those components (Zondervan et al. 2001, 2002, Rost et al. 2002). This may not apply however to coralline macroalgae, such as maerl, which may be susceptible to ocean acidification. The persistence of some species under stressful abiotic conditions would depend on the balance of higher maintenance costs and on the ability they have to change energy allocation to reproduction and somatic growth (Kroeker et al. 2010).

Nutrients, although they are essential for macroalgal growth, are typically present at low concentrations in marine waters unaffected by anthropogenic inputs, which together with grazing pressure (e.g., herbivorous fish), keeps the density of macroalgae at balanced levels (Chung et al. 2007, Diaz-Pulido and McCook 2008).

Other aspects related to hydrodynamics, such as water flow, currents, waves and tides, may also influence species distributional patterns due to hydrodynamically driven processes such as recruitment (Johnson and Brawley 1999, Lotze et al. 2001) and detachment (Gaylord et al. 1994, Milligan and DeWreede 2000, Thomsen and Wernberg 2005). The combination of those factors in the air-water interface and various other gradients, such as light penetration below the surface and spray above the surface result in the typical shore zonation.

It is generally accepted that tides and the influences created by them largely shape the vertical distribution, or zonation, of the different marine macroalgae (e.g., Lewis 1964). Upper shore macroalgae experience the most extreme conditions of emersion and temperature variation, and this is generally the least species rich part of the shore. In the mid-littoral zone, the organisms are tolerant of, or depend on, periodic air exposure and seawater immersion; but are less able to withstand the extremes of hot and cold which may be experienced by top shore algae. Below the mid-littoral zone on the lower shore are other algae which require greater periods of immersion. In the more stable very low shore, or sublittoral fringe, the stable aquatic environment may support dense beds of kelps and greater numbers of algae (Ellis 2003). Above the top of the shore may be the supralittoral fringe, a splash zone with fewer sea-water tolerant organisms including lichens and some small algae or microalgae. To survive mid-littoral zone conditions, macroalgae must maintain photosynthetic activity when exposed to air or, at least, have a sufficient rate of recovery of photosynthesis upon re-immersion to compensate (Bell 1993). Although thallus heating and desiccation influence growth by affecting both emersed and immersed rates of photosynthesis, the rate of recovery is dependent more on the thallus temperature than on desiccation. The survival of a macroalga is threatened when extreme thallus temperatures are encountered during low tide (Bell 1993).

Additionally the survival, distribution and abundance of intertidal macroalgae depends on their ability to withstand the large hydrodynamic forces generated by breaking waves, an ability that may be a function of the combined effects of morphology, size of the plant (Carrington 1990) and the material properties that contribute to frond flexibility (Denny and Gaylord 2002 and references therein). Macroalgae exploit flexibility and reconfiguration, i.e., the alteration of shape, size and orientation as water velocity increases, to reduce the hydrodynamic forces imposed in the wave-swept rocky intertidal zone (Boller and Carrington 2007). In habitats with heavy wave action thalli of intertidal macroalgae are size-limited, and regardless of their morphology their maximum size may be limited by water velocities that occur on exposed coasts (Carrington 1990). In small individuals, flexibility allows the plant to reorient and reconfigure in the flow, thereby assuming a streamlined shape and reducing the applied hydrodynamic force. In large individuals, flexibility allows fronds to 'go with the flow', a strategy that can at times allow the plant to minimise hydrodynamic forces (Denny and Gaylord 2002 and references therein). The combination of these different structural properties of macroalgae with the hydrodynamics and other varied natural factors found on a shore is responsible for differences in macroalgal communities, giving to intertidal rocky shores the typical vertical distribution, or "belts", of macroalgae.

3 Macroalgal Blooms—Causes and Effects on Marine Ecosystems

There are various natural environmental factors which can influence marine macroalgae, as well as anthropogenic pressures. Human interventions directly or indirectly affecting environmental factors (e.g., light, temperature, nutrients, currents, exposure) may lead to detrimental impacts on macroalgal communities, or may at least cause modifications to structural and functional attributes of these marine assemblages. Hydromorphological modifications and nutrient enrichment rank high on the list of anthropogenic pressures, and worldwide, constitute some of the most important management concerns. While macroalgae are naturally present in shallow water marine and transitional soft sediment communities (Abbott and Hollenberg 1976), excess nutrients may lead to the explosive growth of opportunistic macroalgae, particularly in soft sediment areas; blooms also occur in estuaries, or transitional waters, as well as fully saline waters. Opportunist algae are able to store excess nutrients and mobilise them quickly (e.g., Wallentinus 1984); they may also show a greater reproductive capability than other algae (Hoffmann and Ugarte 1985). Excessive loading of nutrients into coastal and transitional waters (or the increase of residence time) and insufficient herbivorous pressure give space to opportunistic species to out-compete slower-growing species (Corredor et al. 1999, Nixon and Fulweiler 2009). Different taxa may show different responses to the same disturbance type (pressure), and perennial species, more sensitive to such kinds of disturbance, may succumb more or less rapidly to competition from more successful, opportunistic forms (Corredor et al. 1999, Diaz-Pulido and McCook 2008, Nixon and Fulweiler 2009).

One of the most serious threats worldwide to environmental health is nutrient enrichment (Bushaw-Newton and Sellner 1999, Gubelit 2012), which can result in a shift in community composition away from long-lived algal species towards short-lived taxa, with the proliferation of opportunistic macroalgae, and consequent impacts on underlying fauna (e.g., macroinvertebrate, fish) and sediments. Blooms and their causes have been addressed world-wide by many workers (e.g., Soulsby et al. 1982, McComb and Humphries 1992, Sfriso et al. 1992, den Hartog 1994, Reise and Siebert 1994, Fletcher 1996a,b, Raffaelli 1998), and their effects are well known. Longer-lived macroalgae are out competed and replaced by fast-growing annual seaweeds, blooms of opportunistic algae may occur on soft sediments, and free floating macroalgae may alternate with phytoplankton in community dominance in some situations (Twilley et al. 1985, Duarte 1995, Corredor et al. 1999, Nixon et al. 2001, Valiela

2006, Gubelit 2012). Early phases of eutrophication may increase overall productivity, with high biomass potentially providing refugia for small fish and invertebrates (Pihl et al. 1996, Raffaelli et al. 1989), but the process can lead to undesirable disturbances (e.g., Raffaelli et al. 1989). While specific blooming taxa may vary globally, they are often green algae such as *Ulva, Cladophora, Chaetomorpha* spp. (e.g., Fletcher 1996a,b, Raffaelli 1998), but red (e.g., *Ceramium, Porphyra* spp.) and brown (e.g., *Pylaiella, Ectocarpus* spp.) may also form blooms to nuisance levels (e.g., Vogt and Schramm 1991).

Macroalgal blooms may occur in sheltered coastal or estuarine (transitional) waters on soft sediments due to nutrient's supply increase (Ryther and Dunstan 1971, Kruk-Dowgiallo 1991, Wilkes 2005, Anderson et al. 1995, Bushaw-Newton and Sellner 1999, Gubelit 2012). Lotze et al. (1999) considered blooms to occur mainly in areas of restricted flushing. Blooms may also result from other environmental factors acting alone or in combination (e.g., temperature, light) (Valiela et al. 1997, Bushaw-Newton and Sellner 1999) that stimulates a metabolic response from a well adapted form, or from the physical concentration of a species in a certain area due to local patterns in water circulation (e.g., hydrodynamic modification) (Bushaw-Newton and Sellner 1999, Gubelit 2012), or even due to biomass increase as a consequence of reduction in grazing pressure (Bushaw-Newton and Sellner 1999, Diaz-Pulido and McCook 2008). Diaz-Pulido and McCook (2008) highlighted that the high abundance of *Sargassum* on inshore reefs has been shown to be due to the lower intensity of grazing by herbivorous fish in these areas, rather than the direct enhancement of algal growth by higher nutrients in coastal waters. However, several studies have shown that nutrient-enriched waters and temperature alone may not be enough for macroalgal blooms to occur, and these only explode when salinity is also at an appropriate level (Dawes et al. 1998, Israel et al. 1999, Kramer and Fong 2000, Eriksson and Bergstrom 2005, Larsen and Sand-Jensen 2006, Chung et al. 2007) or when hydrodynamics are favourable for macroalgal growth or accumulation (Martins et al. 2001).

Moreover, opportunistic macroalgae, such as several members of the Chlorophyta, may come to dominate the benthic communities and to exclude both other macrophytes and less mobile macrofaunal species (Diaz et al. 1990, Corredor et al. 1999, Marques et al. 2003, Lillebø et al. 2005, Patrício et al. 2009). The oxygen produced in daytime by primary producers' growing processes is consumed at night in respiration of microbial, primary producers and by faunal populations and it may lead to a general crash of the community when reduced dissolved oxygen concentrations are reached (Marques et al. 2003). Decomposition starts and the severe anoxia of the sediments creates a hostile exclusion-environment to natural communities. The effects on the survival time of benthic organisms (e.g., macrofauna, macrophytes) under hypoxia is aggravated by the production of hydrogen sulfide in the bottom water layer (Vaquer-Sunyer and Duarte 2010). In particular, as hypoxia progresses, benthic microbial communities shift to sulfate reduction, and sulfide concentrations increase in the environment (Conley et al. 2009). Sulfide is always present in the anoxic layer of marine sediments (Fenchel and Jorgensen 1977) and its diffusion into bottom water is controlled by the oxic sediment depth, which is dictated by the rate of oxygen diffusion and the oxygen consumption in the sediment (Vistisen and Vismann 1997). During hypoxic events, the anoxic layer of the sediments migrates upwards and the hydrogen sulfide produced is released into the bottom water (Vaquer-Sunyer and Duarte 2010). The result is the characteristic foul odour of rotten eggs and the poisoning of the living organisms that have been exposed to the gas (Gamenick et al. 1996).

Macroalgal blooms, triggered by human activities that lead to nitrogen (N) and phosphorus (P) increase, may dominate and cause disruption of the nitrogen and phosphorus cycles (Fong et al. 2004). Opportunistic macroalgae take up available inorganic nutrients from the water as well as intercepting benthic fluxes, and can sequester and store large pools of nutrients in their tissues during bloom development (Peckol and Rivers 1995, Fong and Zedler 2000). The nitrification/denitrification processes can become impaired and lose the major role these mechanisms play in reducing the effects of increased nitrogen loading from human activity (Seitzinger 1988, Nixon et al. 1996). Macroalgae can be a sink (Fong et al. 2004) but also a source (Tyler et al. 2001) of nutrients in the water column since they can alleviate N and P limitations but also load the system during tissue degradation (Corredor et al. 1999). Another important step in the N-cycle of estuaries and sheltered coastal areas that may be facilitated by macroalgae is the transformation of dissolved inorganic nitrogen (DIN) to dissolved organic nitrogen (DON) in the water column. DON is often an important component of the total dissolved N pool in coastal systems (e.g., Fong and Zedler 2000, Tyler et al. 2001). Although there are many negative impacts ascribed to them (e.g., Raffaelli et al. 1998, Valiela et al. 1997, Bolam et al. 2000), macroalgal blooms (e.g., *Ulva* sp.) play an important role in retention and cycling of N in eutrophic coastal areas. The uptake of DIN and its transformation into DON contributes to the processing of N, which increases the supply of DON to the system and may support a microheterotrophic food web (Peckol and Rivers 1995). One important mechanism in the N processing function is the ability of *Ulva* to allocate most of its energy toward nutrient uptake when pulses are large, delaying growth in favour of maximizing uptake and storage. These characteristics, coupled with the ability to accumulate in large masses in the field (Kamer et al. 2001) suggest that *Ulva* sp. plays a significant role in nutrient cycling in shallow coastal systems and estuaries (Fong et al. 2004).

Often macroalgal blooms are accompanied by severe impacts to coastal resources, local economies, and public health (Bushaw-Newton and Sellner 1999). Persistent macroalgal blooms may actually constitute the widespread chronic cause, unrelated to climate change, of seagrass beds and coral reef systems decline (Björk et al. 2008, Short et al. 2011, den Hartog 1994). The growth of epiphytes and the massive proliferation of macroalgae in the water column contributes to a reduction in the amount of light and nutrients accessible to seagrasses, ultimately leading to their decline (van den Hoek 1982, Burkholder et al. 2007, Ralph et al. 2006, Rasmussen et al. 2012). Macroalgal mats may also have influence on higher trophic level organisms in coastal systems. Intertidal flats covered by macroalgal mats were, at the large scale, mostly unattractive to feeding waders (e.g., mortality or abandonment of invertebrates of the algal covered area; interference with foraging behavior) (Tubbs and Tubbs 1980, Cabral et al. 1999), but, at a local scale, e.g., plover peck rates increased due to the development of new opportunistic communities (e.g., increase biomass of amphipods and *Carcinus maenas*) associated to high macroalgal biomass (Cabral et al. 1999). When benthic bio-stabilisers are affected, the sediment stability is also compromised. Sediment erodability depends on the interactions between physical processes, sediment properties and biological processes, which can influence the hydrodynamics and provide some physical protection to the bed (e.g., mussel and seagrass beds), or can enhance cohesiveness and modify the critical erosion threshold (e.g., microphytobenthos). In contrast, bio-destabilisers (e.g., bioturbators such as *Scrobicularia plana, Hydrobia ulvae*) increase surface roughness, reduce the critical erosion threshold and enhance the erosion rate (Widdows and Brisley 2002, Kristensen et al. 2013). As result, an increase in currents, wave action and sediment resuspension is expected, leading to an increase in turbidity and a reduction in the light penetration through the water column, which may affect the primary producers but also the presence of macrobenthos and fish.

For the WFD, it was necessary to decide on appropriate monitoring criteria and to define critical levels of abundance of blooming macroalgae in terms of quality status classes (e.g., Scanlan et al. 2007, Patrício et al. 2007). Consideration also had to be given to the different eutrophication assessment regimes under the WFD, Habitats Directive (Habitats Dir. 1992), Urban Waste Water Treatment Directive (UWWTD 1991), the Nitrates Directive (Nitrates Dir. 1991) and the Oslo-Paris Convention (OSPAR 2003), summarised in, e.g., Scanlan et al. (2007), Elliott and de Jonge (2002).

4 Macroalgae as Transmission Vehicles of Dangerous Substances Inside the Trophic Web

Macroalgae can function as a transmission vehicle of dangerous substances such as heavy metals. Although these may be not visibly affecting macroalgae, they can be accumulated directly in macroalgal tissues, and (through the food chain) pose a health risk to secondary consumers including humans (Angelone and Bini 1992, Chan et al. 2003). Unlike many other pollutants, heavy metals are non-biodegradable substances, which persist in the environment mainly because of the difficulty of remediation, and the high costs associated with the management processes (Kaewsarn and Yu 2001). Heavy metals are among the major environmental hazards due to their affinity for metal sensitive groups, such as thiol groups.

In macroalgae, heavy metals are known to block functional groups of proteins, displace and/or substitute essential metals in the organism, induce conformational changes, denature enzymes and disrupt cells and organelle integrity (Hall 2002). Heavy metals can be responsible for inhibiting growth and photosynthesis, reducing chlorophyll content, affecting reproduction, interfering with cell permeability, causing the loss of K ions, affecting protein synthesis and degradation, and causing oxidation and lipid peroxidation (Sorentino 1979, Strömgren 1980, Rai et al. 1981, Kremer and Markham 1982). The toxicity of heavy metals is also largely dependent on their biological availability (Campbell 1995, Sunda and Huntsman 1998), which is determined by both their physical and chemical states (Langston 1990). Toxicity of heavy metals may be reduced when they are adsorbed to suspended organic matter, thus reducing their ionic fraction in the water column. Both pH and redox potential affect the toxicity of heavy metals by limiting their availability (Peterson et al. 1984). Some macroalgae, such as members of the Phaeophyceae like fucoid algae, can accumulate high levels of various metals in their cell walls (Burdin and Bird 1994, Salgado et al. 2005), which have negatively charged polysaccharides and special compartments (physodes) that enhance their ability to do this (Raize et al. 2004, Salgado et al. 2005). Therefore, heavy-metals can be transmitted through the food chain without causing an evident negative effect on the macroalga itself.

There can also be less obvious impacts of macroalgae on ecosystem health. Some recent studies provided evidence for the possible implication of alien macroalgae in detrimental impacts on the health of native fauna, with macroalgae essentially acting as a vector for viruses. Studying the development of fibropapillomatosis (FP) tumours caused by herpes virus on Hawaiian green turtles, Van Houtan (2010) showed that, after 1950, native Hawaiian algae and seagrasses were displaced by invasive species of macroalgae, especially in locations with elevated nutrient loads, and that in some locations those non-native macroalgae (*Hypnea musciformis* and *Ulva fasciata*) could dominate >90% of green turtle diets (Russell and Balazs 2009). Those macroalgae store the excess of N as arginine (Arg), the only tetra-amine amino acid (Llácer et al. 2008), and pass it on as turtle forage. Dailer et al. (2010) showed that 43% of stored N in those macroalgae originated from discharged sewage, the source of the infection. The implication of Arg in cell inflammation, immune dysfunction and growth of herpes, contributed to put in evidence the importance of Arg in herpes virus promotion and tumour growth in Hawaiian green turtles (Van Houtan 2010).

5 Assessment Concepts, Indicators and Tools (a European Perspective)

Although knowledge on the impact of pollution on the structure of macroalgal dominated assemblages has existed for some decades now, there had been little effort among phycologists to produce assessment indices based on marine macroalgae (Ballesteros et al. 2007a). Prior to the WFD (2000/60/EC), studies using macroalgae attributes (e.g., taxonomic composition, functional and sensitivity features) to provide any absolute quality classification of aquatic systems were scarce. With the WFD implementation, simple indices based on aspects of macroalgal communities, such as the number of species, the percentage of Chlorophyta or the Rhodophyta/Phaeophyta mean Ratio Index, earned extra importance. New macroalgae-based assessment tools were developed so that, when observed results are compared with the water type-specific references values, these methods provide an ecological quality status classification for the study sites. Quality status is generally reported at waterbody level. Each European member state (MS) derived its own monitoring tools, but these often used similar metrics within the overall tool.

Different aspects of macroalgal communities were explored in relation to environmental variables and, among others, included changes in the presence of perennial vs. annual species; the increase in the percentage of opportunistic species and the concomitant reduction in numbers of stress-intolerant taxa; the contribution of the different functional-form groups in the community; and the similarity of species' structure/composition with reference sites. However, the interactions between biological measures and different environmental variables are complex, often non-linear and difficult to interpret. It is important to bear in mind when assessing ecological quality that macroalgae have their own optimal environmental ranges linked to geographical variations (e.g., latitude), and hence different parts of Europe may require different lists of key species (Ramos et al. 2014). Different taxa may also show physiological plasticity according to available resources and physicochemical conditions. Thus, the linkage of the ecological condition to a pressure level, where biological responses may vary not only along an anthropogenic pressure gradient (e.g., of nutrient concentration or toxicity) but also in relation to latitude (or particular regional conditions), constituted a concern in the assessment process. Accounting for such variability contributed to the implementation of the comparability of results between the various European member states' assessment tools.

Member states were required to compare methods to ensure that these evaluated quality status in an equivalent way; this process is known as "intercalibration". Intercalibration aimed to harmonize quality status class boundaries across the different methods, to ensure consistency of assessment across Europe. For simplicity, intercalibration was carried out separately for tools dealing with each particular biological quality element (e.g., for macroalgal rocky shore, macroalgal blooming occurring on soft shores, and for seagrass). The responses of biological quality elements (BQE), in this case macroalgae, against the level of anthropogenic pressure, based on the assumption that similar biological responses are attained for similar environmental disturbances, constituted the common ecological ground in minimizing of bio-geographical differences in the assessment process. It was also necessary to define reference conditions (RC) for each method and member state. RC are the expression of the community under no or only very minor anthropogenic disturbance (see Chapter 2 for further information), and should represent as far as possible, pristine conditions. These reference conditions should conform with status descriptions in the WFD known as "normative definitions".

Some authors give preference to biological data/information obtained from experimental science and predictive models (Orfanidis 2007) for the construction of assessment methods, while others prefer the use of descriptive studies showing the relationships between species abundances and environmental variables (Ballesteros et al. 2007b) to achieve WFD objectives. Despite this evident contradiction, the following steps can be regarded when developing quality assessment methodologies (Borja and Dauer 2008): i) the spatio-temporal scale of the intended application; ii) the selection of the candidate metrics; iii) how the metrics are combined; and iv) the index validation, by testing it using an independent data set, different than the index development data set (calibration data set). In Europe, several marine macroalgae-based assessment tools were developed following the mentioned preferences (Table 1), and their results harmonized during the intercalibration exercise (IC) (2013/480/EU). Where possible, methods were applied to data collected in a different area from the one for which they were developed and, despite differences in the assessment concepts followed by each methodology (Table 1), the ecological quality they provided and the response against the anthropogenic pressure, for most of the cases, were consistently comparable (2013/480/EU). The ecological scientific knowledge, conscientiously integrated during the production phase of the indices, allowed successful intercalibration (required for the full implementation of the WFD in Europe) for the methods assessing rocky shore macroalgae. Methods to assess macroalgal blooming are still being

Table 1. Macroalgae-based assessment methodologies, developed for quality assessment on coastal waters (CW) and transitional waters (TW), and related operational concept, indicators used, region of origin and references. Listed indices were used on the intercalibration exercise and/or proposed as national method to assess the ecological quality of water under the WFD.

Name	Operational concept	Indicators/measures	Origin
BENTHOS – Rocky-shore communities (Pinedo et al. 2007)	changes on the taxonomic composition of macroalgal communities in upper subtidal rocky-shores	taxonomic composition of macroalgae and macroinvertebrates; coverage (horizontal surface)	Mediterranean Sea; Europe (Spain) (CW)
BMI – Blooming Macroalgal Index (Patricio et al. 2007)	development of intertidal opportunistic macroalgal blooms	available intertidal area; areal coverage of opportunistic macroalgae (ha); opportunistic macroalgae cover (%); biomass of opportunistic macroalgae	North East Atlantic; Europe (Portugal) (TW)
CARLIT - cartography of littoral and upper-sublittoral rocky-shore communities (Ballesteros et al. 2007a)	cartography of commonest littoral and upper-sublittoral rocky-shore communities and available information about the value of the communities (sensitivity level) as indicators of water quality, using GIS technology	macroalgal communities distribution (geo-referenced graphical support); sensitivity levels of macroalgal communities (vulnerability and resistance to environmental stress)	Mediterranean Sea; Europe (Spain) (CW)
CCO - Cover, Characteristic species, Opportunistic species on intertidal rocky bottoms (2013/480/EU)	changes on the number and percentage coverage of characteristic macroalgae species and the coverage of opportunistic species in the intertidal community	contribution of each macroalgal community / habitat (referred to as a belt) to the global plant covering of a given intertidal rocky shore; number of characteristic species; total cover of opportunist species in every belt	North East Atlantic; Europe (France) (CW)
CFR – Quality of Rocky Bottoms (Juanes et al. 2008)	changes on the percentage coverage of characteristic macroalgae species and the fraction of opportunistic species in the community throughout the depth gradient, from the intertidal to the shallow subtidal	coverage of "characteristic macroalgae" (C); fraction of opportunistic species (F); richness of "characteristic macroalgae" (R)	Atlantic; Europe (Spain, Cantabric District) (CW)
E-MaQI – Macrophyte Quality Index (Sfriso et al. 2009)	changes in community composition related to environmental degradation in hard and soft substrata	taxonomic composition; species sensitivity scores	Mediterranean Sea; Europe (Italy) (CW & TW)
EEI-c – Ecological Evaluation Index continuous formula (Orfanidis et al. 2001, 2011)	changes in community composition as indicator of the aquatic ecosystem Ecological State (from pristine late-successional to degraded opportunistic species ESG)	taxonomic composition and ESGs using traits combination in relative terms of species morphology, physiology, life strategy and distribution	Mediterranean Sea; Europe (Greece) (CW)

Index	Response	Metrics	Location
EPI- Estonian coastal water phytobenthos Index (macroalgae and angiosperms) (2013/480/EU)	changes on depth distribution of vegetation and of *Fucus vesiculosus*, in particular, and proportion of perennial species biomass in the community	depth distribution of vegetation, depth distribution of *Fucus vesiculosus*, proportion of perennial species	Baltic Sea; Europe (Estonia) (CW)
Exclame (Andral & Derolez 2007)	change on the relative coverage of climax species and variation of species richness of macrophyte (macroalgae and seagrasses) species	total macrophytes cover; relative coverage of climax species ; macrophyte species richness	North East Atlantic; Europe (France) (TW)
Fucus depth limit (macroalgae) (2013/480/EU)	changes in the subtidal lower limit of growing zone of bladder-wrack (*Fucus vesiculosus*) (measured by diving)	lower limit of growing zone of bladder-wrack (*Fucus vesiculosus*)	Baltic Sea; Europe (Finland) (CW)
HPI - Helgoland Phytobenthic Index (Kuhlenkamp et al. 2011)	changes on the taxonomic composition and abundance of macroalgae in intertidal and subtidal rocky-shore communities	species richness; spatial extent of dense "Fucetum", Green algae (total abundance of *Ulva lactuca*); sublittoral depth limit of selected species	North Sea; Europe (Germany) (CW)
MarMAT – Marine Macroalgae Assessment Tool (Neto et al. 2012)	changes on the taxonomic composition, structural and functional attributes of intertidal rocky-shores macroalgal communities	species richness; proportion of Chlorophyta; number of Rhodophyta; number of opportunists/ESG I (ratio); proportion of opportunists; shore description; coverage of opportunists	North East Atlantic; Europe (Portugal) (CW)
MSMDI - Multi Species Maximum Depth Index (Kautsky et al. 2007)	changes on depth distribution (abundance) of common conspicuous perennial species (disturbance sensitive taxa) due to shading from the overgrowth of opportunistic species, increased phytoplankton biomass and increased siltation following eutrophication	depth distribution (abundance) of 3 to 9 conspicuous perennial species (dependent on national type)	Baltic Sea; Europe (Sweden) (CW)
OMA – Opportunistic Macroalgae Tool (Scanlan et al. 2007)	development of intertidal opportunistic macroalgal blooms	AIH - Available Intertidal Habitat (area suitable for algal growth); % cover AIH; AA - Total extent (ha) of the Affected Area; AA/AIH (%); Biomass AIH (Average biomass of algae in the Available Intertidal Habitat; Biomass AA (Average biomass in Affected Area only); % entrained algae (% of quadrats with algae entrained into sediment)	North East Atlantic; Europe (United Kingdom) (CW & TW)
OMAI – Opportunistic Macroalgae-cover/acreage on soft sediment intertidal in coastal waters (2013/480/EU)	development of intertidal opportunistic macroalgal blooms	maximum extent of opportunistic macroalgae (km2); extent of the intertidal area (km2) per waterbody	North East Atlantic; Europe (Germany) (CW)

Table 1. contd....

Table 1. contd.

Name	Operational concept	Indicators/measures	Origin
PAN-EQ-MAT – Pan-Ecological Quality Macroalgae Assessment Tool (Wallenstein et al. 2013)	changes on the taxonomic composition and abundance of macroalgae in intertidal rocky-shore communities	species richness; total cover (%), Opportunist cover (%)	North East Atlantic; Europe (Portugal) (CW)
R-MaQI – Macrophyte Quality Index (Sfriso et al. 2009)	changes in community composition and abundance of species in hard and soft substrata (dichotomic key)	taxonomic composition; species sensitivity scores; Rhodophyceae/Chlorophyceae ratio; environmental conditions (water transparency, salinity, oxygen saturation, sediment grain-size, nutrient concentrations)	Mediterranean Sea; Europe (Italy) (TW)
Rhodophyceae/Chlorophyceae ratio (Giaccone & Catra 2004)	changes on relative coverage of macroalgal groups in intertidal rocky-shore communities	macroalgae taxonomic groups (rhodophyceae, chlorophyceae); coverage (%)	Mediterranean Sea; Europe (Italy) (CW)
RICQI – Rocky Intertidal Community Quality Index (Diez et al. 2012)	changes on the benthic communities based on indicators of the whole assemblage (macroalgae and macrofauna biological elements).	cover of species with different degrees of tolerance to anthropogenic stress (SpBio); morphologically complex algae cover (MCA); species richness (R); community measures related to faunal cover (FC)	North East Atlantic; Europe (Spain, Basque Country) (CW)
RSL – Reduced Species List (Wells et al. 2007)	changes on the taxonomic composition of macroalgal communities in intertidal rocky-shores	species richness; proportion of Chlorophyta; proportion of Rhodophyta; ESG ratio; proportion of opportunists; shore description	North East Atlantic; Europe (United Kingdom) (CW)
RSLA – Rocky Shore Reduced Species List with Abundance (2013/480/EU)	changes on the taxonomic composition of macroalgal communities and abundance of green and brown macroalgal groups at intertidal rocky-shores	species richness; proportion of Chlorophyta; proportion of Rhodophyta; ESG ratio; proportion of opportunists; abundance of greens; abundance of browns; proportion of browns	North East Atlantic; Europe (Norway) (CW)

intercalibrated at the time of writing. Other indices were produced but are not listed here because they were either simple macroalgae indicators, not sufficiently articulated under a mathematical formulation, or were not tested yet for ecological responses against different levels of pressure (for review see www.wiser.eu/results/method-database; Birk et al. 2010, 2012, Blomqvist et al. 2012).

Around 90% of methods listed in Table 1 were developed for coastal waters (CW), some may be successfully applied in the assessment of transitional waters (TW), but others were developed purely for either TWs or CWs. Some apply only to the intertidal, others also include the subtidal zone. Eleven methods were developed for application in the North East Atlantic (NEA) region, six for the Mediterranean Sea (MED), three for the Baltic Sea (BAL) and one for the North Sea (Helgoland, Germany). Five categories of indicators group the macroalgae assemblage measures used by the indices: distribution (e.g., depth limit); abundance (e.g., cover area); taxonomic composition/diversity (e.g., species richness); traits (e.g., sensitivity to pollution). Most include a taxonomic composition indicator in the index formulation (>75%), or they accomplish this feature by assessing taxonomy to a higher level, or as an indicator extracted from community components. The methods that do not include such detailed taxonomic identification usually consider specific groups of macroalgae (e.g., blooming macroalgae) and base the quality assessment on data from those taxa groups only. In this case two approaches are often used; one based on the effect the anthropogenic pressure has on tolerant, fast-growing (opportunistic) species, inducing them to bloom (e.g., methods BMI, OMA, OMAI), and the other, by contrast, based on the decreased expression (biomass and/or diversity) shown by sensitive species (e.g., *Fucus* depth limit index) with the decreasing water quality. Both responses are attributable to the same cause, the degradation of environmental conditions, but focusing on the behavior of taxa with different life strategies and contrasting sensitivities to pollution. Macroalgae are considered good quality indicators (Ballesteros et al. 2007a) and are able both to demonstrate response to strong anthropogenic pressures, and to subtle variations imposed by global changes. Wilkinson et al. (2007) referred to the response of perennial fucoid algae to toxic estuarine pollution, shown by a reduction in the extent of upstream penetration of fucoid algae as water pollution by toxic substances increased. Cossellu and Nordberg (2010) identified that temperature increase due to global climate change, which imposed discrete variations on hydrodynamics, erosion/sedimentation rates, nutrient biogeochemical cycles and increased concentration of nutrients in the water, was responsible for the higher frequency and intensity of filamentous algae blooms in shallow coastal systems of northern Europe.

Although the separation of sites under different degradation levels may sometimes be achieved simply using taxonomic composition data (e.g., BENTHOS), or the ratio between some of those taxonomic groups (e.g., Rhodophyceae/Chlorophyceae ratio), the classification into five quality classes, as required by the WFD, is facilitated and gets more robust when there is a combination of multiple indicators derived from macroalgal assemblages. The WFD requires the use of disturbance sensitive taxa measures to assess quality (WFD 2000/60/EC), along with consideration of taxonomic composition and abundance. It was therefore sensible to allocate the different marine macroalgae taxa to different morpho-functional groups (Carrington 1990, Orfanidis 2001, 2011, Denny and Gaylord 2002, Wells 2002, Balata et al. 2011). These indicators have been recently integrated in assessment tools in order to fulfill, together with taxonomic composition and abundance measures, the WFD legal requirements. Some indices use more complex morpho-functional classifications than others (e.g., EEI-c, RICQI) but simple characteristics, such as the classification of different taxa as annual or perennial (Orfanidis et al. 2001), are used by 17 out of 21 of the listed assessment methodologies (Table 1).

Apart from macroalgae indicators, one index, the RICQI, included macrofaunal indicators in the assessment, and the other six indices (CARLIT, E-MaQI, EEI-c, EPI, EXCLAME, R-MaQI) count also with the contribution of seagrass measures to assess quality (under the WFD angiosperms and macroalgae form separate BQEs in transitional waters but are grouped with macroalgae in coastal waters). Five indices (CARLIT, CCO, CFR, Rhodophyceae/Chlorophyceae ratio, RICQI) base their assessment on community/habitat level (e.g., characteristic communities), with two of them using cartographic (CARLIT) and landscape attributes (CFR subtidal; Guinda et al. 2013) to produce necessary data for the quality evaluation. The depth limit indicator is used in 5 indices (EPI, EXCLAME, *Fucus* Depth Limit, HPI, MSMDI), all applied to subtidal assessment. One recent index, the PAN-EQ-MAT (Wallenstein et al. 2013), based its construction on indices created before for different regions of the Atlantic, and aimed to have an extensive applicability into this larger and inclusive region, the North East Atlantic (NEA, sensu WFD). For simplification, most of the indicators concerning diversity measures, which need information on the taxonomic composition, base their data determination in a reduced species/taxa list (RSL, RTL) adapted to regional conditions (e.g., CCO, RSL, RSLA, MarMAT, PAN-EQ-MAT). These RSL/RTL may be adapted to local macroalgae community compositions thereby allowing the applicability of indices to regions other than the one from which they were originally created. The final assessment results are produced by the combination of the different indicators included in an index, either as the mean value achieved by those indicators or using a different weighting formula for the calculation.

As there are differences found in the calculation formula of each index, there are also differences in data collection and sampling. To reduce costs and time, many of the indices can use nondestructive sampling, such as video records and photographs produced in the field or from remote sensing technology applications. Others (RLS, RSLA) use reduced species lists as surrogates for full species lists (the latter require a much higher level of taxonomic skills).

For the WFD results must be expressed as an Ecological Quality Ratio (EQR) on a zero to one scale, and hence the quality class or Ecological Quality Status (EQS), determined as the deviation of the results from the reference condition (Fig. 1). The five quality classes are: i) High, if the values of the BQE show no, or only very minor, evidence of distortion; ii) Good, if the values of the BQE deviate only slightly from those normally associated with the surface water body type under undisturbed conditions; iii) Moderate, if the values of the BQE show moderate signs of distortion and are significantly more disturbed than under conditions of good status; iv) Poor; and v) Bad (2000/60/EC).

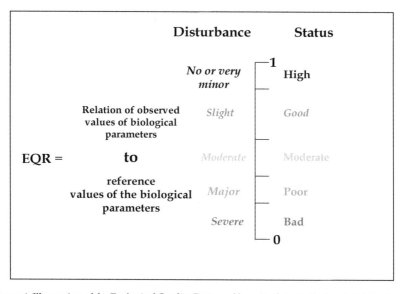

Figure 1. Illustration of the Ecological Quality Ratio and how it relates to the level of disturbance and ecological status during a classification. The class band widths relate to biological changes as a result of disturbance (WFD CIS Guidance Document No. 5, 2003).

Table 2. Description of WFD Normative Definitions for TraC waters (a—coastal waters; b—transitional waters).

a) Coastal Waters

HIGH	All disturbance-sensitive macroalgae associated with undisturbed conditions present. The levels of macroalgal cover are consistent with undisturbed conditions.
GOOD	Most disturbance-sensitive macroalgae associated with undisturbed conditions are present. The level of macroalgal cover shows slight signs of disturbance.
MODERATE	Macroalgal cover is moderately disturbed and may be such as to result in an undesirable disturbance in the balance of organisms present in the water body.
POOR	Major alterations to the values of the biological quality elements for the surface water body type. Relevant biological communities deviate substantially from those normally associated with the surface water body type under undisturbed conditions.
BAD	Severe alterations to the values of the biological quality elements for the surface water body type. Large portions of the relevant biological communities normally associated with the surface water body type under undisturbed conditions are absent.

b) Transitional Waters

HIGH	The composition of macroalgal taxa is consistent with undisturbed conditions. There are no detectable changes in macroalgal cover due to anthropogenic activities.
GOOD	There are slight changes in the composition and abundance of macroalgal taxa compared to the type-specific communities. Such changes do not indicate any accelerated growth of phytobenthos or higher forms of plant life resulting in undesirable disturbance to the balance of organisms present in the water body or to the physico-chemical quality of the water.
MODERATE	The composition of macroalgal taxa differs moderately from type-specific conditions and is significantly more distorted than at good quality. Moderate changes in the average macroalgal abundance are evident and may be such as to result in an undesirable disturbance to the balance of organisms present in the water body.
POOR	Major alterations to the values of the biological quality elements for the surface water body type. Relevant biological communities deviate substantially from those normally associated with the surface water body type under undisturbed conditions.
BAD	Severe alterations to the values of the biological quality elements for the surface water body type. Large portions of the relevant biological communities normally associated with the surface water body type under undisturbed conditions are absent.

Note: Members States were required to produce expanded versions of these to describe quality status in relation to specific methods.

6 Case Studies

This section shows the potential of some macroalgae-based assessment tools in operational assessment programmes of rocky shore coastal systems. All the methods were developed for WFD monitoring, and their responses were tested against site pressure levels and compared with other methodologies within the aims of the IC exercise performed between EU member states (MS).

The nutrient enrichment or the elevated concentrations of other substances in water, of anthropogenic origin, is a worldwide concern in shallow coastal environments, where the resulting symptoms of eutrophication have been recognised as threats to environmental health for many years (Norkko and Bonsdorff 1996, Valiela et al. 1997, Raffaelli et al. 1998, Sfriso et al. 2001). In coastal rocky shores from the NEA region, the RSLA, MarMAT (Neto et al. 2012) and CFR (Juanes et al. 2008, Guinda et al. submitted) were calculated and the results compared with different pressure levels influencing the sampling sites (Juanes et al. 2012). This procedure aimed to verify the extent to which marine macroalgae were able to provide a comprehensive response against different pressure levels and types, ensuring that either individual macroalgae-based indicators or multimetric methodologies could be integrated satisfactorily in operational environmental assessments. The pressures tested have an anthropogenic origin (agriculture, industrial and urban) and affected the selected sites under a distributional gradient and for different time-periods. Sampling sites were located along the European coast, from Portugal, in the south, through north Spain, to the far north in Norwegian fjord systems, and allowed the testing of the responses of the assessment tools developed for each area.

6.1 Eutrophication Effects on Algal Communities in a Fjord at the Arctic Circle

According to the WFD, water quality across Europe should be characterized by use of different BQEs of which macroalgal indices are some. Hence, a new rocky shore macroalgae index was developed in Norway based on the "Reduced Species List" used in the United Kingdom and Ireland (Wells et al. 2007), and which included species that were common along the Norwegian West Coast up to the Arctic Circle and also their abundance. Algal community data and pressure data were collected from four fjords in Norway—two close to the Arctic Circle and two on the West Coast of Norway (Fig. 2).

The index developed, "RSLA" (Reduced Species List with Abundance), is a multi-metric index and includes different sub-metrics such as number of species (from a pre-made species list of the most common species from this coastline), ecological status groups (ESG, Orfanidis et al. 2001), number of opportunists, proportion of red-, green- and brown algae, summarized abundance for the same groups, as well as abundance of some dominant wracks. These were all tested against pressure data (i.e.,

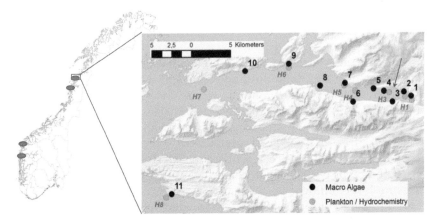

Figure 2. Stations for macro algal collections. Four stations (red dots) on the left figure, indicates areas of data collection for development of the RSLA-index. Map on right shows where macro alga (black dots) and hydrograhical/hydrochemitry sampling (blue dots) were carried out in Glomfjord and the reference station in Tjongsfjorden. The arrow indicates the discharge site for nitrogen and phosphorous compounds from a fertilizer plant.

Color image of this figure appears in the color plate section at the end of the book.

nutrients concentration in seawater near the sampling sites). The number of species found at a site (species richness), was adjusted according to the expected number to be found at that specific site, which is based on the shoreline's suitability to sustain a certain number of species. Characteristics to be considered are: the shoreline's toughness, slope, aspect, cracks and crevices, boulders, out-springs, piers and so on. For all the descriptors a score between –2 and 4 is given and summarized to give the site a score value. An exponential formula based on empiric data from UK shorelines was used to calculate a "de-shoring" factor (Wells 2007) which can be multiplied by the actual number of species recorded on the site, resulting in a normalized species number to the shoreline description.

The abundance data were registered according to a semi-quantitative scale from 1 to 5 (ISO/DIS19493:2006), where 1 is a single finding and the numbers 2 to 6 represent an increasing percentage cover up to 100%.

Multivariate analyses were performed between the nutrient data collected near the registration sites, and the calculated parameters for the algae found during the semi-quantitative registrations at the respective sites. The values for each parameter, EQR, were calculated as ratios against a reference value according to the description in the WFD (2000/60/EC) and Table 3 shows the correlation coefficients found among averaged nutrient values and the calculated parameters.

Table 3. Correlation for each parameter against nutrients.

	Total P	PO4	Total_N	NO3	NH4
Species richness normalized	–0.7609	–0.7388	–0.8056	–0.7494	–0.8693
Proportion of Rhodophyta	–0.3653	–0.3663	–0.4029	–0.4178	–0.0585
Proportion of Chlorophyta	0.8568	0.8502	0.8476	0.8429	0.8007
Proportion of Phaeophyta	–0.6385	–0.6256	–0.6251	–0.5627	–0.7429
Proportion of opportunists	0.5360	0.4834	0.5419	0.4538	0.9607
ESG Ratio	–0.1009	–0.1270	–0.0642	–0.1782	0.2327
Abundance of Chlorophyta	0.4999	0.5244	0.4797	0.4873	0.1521
Abundance of Phaeophyta	–0.8067	–0.8209	–0.8836	–0.8372	–0.7903

The species richness, proportion of red and browns, as well as the abundance of browns showed good negative correlations with increasing nutrient pressure as opposed to the proportion of greens and opportunists, and abundance of greens, which showed strong positive correlation with increasing nutrient concentrations in the water. ESG ratios did not show any good correlations, especially when the number of species at a site was low. Still, all the parameters were included in the index RSLA; however, some parameters were excluded from the final combined calculations of the water quality status when species number was below fourteen. The status for water quality was calculated as an average of all the normalized EQR-values from the included parameters.

To test the RSLA, a location in Norway, Glomfjord—a Norwegian fjord at the Arctic Circle (N 66° 48.58, E 13° 56.18) adjacent to Norway's largest glacier Svartisen—was chosen. A fertilizer production plant was established in the inner part of the fjord (Fig. 1) as early as 1947 with production of ammonia in 1947 and granulated mineral fertilizer (nitrogen, phosphorous, potassium) in 1955. Today they also produce nitric acid and calcium nitrate in two other plants at the same location. Over the years large quantities of nitrate and phosphate as well as other essential nutrients for primary production has been discharged into the fjord. Prior to 1985, the annual discharge was approximately 1900 and 260 tons of nitrogen and phosphorous, respectively, at about 7 m depth. However, during the late 1980s an extensive cleanup program was implemented and over a 5-year period, the discharges were reduced to an ongoing annual discharge of approximately 600 tons of nitrogen and 35 tons of phosphorous (Fig. 3). Hence, the output of fertilizers to the fjord has resulted in eutrophication problems in the surface layers.

Norway has, since 1993, characterized the water quality of seawater based on several physical, physicochemical and hazardous substances as well as on some biological parameters that include nutrient levels

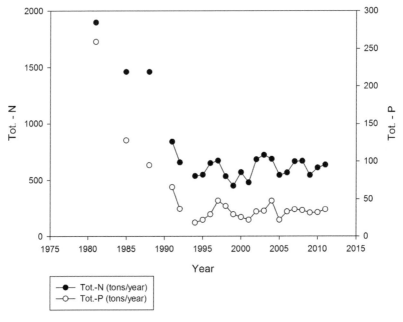

Figure 3. Discharge of total nitrogen and phosphorous compounds into Glomfjord from 1981 to 2011.

(Molvær et al. 1997). The quality status classes are similar to the ones chosen for the WFD, i.e., five status classes from "very good" to "very bad". The class boundaries for nutrient levels are based either on average summer or winter values of nutrients like Total P, PO_4-P, Total N, NO_3-N + NO_2-N and NH_3-N, collected from 0, 5 and 10 m depths.

Nutrient samples have been taken at several stations in Glomfjord (Fig. 1) periodically since 1981 and based on these values the status of the fjord has been classified as shown in Table 4. The average values show that the water quality is poor closest to the outlet and improves outwards along the fjord. All parameters show "very bad" and "bad" status closest to the outlet (station H3, Table 4, Fig. 2) and for most of the parameters, the water quality at the outmost station in Glomfjord are classified as either "good" or "very good" except for phosphate that showed somewhat elevated concentrations. At three stations, i.e., H3, H5 and H6, two extra hydrographical stations were placed perpendicular to the length of the fjord (not shown in Fig. 1), to establish if there were differences in nutrient condition on the northern and southern side of the fjord due to the Coriolis force. This was found to be the case as the Coriolis force resulted in a predominant inflow of water on the southern side of the fjord and the outflow of water happened on the northern side of the fjord. Hydrographical data showed that the currents

Table 4. Average nutrient levels (µg/l) at the hydrographical (H) stations in Glomfjord. Status classes are; 1-Very Good, 2 – Good, 3 – Moderate, 4 – Bad and 5 – Very Bad according to Molvær et al (1979). (n=20 for all parameters except for ammonium were n =10.)

Hydrographical station	Tot. P	sd	Status class	PO4-P	sd	Status class	Tot. N	sd	Status class	NO3	sd	Status class	NH4	sd	Status class
H1	5.1	±2.1	1	4.3	±2.7	2	195	±66	1	77	±22	4	39	±32	2
H2	44.5	±37.1	4	41.6	±37.6	4	350	±276	3	114	±108	4	96	±156	3
H3	85.7	±94.8	5	83	±95.2	5	848	±696	5	301	±278	5	317	±447	4
H4	66	±34.9	5	57.5	±34.9	5	760	±345	4	220	±109	4	347	±293	5
H5	51.3	±20.6	4	45.7	±20.9	4	605	±325	4	198	±86	4	177	±181	3
H6	26.1	±12.3	3	27	±16.4	4	300	±125	2	87	±66	4	56	±46	3
H7	13.4	±8.1	2	10.2	±5.7	3	178	±11	1	23	±19	2	30	±18	2

pattern was a bit zigzag in the inner part of the fjord near form H3 to H5. The Coriolis force resulted in differences in nutrient condition between the two sides of the fjord, as the nutrient levels on the northern side were higher than on the southern side of the fjord at similar distance from the head of the fjord and outwards. The stations closest to the discharge point (Fig. 2) shows the highest nutrient levels and the concentration decreased out along the fjord to almost reference values at station H7.

Macroalgae have been collected at 11 sites in Glomfjord during 3 periods; in 1981 and 1982, in 1991 and 1992, and in 2011. The macroalgae have been registered in a way that made it possible to compare datasets and use them as input to the newly developed RSLA index for Norway. The results from the calculation shows that the EQR-values for RSLA coincide with the classification based on nutrients and shows that the stations closest to the outlet from the fertilizer plant had lowest EQR-values. The head of the fjord showed almost as good conditions as the mouth of the fjord and the northern side of the fjord was more affected by the discharge than the southern side (Fig. 4) most probably due to the Coriolis force.

The eutrophication situation seemed to worsen over the ten year period from 1981/1982 to 1991/1992 as the algae communities closest to the fertilizer plant deteriorated from "moderate" to "bad" status even though the discharge from the fertilizer plant was significantly reduced. The effects of the discharge seemed to also affect one station on the southern side of the fjord during these two periods. During the late 1990's, glacier water used in a very large hydro power plant was transferred from Glomfjord to an adjacent fjord, Holandsfjorden, and this may have caused a shift in the fjord's surface water circulation pattern which again can explain why in 2011 an improvement was seen on the southern side of the fjord. However, on the northern side the deterioration still continued. One explanation can be that the effect of the Coriolis force will lead to the outgoing current being more restricted to the northern side of the fjord as the zigzag pattern

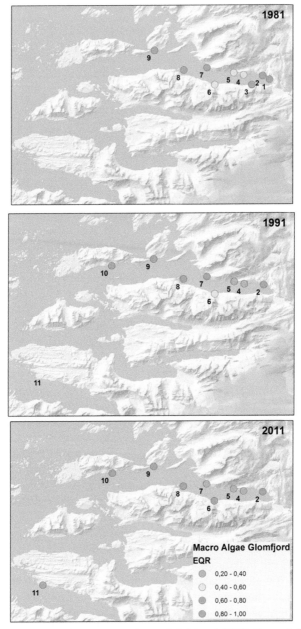

Figure 4. Status classes as EQR values according to definitions in WFD. Orange is "bad", yellow is "moderate", green is "good" and blue is "high". Note that not all stations were monitored among the three periods and reference station 11 was only registered in 2011.

Color image of this figure appears in the color plate section at the end of the book.

of the current in the middle of the fjord, as seen in 1980's, seemed to have been weakened, resulting in a more frequent supply of nutrients to the organisms on the northern shore line. Both station 8 and 6 have improved on the southern side as opposed to stations 7 and 10 on the northern side that have shown signs of deterioration. However, at station 7, 9 and 10, an increase in barnacle settling and grazing may complicate the interpretations based on changes in EQR-values, current patterns and discharge cleanup that has been carried out in Glomfjord around the millennium. Even so RSLA seems to show a good response to these changes and can be a useful tool in coastal water management.

6.2 Eutrophication Effects on Coastal Macroalgae Communities of Southern Europe

This case study took place in southern European Atlantic coast, in Portugal. The study aimed to apply an assessment methodology, the Marine Macroalgae Assessment Tool (MarMAT) (Neto et al. 2012), developed for classification of the water quality based on rocky intertidal macroalgae assemblages, at the same time that the comparison against pressure levels could be done.

Sampling of biological data was conducted at intertidal rocky shores along the northern coast of Portugal (Fig. 5), mainly during summer and spring, and the anthropogenic pressure was quantified based on the number of inhabitants, the industrial land use and the agriculture/forest/fishing surface area [sensu the Land Uses Simplified Index (LUSI); Royo et al. 2009] influencing the study sites (Table 5). Macroalgae data resulted from non-destructive quantitative assessments, restricted to one shore sample (composed by 21 photograph-replicates) collected during a single low tide event (Neto et al. 2012). Taxonomic composition and the substratum area covered by macroalgae were determined for further analysis and for the calculation of the EQR for each shore. The total pressure scores were compared with the EQR from sampling sites to validate the response of MarMAT against the anthropogenic pressures.

The results achieved illustrate well the capacity of marine macroalgae communities to respond to nutrient enrichment. In this case, macroalgae were able to integrate efficiently the effects of different levels of

Table 5. Criteria used to assess anthropogenic pressures. Indicators of anthropogenic pressure and years considered for the assessment.

	SCORES			
	1	2	3	4
No. Inhabitants x 1000 (2008)	<350	<700	<1050	<1400
Industrial Land Use (ha) (2008)	<1250	<2500	<3750	<5000
Agriculture/Forest/Fishing Surface Area (ha) (1999)	<4500	<9000	<13500	<18000

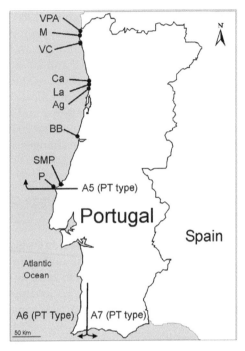

Figure 5. Sampling sites of the A5 coastal waters, Portuguese (PT) typology (EU NEA1) included in the study: Vila Praia de Âncora (VPA), Montedor (Mo), Viana do Castelo (VC), Cabedelo (Ca), Lavadores (La) and Aguda (Ag), Buarcos Bay (BB), São Martinho do Porto (SMP) and Peniche (P).

environmental pressure (Fig. 6). The species richness and the number of taxa from Rhodophyta and Phaeophyceae macroalgal groups were negatively correlated with the total pressure affecting the study site. With a positive correlation, this was also true for the % of Chlorophyta taxa, the % of opportunistic taxa and the coverage of opportunists (%) in the community, that increased their presence in parallel to pressure's increase.

The high correlations found between environmental pressures and the results achieved by marine macroalgae indicators, both individually (e.g., no. of species, no. of Rhodophyta, coverage of opportunists) and when combined in a multimetric assessment tool (MarMAT), are clear evidence of the capacity of macroalgal communities to describe the ecological quality of the environment they live in. The response of the MarMAT against the anthropogenic pressures was also in accordance with expectations (Fig. 7), consistently providing the worst quality classifications at sites reporting higher total anthropogenic pressure values (Fig. 8).

However, in practice, the correct application of MarMAT, and other assessment methods, requires attention. The reference condition (RC) and

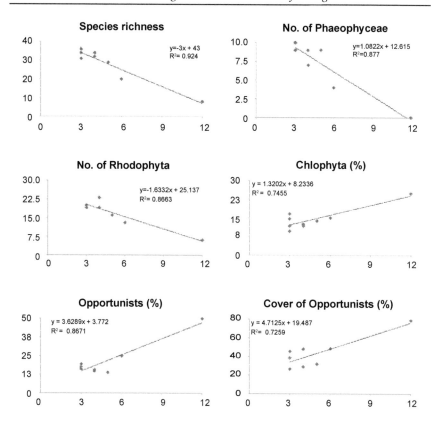

Figure 6. Correlation between macroalgae indicators (vertical axes) and the total pressure scores (horizontal axes) obtained for sampling site.

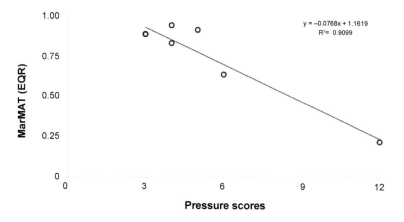

Figure 7. Correlation between EQRs reported by the MarMAT and the total anthropogenic pressures quantified in the sampled sites.

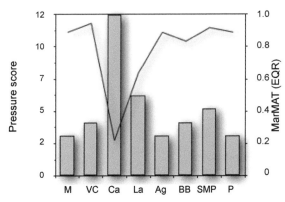

Figure 8. Plot of EQRs reported by the MarMAT against the total pressure score defined for the sample sites. Grey bars = pressure scores; Black line = EQR; Montedor (M); Viana do Castelo (VC); Cabedelo (Ca); Lavadores (La); Aguda (Ag); Buarcos Bay (BB); São Martinho do Porto (SMP); Peniche (P).

the downscale evolution of macroalgae indicator values towards the lower quality condition must be understood, and the geographic differences should be identified and somehow made capable of being expressed in the assessment results. The assessment methods should ideally be easy to use and the quality result fast to produce. The quality classes should then be defined in relation to the RC, as the deviation in quality from that reference point. To make the methods easy to use, many produce a reduced taxa (or species) list (RTL, RSL) or defined characteristic communities that should be used as a RC (see Chapter 2 for further information). These can be selected from historical taxa records, and reduce the number of entries in the list by grouping taxa with similar characteristics, e.g., calcareous algae rather than specific taxa (Wells et al. 2007, Gaspar et al. 2012), or defining the composition of a few characteristic dominant assemblages (Juanes et al. 2008). In addition, inclusion of the geographical adaptation of the reference conditions and the boundaries is important to improve the reporting accuracy of assessment methods. Because the differences observed in macroalgal communities are attributable to various environmental factors (Ballesteros et al. 2007a), such as nutrients, temperature or currents, it is clear that the human presence in coastal areas may influence the variation of those parameters.

The MarMAT successfully captured the total anthropogenic pressure calculated for the sites (Fig. 7). This validates the MarMAT methodology in terms of the requirements listed in the WFD for the behaviour of assessment tools. Another important feature that assessment tools must incorporate is the ability to report the five quality classes (bad to high). The MarMAT results indicated that the Cabedelo site (Ca) had a bad EQS (EQR = 0.19) and that the Montedor, Viana do Castelo, and Aguda sites had high EQS (e.g.,

EQR = 0.94) (Fig. 8). In addition, a classification of high EQS was obtained whenever the total anthropogenic pressure was low along the length of the study area. This indicates that the list of taxa, the RTL, is balanced and does not restrict the outputs of the tool.

The class boundaries adopted were initially equidistant (0.20) but they were improved in the IC exercise performed by MSs. Boundaries were adjusted through comparison with other methodologies (e.g., the CFR). This procedure ensured compliance regarding EQS assessments in contiguous coastal areas and the WFD.

6.3 Coastal Discharge Effects on Algal Communities in the Southern Gulf of Biscay

The contribution of the rocky shore macroalgae communities ("coastal reefs") to the specific richness and, consequently, to the structural and functional biodiversity of the marine environment seems to be very significant, as recognized by their proposal as natural habitats of European interest, whose conservation requires the designation of Special Areas for Conservation at European level (92/43/CE Habitat Directive, habitat code 1170: reefs). In the same way, benthic habitats remain as an important part of the "coastal water bodies" established for the implementation of the WFD (2000/60/CE).

Coastal reefs, including both the intertidal and the subtidal environments, constitute an ecological unit, whose conservation management must be focused from an integrated point of view. Urban and industrial discharges through surface or deep water submarine outfalls are widespread on all coasts, with a potential impact on rocky shore communities, whose extent will mainly depend on physical characteristics (e.g., residence time, exposure, tidal level) and intrinsic features of the exposed communities (e.g., vulnerability, singularity). Establishing causal relationships between stressors and effects at the individual, species or community level on marine systems is a difficult task that requires the use of multiple lines of evidence (Adams 2005), indicators or indices that measure the extent of impact of an activity on part of the ecosystem.

The "Quality of Rocky Bottoms" index (CFR by its Spanish acronym) (Juanes et al. 2008, Guinda et al. submitted) is a multimetric method used for the assessment of macroalgae communities at intertidal and subtidal areas. It was developed according to the WFD basic requirements (2013/480/ EU; Juanes et al. 2012) in order to carry out general management and monitoring works. The method is based on a community approach which avoids exhaustive taxonomical identifications, allowing its easy and fast application over extensive coastal water bodies by either direct or remote assessment methods (e.g., ROVs, Guinda et al. 2013). At present, it is the only intercalibrated method along the NE Atlantic region available for the macroalgae ecological assessment of the whole water body in both intertidal and subtidal habitats.

In this case study, data from different surveys (2006–2012) regarding the specific effects of urban and industrial point discharges on intertidal and subtidal zones were compiled and analysed, in terms of the ecological quality of macroalgae communities, using the continuous scoring CFR index (Guinda et al. submitted) calculated by the specific software recently developed by IH Cantabria (www.cfr.ihcantabria.com).

The data were collected at 7 coastal sites located at the southern Gulf of Biscay (N Spain, Fig. 9), including several intertidal and subtidal stations placed along theoretical contaminant gradients generated by specific point organic discharges (mainly urban), from small to large cities (from W to E: Tapia de Casariego, S Vicente de la Barquera, Santander and Castro Urdiales), and industrial discharges (Usgo and Ontón). According to the IC decision (2013/480/EU), all sites belongs to the CW type 1/26-A2, characterized by open oceanic and shallow coasts, exposed or sheltered,

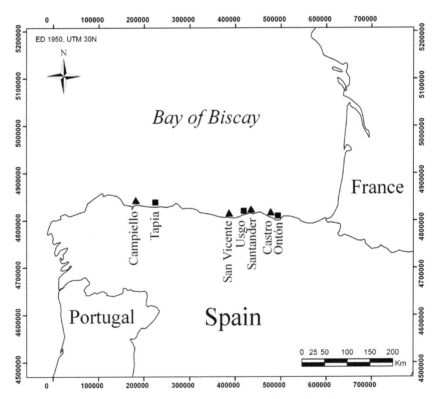

Figure 9. Sampling sites included in the case study according to their discharge type: urban (triangles) or industrial (squares).

with euhaline and temperate waters (mainly, >13°C) and high irradiance (mainly, >29 M/m² day).

The sampling procedure consisted of non-destructive visual estimations of the three metrics which integrate the CFR index: Coverage of "Characteristic Macroalgae" (C), Fraction of Opportunistic Macroalgae (F) and Richness of "Characteristic Macroalgae" (R). Sampling surfaces were station-specific, extending from the mid-littoral to the low littoral at intertidal zones, or covering broad underwater areas in subtidal zones (e.g., 10–100 m²). Stations were located close to different types of anthropogenic pressures both in impacted and control areas. In all cases, the pressure level (PL) was estimated according to the semiquantitative scale in 5 categories (0–4) used for urban and industrial discharges during the IC process of macroalgae in the Northeast Atlantic Geographical Intercalibration Group (2013/480/EU) (Table 6). This system was based on i) the type of discharge (urban/industrial pollution), ii) the distance to the contaminant source and iii) its magnitude, in terms of inhabitant equivalents (IE) or physicochemical risk (IPPC industries).

The values of the three metrics were transformed into scores, ranging between 0 and 1, by applying the equations and the reference conditions for each of the metrics, according to the typology of the survey area (intertidal/subtidal), indicated in Guinda et al. (submitted). Finally, the CFR index was calculated according to the Equation 1, which consists on a weighted sum of the scores obtained by the three metrics:

$$CFR = 0.45 \cdot C + 0.35 \cdot F + 0.2 \cdot R \qquad \text{(Eq. 1)}$$

The resulting CFR index value (0–1) was equivalent to the EQR for classification of the five quality classes of the WFD according to the intercalibrated thresholds (2013/480/EU).

Main results obtained from this exercise showed, first, the important differences in correlation between the pressure level of urban discharges and the values obtained by each independent metric included in the CFR index

Table 6. Pressure assessment system used by the Northeast Atlantic Geographical Intercalibration Group for macroalgae (2013/480/EU) (P.L. = 0: No pressure, P.L. = 1: Low pressure, P.L. = 2: Moderate pressure, P.L. = 3: High pressure, P.L. = 4: Very high pressure).

DISCHARGE TYPE	MAGNITUDE (inh-eq)	DISTANCE (m)			
		>500	500-100	100-50	<50
Organic	<2000	0	0	1	2
Organic/Temp	2000- 10000	0	1	2	3
Organic/Temp	10k-150k	1	2	3	4
Organic/IPPC industries	>150k	2	3	4	4

(Fig. 10). The best correlations corresponded to the fraction of opportunistic species (R^2, 0.729) and to the coverage of characteristic macroalgae (R^2, 0.457) while the richness of characteristic macroalgae presented a more variable response for certain stations. The increase of nutrients around this type of discharge may have a quick and subtle response in terms of abundance of opportunistic species, but within the first 500 m. Effects on coverage of conspicuous populations of seaweeds seem to be located a shorter distances from the discharge. In contrast, once the integration of metrics is done, the global CFR index has correctly reflected the defined pressure gradient to urban discharges.

A similar trend in both the standardized scored metrics and the EQR values for the CFR index was observed in case of industrial discharges (Fig. 11). However, in this case significant correlations between each single metric and the pressure at the station level were obtained for the three indicators, with higher R^2 values for Coverage (0.772) and Richness of Characteristic macroalgae (0.418). Conspicuous growth of opportunistic macroalgae concentrates in a closer area around the discharge (less than 100 m) associated to reduction in number and abundance of characteristic species. Furthermore, the quality of macroalgae communities based on the CFR values for each station reflects the pressure gradient provoked by industrial discharges, which main effects seem to be very concentrated around the discharge.

Considering the global response of macroalgae communities to all different discharges, which takes place together within the same water bodies (Fig. 12), a significant correlation between the EQR and the maximum pressures was estimated at the station level. However, there is still a great amount of variability which can be attributed to different causes. For instance, natural factors (presence of sediments, unsuitable substrates, coastal geomorphology, local currents, etc.), sampling uncertainties in very exposed stations or overestimation of the pressure level, set by expert judgement. Discrimination between natural and anthropogenic causes of variability is one of the main challenges within the WFD for different BQEs. In this sense, the results obtained in this study demonstrate that, despite the variability observed in some cases, the CFR index has responded adequately to the anthropogenic pressures addressed.

In **summary**, it is widely accepted that the use of biotic indices for the assessment of the ecosystem health constitutes an enormous simplification of the environmental reality, which is subject to many sources of variability. However, the regulatory frameworks (e.g., WFD) encourage their use in order to produce comprehensible and standardized results at different geographical regions. The elevated costs associated to the extensive monitoring of all the European water bodies require site-specific assessments that reveal those significant pressures. Along the European

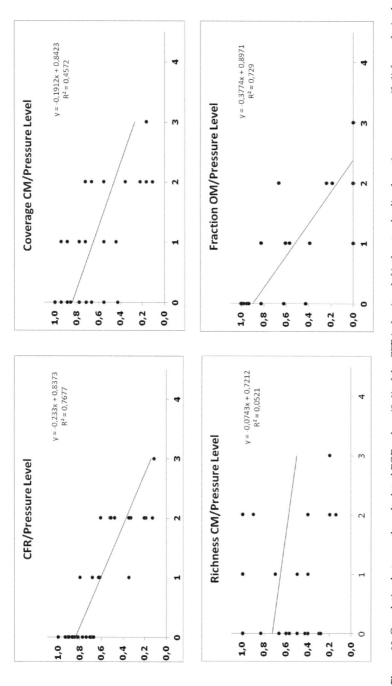

Figure 10. Correlation between the calculated EQR values (0–1) of the CFR index (top left), the standardized continuous scores (0–1) for each single indicator: Coverage of Characteristic Macroalgae (C, top right), Fraction of Opportunistc Macroalgae (F, bottom right) and Richness of Characteristic Macroalgae (R, bottom left) and the maximum Pressure Level (0–4) estimated for each sampling station exposed to point urban discharges.

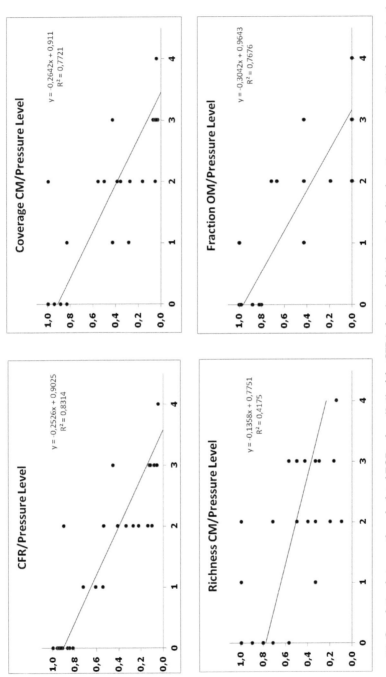

Figure 11. Correlation between the calculated EQR values (0–1) of the CFR index (top left), the standardized continuous scores (0–1) for each single indicator: Coverage of Characteristic Macroalgae (C, top right), Fraction of Opportunistic Macroalgae (F, bottom right) and Richness of Characteristic Macroalgae (R, bottom left) and the maximum Pressure Level (0–4) estimated for each sampling station exposed to a point industrial discharge.

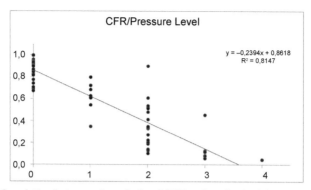

Figure 12. Correlation between the calculated EQR values (0–1) of the CFR index and the maximum Pressure Level (0–4) estimated for each sampling station exposed to both point urban and industrial discharge.

coast, macroalgal assemblages present at different biogeographical regions responded comparatively well to the anthropogenic pressure affecting each of the areas as proved by the similar results obtained by the single indicators (e.g., species richness, % opportunists) for similar pressure levels. At the same time, different assessment indices (RSLA, MarMAT and CFR) were able to provide results significantly correlated to the pressure affecting each of the sites. The capacity to react to environmental conditions allows macroalgae to be seen as functional natural loggers, able to transmit (though physiological signals) the accumulated characteristics of the water column. In this context, it's possible to create tools to measure ecological quality based on the response of macroalgae indicators and, after a well succeed IC exercise, performed at the European level; the RSLA, MarMAT and CFR indices (as well as all the other intercalibrated indices) have proven to be appropriate methodologies to carry out extensive monitoring of intertidal rocky shores. This possibility, in the case of CFR, is also applicable to subtidal environments.

Acknowledgments

Fundação para a Ciência e a Tecnologia supported João M. Neto through the programs POPH and QREN (FSE and national funds of MEC). This research was partially funded by the Spanish Ministry of Science and Innovation (VIGES2 project, CTM2008-04649) and by a research grant from the Environment Department of the Regional Government of Cantabria.

Keywords: North East Atlantic, Water Framework Directive, rocky shore, pollution sensitive, opportunistic, morpho-functional traits, assessment tools, pressure response, physiological mechanisms, biogeographical variation, reference conditions, intercalibration, ecological quality status

References Cited

Abbott, I.A. and G.J. Hollenberg. 1976. Marine Algae of California Stanford University Press, Stanford, California. 827.

Adams, S.M. 2005. Assessing cause and effect of multiple stressors on marine systems. Marine Pollution Bulletin 51: 649–657.

Anderson, D.M. 1995. ECOHAB, the ecology and oceanography of harmful algal blooms: A national research agenda. Woods Hole, MA: Woods Hole Oceanographic Institution. 66 pp.

Andral, B. and V. Derolez. 2007. Mise en oeuvre du Contrôle de surveillance. Résultats de la campagne 2006, District Rhône et Côtier Mediterraneens (Directive Cadre Eau). IFREMER, Décembre 2007 RST. DOP/LER-PAC/07-28.

Angelone, M. and C. Bini. 1992. Trace elements concentrations in soils and plants of Western Europe. pp. 19–60. In: Adriano, D.C. (ed.). Biogeochemistry of Trace Metals. Lewis Publishers, Boca Raton, FL.

Angus, S. 2014. The implications of climate change for coastal habitats in the Uists, Outer Hebrides. Ocean & Coastal Management 94: 38–43.

Angus, S. and A. Rennie. 2014. An Ataireachd Aird: The storm of January 2005 in the Uists, Scotland. Ocean & Coastal Management 94: 22–29.

Arévalo, R., S. Pinedo and E. Ballesteros. 2007. Changes in the composition and structure of Mediterranean rocky-shore communities following a gradient of nutrient enrichment: Descriptive study and test of proposed methods to assess water quality regarding macroalgae. Marine Pollution Bulletin 55: 104–113.

Balata, D., L. Piazzi and F. Rindi. 2011. Testing a new classification of morphological functional groups of marine macroalgae for the detection of responses to stress. Marine Biology 158: 2459–2469 (DOI 10.1007/s00227-011-1747-y).

Ballesteros, E., X. Torras, S. Pinedo, M. García, L. Mangialajo and M. de Torreset. 2007a. A new methodology based on littoral community cartography dominated by macroalgae for the implementation of the European Water Framework Directive. Marine Pollution Bulletin 55: 172–180.

Ballesteros, E., S. Pinedo and R. Arévalo. 2007b. Comments on the development of new macroalgal indices to assess water quality within the Mediterranean Sea: A reply. Marine Pollution Bulletin 54: 628–630.

Bartsch, I., C. Wiencke and T. Laepple. 2012. Global seaweed biogeography under a changing climate: the prospected effects of temperature. pp. 383–406. In: Wiencke, C. and K. Bischof (eds.). Seaweed Biology, Ecological Studies 219. Springer-Verlag, Berlin, Heidelberg (DOI 10.1007/978-3-642-28451-9_18).

Bell, E.C. 1993. Photosynthetic response to temperature and desiccation of the intertidal alga *Mastocarpus papillatus* Kützing. Marine Biology 117: 337–346.

Birk, S. and N. Willby. 2010. Towards harmonization of ecological quality classification: establishing common grounds in European macrophyte assessment for rivers. Hydrobiologia 652: 149–163.

Birk, S., W. Bonne, A. Borja, S. Brucet, A. Courrat, S. Poikane, A. Solimini, W. van de Bund, N. Zampoukas and D. Hering. 2012. Three hundred ways to assess Europe's surface waters: an almost complete overview of biological methods to implement the Water Framework Directive. Ecological Indicators 18: 31–41.

Björk, M., F. Short, E. Mcleod and S. Beer. 2008. Managing Seagrasses for Resilience to Climate Change. IUCN, Gland, Switzerland, 56 pp.

Blomqvist, M., D. Krause-Jensen, P. Olsson, S. Qvarfordt and S.A. Wikström. 2012. Potential eutrophication indicators based on Swedish coastal macrophytes. Deliverable 3.2-1, WATERS Report no. 2012:2. Havsmiljöinstitutet, Sweden (http://www.waters.gu.se/rapporter).

Bolam, S.G., T.F. Fernandes, P. Read and D. Raffaelli. 2000. Effects of macroalgal mats on intertidal sandflats: an experimental study. Journal of Experimental Marine Biology and Ecology 249: 123–137.

Boller, M.L. and E. Carrington. 2007. Interspecific comparison of hydrodynamic performance and structural properties among intertidal macroalgae. The Journal of Experimental Biology 210: 1874–1884 (doi:10.1242/jeb.02775).

Borja, A. and D.M. Dauer. 2008. Assessing the environmental quality status in estuarine and coastal systems: comparing methodologies and indices. Ecological Indicators 8: 331–337.

Burdin, K.S. and K.T. Bird. 1994. Heavy metal accumulation by car-rageenan and agar producing algae. Botanica Marina 37: 467–470.

Burkholder, J., D.A. Tomasko and B.W. Touchette. 2007. Seagrasses and eutrophication. Journal of Experimental Marine Biology and Ecology 350: 46–72.

Bushaw-Newton, K.L. and K.G. Sellner. 1999 (online). Harmful Algal Blooms. *In*: NOAA's State of the Coast Report. Silver Spring, MD: National Oceanic and Atmospheric Administration (http//state-of-coast.noaa.gov/bulletins/html/hab_14/hab.html).

Cabral, J.A., M.Â. Pardal, R.J. Lopes, T. Múrias and J.C. Marques. 1999. The impact of macroalgal blooms on the use of the intertidal area and feeding behaviour of waders (Charadrii) in the Mondego estuary (west Portugal). Acta Oecologica 20: 417–427.

Campbell, P.G.C. 1995. Interactions between metals and aquatic organisms. pp. 45–102. *In*: Tessier, A. and D.R. Turner (eds.). Metal Speciation and Bioavailability in Aquatic Systems. John Wiley, New York.

Carrington, E. 1990. Drag and dislidgment of an intertidal macroalga: consequences of morphological variation in *Mastocarpus papillatus* Kützing. Journal of Experimental Marine Biology and Ecology 139: 185–200.

Chan, S.M., W. Wang and I. Ni. 2003. The uptake of Cd, Cr, and Zn by the Macroalga *Enteromorpha crinita* and subsequent transfer to the marine Herbivorous Rabbitfish, *Sigunus canaliculatus*. Archives of Environmental Contamination and Toxicology 44: 298–306.

Chung, C., R.L. Hwang, S.H. Lin, T.M. Wu, J.Y. Wu, S.W. Su, C.S. Chen and T.M. Lee. 2007. Nutrients, temperature, and salinity as primary factors influencing the temporal dynamics of macroalgal abundance and assemblage structure on a reef of Du-Lang Bay in Taitung in southeastern Taiwan. Botanical Studies 48: 419–433.

Conley, D.J., J. Carstensen, R. Vaquer-Sunyer and C.M. Duarte. 2009. Ecosystems thresholds with hypoxia. Hydrobiologia 629: 21–29 (DOI: 10.1007/s10750-009-9764-2).

Corredor, J.E., R. Howarth, R.R. Twilley and J.M. Morell. 1999. Nitrogen cycling and anthropogenic impact in the tropical interamerican seas. Biogeochemistry 46: 163–178.

Cossellu, C. and K. Nordberg. 2010. Recent environmental changes and filamentous algal mats in shallow bays on the Swedish west coast—A result of climate change? Journal of Sea Research 63: 202–212.

Dailer, M., R.S. Knox, J.E. Smith, M. Napier and C.M. Smith. 2010. Using delta-15 N values in algal tissue to map locations and potential sources of anthropogenic nutrient inputs on the island of Maui, Hawaii, USA. Marine Pollution Bulletin 60: 655–671.

Dawes, C.J., J. Orduna-Rojas and D. Robledo. 1998. Response of the tropical red seaweed Gracilaria cornea to temperature, salinity and irradiance. J. Appl. Phycol. 10: 419–425.

Denny, M. and B. Gaylord. 2002. The mechanics of wave-swept algae. The Journal of Experimental Biology 205: 1355–1362.

den Hartog, C. 1994. Suffocation of a littoral *Zostera* bed by *Enteromorpha radiata*. Aquatic Botany 47: 21–28.

Diaz, M.R., J.E. Corredor and J.M. Morell. 1990. Inorganic nitrogen uptake by *Microcoleus lynbyaceus* mat communities in a semi-eutrophic marine community. Limnology and Oceanography 35: 1788–1795.

Diaz-Pulido, G. and L. McCook. 2008. Macroalgae (Seaweeds). 44 pp. *In*: Chin, A. (ed.). The State of the Great Barrier Reef On-line, Great Barrier Reef Marine Park Authority,

Townsville (http://www.gbrmpa.gov.au/corp_site/info_services/publications/sotr/downloads/SORR_Macroalgae.pdf).

Díez, I., M. Bustamante, A. Santolaria, J. Tajadura, N. Muguerza, A. Borja, I. Muxika, J.I. Saiz-Salinas and J.M. Gorostiaga. 2012. Development of a tool for assessing the ecological quality status of intertidal coastal rocky assemblages, within Atlantic Iberian coasts. Ecological Indicators 12: 58–71.

Duarte, C.M. 1995. Submerged aquatic vegetation in relation to different nutrient regimes. Ophelia 41: 87–112.

Elliott, M. and V.N. de Jonge. 2002. The management of nutrients and potential eutrophication in estuaries and other restricted water bodies. Hydrobiologia 475–476: 513–524.

Ellis, D.V. 2003. Rocky shore intertidal zonation as a means of monitoring and assessing shoreline diversity recovery. Marine Pollution Bulletin 46: 305–307.

Eriksson, B.K. and L. Bergstrom. 2005. Local distribution patterns of macroalgae in relation to environmental variables in the northern Baltic Proper. Estuarine, Coastal and Shelf Science 62: 109–117.

Fenchel, T. and B.B. Jorgensen. 1977. Detritus food chains of aquatic ecosystems: The role of bacteria. pp. 1–58. *In*: Alexander, M. (ed.). Advances in Microbial Ecology, Vol. 1. Plenum Press.

Fletcher, R.L. 1996a. The occurrence of "green tides"—a review. pp. 7–43. *In*: Schramm, W. and P.H. Nienhuis (eds.). Marine Benthic Vegetation: Recent Changes and the Effects of Eutrophication. Springer, Berlin, Heidelberg, New York.

Fletcher, R.L. 1996b. The British Isles. pp. 150–223. *In*: Schramm, W. and P.H. Nienhuis (eds.). Marine Benthic Vegetation: Recent Changes and the Effects of Eutrophication. Springer, Berlin, Heidelberg, New York.

Fong, P. and J.B. Zedler. 2000. Sources, sinks, and fluxes of nutrients (N + P) in a small highly-modified estuary in southern California. Urban Ecosystems 4: 125–144.

Fong, P., R.M. Donohoe and J.B. Zedler. 1994. Nutrient concentration in tissue of the macroalga Enteromorpha as a function of nutrient history: an experimental evaluation using field microcosms. Marine Ecology Progress Series 106: 273–281.

Fong, P., J.J. Fong and C.R. Fong. 2004. Growth, nutrient storage, and release of dissolved organic nitrogen by *Enteromorpha intestinalis* in response to pulses of nitrogen and phosphorus. Aquatic Botany 78: 83–95.

Fugita, R.M. 1985. The role of nitrogen status in regulating transient ammonium uptake and nitrogen storage by macroalgae. Journal of Experimental Marine Biology and Ecology 92: 283–301.

Galloway, J.N., J.D. Thornton, S.A. Norton, H.L. Volcho and R.A. McLean. 1982. Trace metals in atmospheric deposition: a re-view and assessment. Atmospheric Environment 16: 1677.

Gamenick, I., A. Jahn, K. Vopel and O. Giere. 1996. Hypoxia and sulphide as structuring factors in a macrozoobenthic community on the Baltic Sea shore: colonisation studies and tolerance studies. Marine Ecology Progress Series 144: 75–85.

Gaspar, R., L. Pereira and J.M. Neto. 2012. Ecological reference conditions and quality states of marine macroalgae sensu Water Framework Directive: An example from the intertidal rocky shores of the Portuguese coastal waters. Ecological Indicators 19: 24–38.

Gattuso, J.P. and R.W. Buddemeier. 2000. Ocean biogeochemistry: calcification and CO2. Nature 407: 311–313.

Gaylord, B., C.A. Blanchette and M.W. Denny. 1994. Mechanical consequences of size in wave swept algae. Ecological Monographs 64: 287–313.

Gorostiaga, J.M. and I. Díez. 1996. Changes in the sublittoral benthic marine macroalgae in the polluted area of Abra de Bilbao and proximal coast (Northern Spain). Marine Ecology Progress Series 30: 157–167.

Graham, M.H. 2004. Effects of local deforestation on the diversity and structure of Southern California giant kelp forest food webs. Ecosystems 7: 341–357.

Gubelit, Y.I. 2012. Factors, influencing on macroalgal communities in the Neva estuary (eastern Baltic Sea). Baltic International Symposium (BALTIC) (DOI: 10.1109/BALTIC.2012.6249173).

Guinda, X., A. Gracia, A. Puente, J.A. Juanes, Y. Rzhanov and L. Mayer. 2013. Application of landscape mosaics for the assessment of subtidal macroalgae communities using the CFR index. Deep-Sea Research Part II: Topical Studies in Oceanography (accepted) (DOI 10.1016/j.dsr2.2013.09.037).

Guinda, X., J.A. Juanes and A. Puente. 2014. A validated method for the assessment of macroalgae according to the European Water Framework Directive. Journal of Marine Environmental Research (submitted).

Hall, J.L. 2002. Cellular mechanisms for heavy metal detoxification and tolerance (review). Journal of Experimental Botany 53(366): 1–11.

Harley, C.D.G., K.M. Anderson, K.W. Demes, J.P. Jorve, R.L. Kordas, T.A. Coyle and M.H. Graham. 2012. Effects of climate change on global seaweed communities. Journal of Phycology 48: 1064–1078 (DOI: 10.1111/j.1529-8817.2012.01224.x).

Hoffmann, A.J. and R. Ugarte. 1985. The arrival of propagules of marine macroalgae in the intertidal zone. Journal of Experimental Marine Biology and Ecology 92(1): 83–95.

Israel, A., M. Martinez-Goss and M. Friedlander. 1999. Effect of salinity and pH on growth and agar yield of *Gracilaria tenuistipitata* var. liui in laboratory and outdoor cultivation. Journal of Applied Phycology 11: 543–549.

Johnson, L.E. and S.H. Brawley. 1999. Dispersal and recruitment of a canopy-forming intertidal alga: The relative roles of propagule availability and post-settlement processes. Oecologia 117: 517–526.

Juanes, J.A., X. Guinda, A. Puente and J.A. Revilla. 2008. Macroalgae, a suitable indicator of the ecological status of coastal rocky communities in the NE Atlantic. Ecological Indicators 8(4): 351–359.

Juanes, J.A., X. Guinda, J.M. Neto, A. Pedersen, R. Melo, R. Nogueira-Mendes, R. Gaspar, A. de Ugarte, A. Borja, I. Muxika, I. Hernández, R. Bermejo, R. Buchet, E. Ar Gall, M. Le Duff, R. Wilkes, C. Scanlan, M. Best, W. Heiber and I. Bartsch. 2012. Intercalibration of coastal intertidal macroalgae assessment methods in the NE Atlantic region within the European Water Framework Directive context, 50th ECSA Conference: Today's science for tomorrow's Management, Venice (Italy).

Kaewsarn, P. and Q. Yu. 2001. Cadmium removal from aqueous solutions by pre-treated biomass of marine alga *Padina* sp. Environmental Pollution 122: 209–213.

Kamer, K. and P. Fong. 2000. A fluctuating salinity regime mitigates the negative effects of reduced salinity on the estuarine macroalga, *Enteromorpha intestinalis* (L.) link. Journal of Experimental Marine Biology and Ecology 254: 53–69.

Kamer, K., K.A. Boyle and P. Fong. 2001. Macroalgal dynamics in a highly eutrophic southern California estuary. Estuaries 24: 623–635.

Kautsky, L., C. Wibjörn and H. Kautsky. 2007. Bedömningsgrunder för kust och hav enligt krav i ramdirektivet vatten – makroalger och några gömfröiga vattenväxter. Rapport till Naturvårdsverket 2007-05-25. 50 pp.

Kremer, B.D. and J.W. Markham. 1982. Primary metabolic effects of cadmium in the brown alga *Laminaria saecharina*. Z. Pflanzenphysiol. 108: 125–130.

Kristensen, E., J.M. Neto, M. Lundkvist, L. Frederiksen, M.A. Pardal, T. Valdemarsen and M.R. Flindt. 2013. Influence of benthic macroinvertebrates on the erodability of estuarine cohesive sediments: Density- and biomass-specific responses. Estuarine, Coastal and Shelf Science 134: 80–87.

Kroeker, K.J., R.L. Kordas, R.N. Crim and G.G. Singh. 2010. Meta-analysis reveals negative yet variable effects of ocean acidification on marine organisms. Ecol. Lett. 13: 1419–34.

Kruk-Dowgiallo, L. 1991. Long-term changes in the structure of underwater meadows of the Puck lagoon. Acta Ichthyol. Piscator, Supplement 22: 77–84.

Kuhlenkamp, R., P. Schubert and I. Bartsch. 2011. Macrophyte Monitoring Helgoland— Report 17 (Water Framework Directive Monitoring Component Macrophytobenthos N5 Helgoland, EQR Evaluation 2010). Bremerhaven, Germany (LLUR-SH: LLUR-AZ 0608.451013).

Langston, W.J. 1990. Toxic effect of metals and the incidence of metal pollution in marine ecosystems. pp. 101–122. *In*: Furness, R.W. and P.S. Rainbow (eds.). Heavy Metals in the Marine Environment. CRC Press, Boca Raton, FL.

Larsen, A. and K. Sand-Jensen. 2006. Salt tolerance and distribution of estuarine benthic macroalgae in the Kattegat- Baltic Sea area. Phycologia 45: 13–23.

Lewis, J.R. 1978. The Ecology of Rocky Shores. Hodder and Stoughton, London.

Lillebø, A.I., J.M. Neto, I. Martins, T. Verdelhos, S. Leston, P.G. Cardoso, S.M. Ferreira, J.C. Marques and M.A. Pardal. 2005. Management of a shallow temperate estuary to control eutrophication: the effect of hydrodynamics on the system nutrient loading. Estuarine Coastal and Shelf Science 65: 697–707.

Llácer, J.L., I. Fita and V. Rubio. 2008. Arginine and nitrogen storage. Current Opinion in Structural Biology 18: 673–681.

Lotze, H.K. and W. Schramm. 2000. Can ecophysiological traits explain species dominance patterns in macroalgal blooms? Journal of Phycology 36: 287–295.

Lotze, H.K., W. Schramm, D. Schories and B. Worm. 1999. Control of macroalgal blooms at early developmental stages: *Pilayella littoralis* versus *Enteromorpha* spp. Oecologia 119: 46–54.

Lotze, H.K., B. Worm and U. Sommer. 2000. Propagule banks, herbivory and nutrient supply control population development and dominance patterns in macroalgal blooms. Oikos 89: 46–58.

Marques, J.C., S.N. Nielsen, M.A. Pardal and S.E. Jørgensen. 2003. Impact of eutrophication and river management within a framework of ecosystem theories. Ecological Modelling 166: 147–168.

Martins, I., J.M. Oliveira, M.R. Flindt and J.C. Marques. 1999. The effect of salinity on the growth rate of the macroalgae *Enteromorpha intestinalis* (Chlorophyta) in the Mondego estuary (west Portugal). Acta Oecologica 20(4): 259–265.

Martins, I., M.A. Pardal, A.I. Lillebø, M.R. Flindt and J.C. Marques. 2001. Hydrodynamics as a major factor controlling the occurrence of green macroalgal blooms in a eutrophic estuary: a case study on the influence of precipitation and river management. Estuarine, Coastal and Shelf Science 52: 165–177.

McComb, A.J. and R. Humphries. 1992. Loss of nutrients from catchments and their ecological impacts in the Peel-Harvey estuarine system, Western Australia. Estuaries 15: 529–537.

Milligan, K.L.D. and R.E. DeWreede. 2000. Variations in holdfast attachment mechanics with developmental stage, substratum-type, season, and wave-exposure for the intertidal kelp species *Hedophyllum sessile* (C. Agardh) Setchell. Journal of Experimental Marine Biology and Ecology 254: 189–209.

Molvær, J., J. Knutzen, J. Magnusson, B. Rygg, J. Skei and J. Sørensen. 1997. Classification of environmental quality in fjords and coastal waters: a guide. Norwegian State Pollution Control Authority (SFT) publication 97:03. Oslo, Norway 36 pp.

Müller, R., T. Laepple, I. Bartsch and C. Wiencke. 2009. Impact of oceanic warming on the distribution of seaweeds in polar and cold-temperate waters. Botanica Marina 52: 617–638 (DOI 10.1515/BOT.2009.080).

Naldi, M. and P. Viaroli. 2002. Nitrate uptake and storage in the seaweed *Ulva rigida* C. Agardh. in relation to nitrate availability and thallus nitrate content in a eutrophic coastal lagoon (Sacca di Goro, Po River Delta, Italy). Journal of Experimental Marine Biology and Ecology 269(1): 65–83.

Neto, J.M., R. Gaspar, L. Pereira and J.C. Marques. 2012. Marine Macroalgae Assessment Tool (MarMAT) for intertidal rocky shores. Quality assessment under the scope of the European Water Framework Directive. Ecological Indicators 19: 39–47.

Nixon, S., B. Buckley, S. Granger and J. Bintz. 2001. Responses of very shallow marine ecosystems to nutrient enrichment. Human and Ecological Risk Assessment: An International Journal 7: 1457–1481 (DOI: 10.1080/20018091095131).

Nixon, S.W. and R.W. Fulweiler. 2009. Nutrient pollution, eutrophication, and the degradation of coastal marine ecosystems. pp. 25–60. *In*: Duarte, Carlos M. (ed.). Global Loss of Coastal Habitats Rates, Causes and Consequences. Fundación BBVA.

Nixon, S.W., J.W. Ammerman, L.P. Atkinson, V.M. Beroundsky, G. Billen, W.C. Boicourt, W.R. Boynton, T.M. Church, D.M. Ditoro, R. Elmgren, J.H. Garber, A.E. Giblin, R.A. Jahnke, N.J.P. Owens, M.E.Q. Pilson and S.P. Seitzinger. 1996. The fate of nitrogen and phosphorus at the land-sea margin of the North Atlantic Ocean. Biogeochemistry 35: 141–180.

Norderhaug, K.N., H. Christie, J.H. Fossa and S. Fredriksen. 2005. Fish-macrofauna interactions in a kelp (*Laminaria hyperborea*) forest. Journal of the Marine Biology Association of the U.K. 85: 1279–1286.

Norkko, A. and E. Bonsdorff. 1996. Population responses of coastal zoobenthos to stress induced by drifting algal mats. Marine Ecology Progress Series 140: 141–151.

Orfanidis, S. 2007. Comments on the development of new macroalgal indices to assess water quality within the Mediterranean Sea. Marine Pollution Bulletin 54: 626–627.

Orfanidis, S., P. Panayotidis and N. Stamatis. 2001. Ecological evaluation of transitional and coastal waters: a marine benthic macrophytes-based model. Mediterranean Marine Research 2(2): 45–65.

Orfanidis, S., P. Panayotidis and K.I. Ugland. 2011. Ecological Evaluation Index con-tinuous formula (EEI-c) application: a step forward for functional groups, the formula and reference condition values. Mediterr. Mar. Sci. 12(1): 199–231.

Patrício, J., J.M. Neto, H. Teixeira and J.C. Marques. 2007. Opportunistic macroalgae metrics for transitional waters. Testing tools to assess ecological quality status in Portugal. Marine Pollution Bulletin 54: 1887–1896.

Patrício, J., J.M. Neto, H. Teixeira, F. Salas and J.C. Marques. 2009. The robustness of ecological indicators to detect long-term changes in the macrobenthos of estuarine systems. Marine Environmental Research 68: 25–36.

Peckol, P. and J.S. Rivers. 1995. Physiological responses of the opportunistic macroalgae *Cladophora vagabunda* (L.) van den Hoek and *Gracilaria tikvahiae* (McLachlan) to environmental disturbances associated with eutrophication. Journal of Experimental Marine Biology and Ecology 190: 1–16.

Pedersen, A., G. Kraemer and C. Yarish. 2004. The effects of temperature and nutrient concentrations on nitrate and phosphate uptake in different species of *Porphyra* from Long Island Sound (USA). Journal of Experimental Marine Biology and Ecology 312: 235–252.

Pedersen, M.F. and J. Borum. 1996. Nutrient control of algal growth in estuarine waters. Nutrient limitation and the importance of nitrogen requirements and nitrogen storage among phytoplankton and species of macroalgae. Marine Ecology Progress Series 142: 261–272.

Peterson, H.G., F.P. Healey and R. Wagemann. 1984. Metal toxicity to algae: a highly pH dependant phenomenon. Canadian Journal of Fisheries and Aquatic Sciences 41: 974–979.

Pihl, L., G. Magnusson, I. Isaksson and I. Wallentinus. 1996. Distribution and growth dynamics of ephemeral macroalgae in shallow bays on the Swedish west coast. Journal of Sea Research 35: 169–180.

Pinedo, S., M. Garcia, M.P. Satta, M. De Torres and E. Ballesteros. 2007. Rocky-shore communities as indicators of water quality: a case study in the Northwestern Mediterranean. Marine Pollution Bulletin 55: 126–135.

Raffaelli, D., S. Hull and H. Milne. 1989. Longterm changes in nutrients, weed mats and shorebirds in an estuarine system. Cahiers de Biologie Marine 30: 259–270.

Raffaelli, D.G., J.A. Raven and L.J. Poole. 1998. Ecological impact of green macroalgal blooms. Oceanography and Marine Biology 36: 97–125.

Rai, L., J.P. Gaur and H.D. Kumar. 1981. Physiology and heavy metal pollution. Biological Reviews 56: 99–151.

Raize, O., Y. Argaman and S. Yannai. 2004. Mechanisms of biosorption of different heavy metals by brown marine macroalgae. Biotechnology and Bioengineering 87: 451–458 (DOI: 10.1002/bit.20136).

Ralph, P.J., D. Tomasko, K. Moore, S. Seddon and C.M.O. Macinnis-Ng. 2006. Human impacts on seagrasses: eutrophication, sedimentation, and contamination. pp. 567–593. *In*: Larkum,

A.W.D., R.J. Orth and C.M. Duarte (eds.). Seagrasses: Biology, Ecology and Conservation. Springer, The Netherlands.

Ramos, E., J.A. Juanes, C. Galván, J.M. Neto, R. Melo, A. Pedersen, C. Scanlan, R. Wilkes, E. van den Bergh, M. Blomqvist, H.P. Karup, W. Heiber, J. Reitsma, M.C. Ximenes, A. Silió, F. Méndez and B. González. 2012. Coastal waters classification based on physical attributes along the NE Atlantic region. An approach for rocky macroalgae potential distribution. Estuarine, Coastal & Shelf Science 112: 105–114.

Ramos, E., A. Puente, J.A. Juanes, J.M. Neto, A. Pedersen, I. Bartsch, C. Scanlan, R. Wilkes, E. Van den Bergh, E. Ar Gall and R. Melo. 2014. Biological validation of physical coastal waters classification along the NE Atlantic region based on rocky macroalgae distribution. Estuarine, Coastal and Shelf Science (in press).doi: 10.1016/j.ecss.2014.05.036.

Rasmussen, J.R., B. Olesen and D. Krause-Jensen. 2012. Effects of filamentous macroalgae mats on growth and survival of eelgrass, Zostera marina, seedlings. Aquatic Botany 99: 41–48.

Reise, K. and I. Siebert. 1994. Mass occurrence of green algae in the German Wadden Sea. German Journal of Hydrography (Suppl.)1: 171–180.

Rönnbäck, P., N. Kautsky, L. Pihl, M. Troell, T. Söderqvist and H. Wennhage. 2007. Ecosystem goods and services from Swedish coastal habitats: identification, valuation, and implications of ecosystem shifts. Ambio 36: 534–544.

Rost, B., I. Zondervan and U. Riebesell. 2002. Light-dependent carbon isotope fractionation in the coccolithophorid *Emiliania huxleyi*. Limnology Oceanography 47: 120–128.

Royo, C.L., C. Silvestri, G. Pergent and G. Casazza. 2009. Assessing human-induced pressures on coastal areas with publicly available data. Journal of Environmental Management 90: 1494–1501.

Runcie, J.W., R.J. Ritchie and A.W.D. Larkum. 2003. Uptake kinetics and assimilation of inorganic nitrogen by *Catenella nipae* and *Ulva lactuca*. Aquatic Botany 76: 155–174.

Russell, D.J. and G.H. Balazs. 2009. Dietary shifts by green turtles (*Chelonia mydas*) in the Kaneohe Bay region of the Hawaiian islands: a 28 year study. Pacific Science 63: 181–192.

Ryther, J.H. and W.M. Dunstan. 1971. Nitrogen, phosphorous and eutrophication in the coastal marine environment. Science 171: 1008–1013.

Salgado, L.T., L.R. Andrade and G.M. Amado Filho. 2005. Localization of specific monosaccharides in cells of the brown alga *Padina gymnospora* and the relation to heavy-metal accumulation. Protoplasma 225: 132–128.

Scanlan, C.M., J. Foden, E. Wells and M.B. Best. 2007. The monitoring of opportunistic macroalgal blooms for the Water Framework Directive. Marine Pollution Bulletin 55: 162–171.

Seitzinger, S.P. 1988. Denitrification in freshwater and coastal marine ecosystems: ecological and geochemical significance. Limnology and Oceanography 33(4): 702–724.

Sfriso, A., B. Pavoni, A. Marcomini and A.A. Orio. 1992. Macroalgal nutrient cycles and pollutants in the Lagoon of Venice. Estuaries 15(4): 517–528.

Sfriso, A., T. Birkemeyer and P.F. Ghetti. 2001. Benthic macrofauna changes in areas of Venice lagoon populated by seagrasses or seaweeds. Marine Environmental Research 52: 323–349.

Sfriso, A., C. Facca and P.F. Ghetti. 2009. Validation of the Macrophyte Quality Index (MaQI) set up to assess the ecological status of Italian marine transitional environments. Hydrobiologia 617: 117–141.

Short, F.T., B. Polidoro, S.R. Livingstone, K.E. Carpenter, S. Bandeira, J.S. Bujang, H.P. Calumpong, T.J.B. Carruthers, R.G. Coles, W.C. Dennison, P.L.A. Erftemeijer, M.D. Fortes, A.S. Freeman, T.G. Jagtap, A.H.M. Kamal, G.A. Kendrick, W.J. Kenworthy, Y.A. La Nafie, I.M. Nasution, R.J. Orth, A. Prathep, J.C. Sanciangco, B. van Tussenbroek, S.G. Vergara, M. Waycott and J.C. Zieman. 2011. Extinction risk assessment of the world's seagrass species. Biological Conservation 144: 1961–1971.

Sorentino, C. 1979. The effect of heavy metals on phytoplankton—a review. Phykos 18: 149–161.

Soulsby, P.G., D. Lowthion and M. Houston. 1982. Effects of macroalgal mats on the ecology of intertidal mudflats. Marine Pollution Bulletin 13: 162–166.

Strömgren, T. 1980. Combined effects of copper, zinc and mercury on the increase in length of *Ascophyllum nodosum*. Journal of Experimental Marine Biology and Ecology 48: 225–231.

Sunda, W.G. and S.A. Huntsman. 1998. Processes regulating cellular metal accumulation and physiological effects: phytoplankton as model systems. Sci. Total Environ. 219: 165–181.

Thomsen, M.S. and T. Wernberg. 2005. Miniview: What affects the forces required to break or dislodge macroalgae? European Journal of Phycology 40(2): 139–148.

Tubbs, C.R. and J.M. Tubbs. 1980. Waders and shelduck feeding distribution in Langstone Harbour, Hampshire. Bird Study 27: 239–248.

Twilley, R.R., W.M. Kemp, K.W. Staver, J. Stevenson and W. Boynton. 1985. Nutrient enrichment of estuarine submersed vascular plant communities. 1. Algal growth and effects on production of plants and associated communities. Marine Ecology Progress Series 179–191.

Tyler, A.C., K.J. McGlathery and I.C. Anderson. 2001. Macroalgae mediation of dissolved organic nitrogen fluxes in a temperate coastal lagoon. Estuarine, Coastal and Shelf Science 53: 155–168.

Valiela, I. 2006. Global Coastal Change. Blackwell Publishing, Oxford.

Valiela, I., J. McClelland, J. Hauxwell, P.J. Behr, D. Hersh and K. Foreman. 1997. Macroalgal blooms in shallow estuaries: controls and ecophysiological and ecosystem consequences. Limnology and Oceanography 42: 1105–1118.

van den Hoek, C. 1982. The distribution of benthic marine algae in relation to the temperature regulatiog of their life histories. Biological Journal of the Linnean Society 18: 81–144.

Van Houtan, K.S., S.K. Hargrove and G.H. Balazs. 2010. Land Use, Macroalgae, and a Tumor-Forming Disease in Marine Turtles. PLoS ONE 5(9): e12900 (DOI: 10.1371/journal.pone.0012900).

Vaquer-Sunyer, R. and C.M. Duarte. 2010. Sulfide exposure accelerates hypoxia-driven mortality. Limnology and Oceanography 55(3): 1075–1082 (DOI:10.4319/lo.2010.55.3.1075).

Vistisen, B. and B. Vismann. 1997. Tolerance to low oxygen and sulfide in *Amphiura filiformis* and *Ophiura albida* (Echinodermata: Ophiuroidea). Marine Biology 128: 241–246 (DOI:10.1007/s002270050088).

Vogt, H. and W. Schramm. 1991. Conspicuous decline of Fucus in Kiel Bay (western Baltic): what are the causes? Marine Ecology Progress Series 69: 189–194.

Wallenstein, F.M., A.I. Neto, R.F. Patarra, A.C.L. Prestes, N.V. Álvaro, A.S. Rodrigues and M. Wilkinson. 2013. Indices to monitor coastal ecological quality of rocky shores based on seaweed communities: simplification for wide geographical use. Journal of Integrated Coastal Zone Management 13(1): 15–25.

Wallentinus, I. 1984. Comparisons of nutrient uptake rates for Baltic macroalgae with different thallus morphologies. Marine Biology 80: 215–225.

Wells, E. 2002. Seaweed species biodiversity on rocky intertidal seashores in the British Isles. Ph.D. Thesis, Heriot Watt University, Edinburgh.

Wells, E., M. Wilkinson, P. Wood and C. Scanlan. 2007. The use of macroalgal species richness and composition on intertidal rocky seashores in the assessment of ecological quality under the European Water Framework Directive. Marine Pollution Bulletin 55: 151–161.

Widdows, J. and M. Brisley. 2002. Impact of biotic and abiotic processes on sediment dynamics and the consequences to the structure and functioning of the intertidal zone. Journal of Sea Research 48: 143–156.

Wilkes, R. 2005. A desk and field study of biomass estimation techniques for monitoring green tides in the Irish environment. A report to the Irish Environmental Protection Agency, Environmental Research, Technological Development & Innovation Programme.

Wilkinson, M., P. Wood, E. Wells and C. Scanlan. 2007. Using attached macroalgae to assess ecological status of British estuaries for the European Water Framework Directive. Marine Pollution Bulletin 55: 136–150.

Zondervan, I., R.E. Zeebe, B. Rost and U. Riebesell. 2001. Decreasing marine biogenic calcification: a negative feedback on rising atmospheric pCO2. Global Biogeochemical Cycles 15: 507–516.

Zondervan, I., B. Rost and U. Riebesell. 2002. Effect of CO_2 concentration on the POC/PIC ration in the coccolithophore Emiliania huxleyi grown under light-limiting conditions and different day lengths. Journal of Experimental Marine Biology and Ecology 272: 55–70.

CHAPTER 4

Understanding Biological Invasions by Seaweeds

Fátima Vaz-Pinto,[1,a,]* *Ivan F. Rodil,*[1,b] *Frédéric Mineur,*[2,3]
Celia Olabarria[4] *and Francisco Arenas*[1,c]

1 Introduction

Every single day thousands of marine species are transported by human activities around the Earth, far away from their native location to new areas where they become non-indigenous species (NIS). However, whether this introduced species will be successful in its new host habitat is a dynamic, very complex process entailing several transitional stages (Williamson 2006). During the process of invasion, NIS must overcome several ecological barriers at different stages of invasion before they are able to inflict ecological or economic harm. Colautti and MacIssac (2004) suggested a well defined framework linking the invasion process with different filters that have to be overcome to achieve subsequent stages. Throughout this chapter, an overview of the stages involved in the process of invasion of marine macroalgae will be presented, followed by a set of management

[1] Laboratory of Coastal Biodiversity, CIIMAR—University of Porto, Rua dos Bragas 289, 4050-123 Porto, Portugal.
[a] Email: f_vazpinto@yahoo.com
[b] Email: ibantxorodil@gmail.com
[c] Email: farenas@ciimar.up.pt
[2] School of Biological Sciences, Queen's University of Belfast, UK.
[3] Phycology Research Group, Ghent University, Belgium.
 Email: f.mineur@qub.ac.uk
[4] Ecology and Animal Biology Department, University of Vigo, 36310 Vigo (Pontevedra), Spain.
 Email: colabarria@uvigo.es
* Corresponding author

strategies that ought to be implemented at international level. Also some very remarkable European invasion case studies are described in detail to illustrate some of the processes and patterns described earlier in the chapter.

2 Vectors of Transport: Starting the Invasion Pathway

The number of marine macroalgal, i.e., seaweed, species introduced into the European region (Atlantic Ocean and Mediterranean Sea) has been increasing exponentially during the last two centuries. Concomitantly, rates at which introduced species were spreading in Europe have also been increasing (Mineur et al. 2010a). Although climate change could explain increased spread rates in a poleward direction (Sorte et al. 2010), other environmental changes and the increased occurrence and efficiency of human-mediated vectors of transport is more likely to account for this trend in a general manner.

Several vectors have been reported in the literature to be involved in the introduction and subsequent spread of seaweeds: hull fouling, ballast water, shellfish farming, aquaculture, scientific research and fishing gear. The Suez Canal, allowing the range extension of Red Sea species to the Mediterranean Sea (Lessepsian migration) (Por 1978), is usually considered as a vector. It does not fit the traditional definition (no transport; only natural dispersal), but has effectively the same result (removal of a natural barrier). The relative importance of the different vectors for seaweed introductions has been widely reviewed over the past decade (e.g., Ribera and Boudouresque 1995, Boudouresque and Verlaque 2002a, Ribera Siguan 2003, Hewitt et al. 2007, Williams and Smith 2007).

Seaweeds are a major component of fouling communities and any object immersed in the sea for a certain period will be eventually colonized by fouling organisms (Callow and Callow 2002). Therefore, the hull of a boat or any other floating structure, if not properly treated, will get covered by algae, including invaders (e.g., Godwin 2003), and will potentially carry them over long distances. Hence, most seaweed introductions, often observed in areas where there is maritime traffic, are attributed to hull fouling. So far, macroalgal surveys on boat hulls have mostly showed opportunistic and cosmopolitan taxa such as *Ulva* (Chlorophyta) and ectocarpoid species (Ochrophyta, Phaeophyceae) on both commercial cargo ships and recreational yachts (Mineur et al. 2007a, Mineur et al. 2008). Potential introductions through this vector would occur in punctual events, such as the movements of large floating structures (see Mineur et al. 2012a for a review). Various components found on boats, such as rope, buoys or nets have also the capacity of sheltering numerous propagules of different seaweeds.

Ballast water, since the seminal work of Carlton (1985), is also increasingly invoked for seaweed introductions. Seaweeds (at various life history stages) have been sampled in ballast water (Carlton and Geller 1993, Smith et al. 1999, Gollasch et al. 2002, David et al. 2007, Flagella et al. 2007). Nevertheless, only cosmopolitan species (or undetermined taxa) have been found in these studies, and there is no real evidence for the transport of non-indigenous seaweed species in ballast water.

Quite early on in the research on invasive species, it was suspected that shellfish transfers were responsible of the introduction of many marine species (Elton 1958). Over the last three decades, a number of seaweed introductions in Europe have been related to imports of oysters *Crassostrea gigas* (Verlaque 2001). The involvement of this vector has been relatively unambiguous in numerous cases (e.g., Cabioc'h and Magne 1987, Verlaque 2001, Verlaque et al. 2005, Mineur et al. 2012b), as several NIS, native to the North Pacific, were found at European oyster farming sites following imports of oysters from the Pacific. The efficiency of this vector in transporting a wide range of seaweed species, including invasives, has been demonstrated experimentally (Mineur et al. 2007b).

The aquarium trade is a large industry in European countries and involves the continual import of macroalgal species, either intentionally (e.g., Stam et al. 2006) or accidentally, associated with "live rocks" (Padilla and Williams 2004). Aquarium hobbyists practice this multibillion-dollar activity in virtually any home or office (or even large yacht) and the risks of accidental (or intentional) release into the sea, especially in urbanized coastal areas, are therefore very high. Different genera of seaweed species are sold through shops, or by individuals through auction websites, but survey efforts have mostly been focusing on the genus *Caulerpa* (Chlorophyta) (e.g., Walters et al. 2006).

Seaweeds can also be used as packing material for fishing bait. Two Atlantic species, *Fucus spiralis* (Ochrophyta, Phaeophyceae) and *Polysiphonia fucoides* (Rhodophyta), are believed to have been introduced into lagoons of the Mediterranean coast of France by this means (Sancholle 1988, Verlaque and Riouall 1989). *Fucus* sp. and *Ascophyllum nodosum* (Ochrophyta, Phaeophyceae), native to the Atlantic, are also commonly used as decoration for seafood plates in sea-front restaurants in the French Riviera, where this species are not naturally present.

Scientific research is also responsible for some introductions. For instance, in Helgoland Island (Germany), field experiments were regularly undertaken with species non-indigenous to the island (e.g., Munda 1977, Bolton 1983). One experiment led to the establishment of *Mastocarpus stellatus* (Rhodophyta), a species native to the North East Atlantic but not present in Helgoland (Munro et al. 1999). Also, the first observations of numerous NIS within a region (or larger area) have been made in tanks (inside or outdoors) at marine stations. This is the case for *Antithamnionella spirographidis* (Rhodophyta) found at Naples and Banyuls (Funk 1923, Feldmann 1942), *Codium fragile* subsp. *fragile* (Chlorophyta) (Feldmann 1956) and *Colpomenia peregrina* (Ochrophyta, Phaeophyceae) (Mendez Domingo 1957) at Banyuls, and the sporophyte phase of *Bonnemaisonia hamifera* (Chemin 1930) and *Goniotrichopsis sublittoralis* (Magne 1992) at Roscoff (Rhodophyta). Marine organisms are commonly kept alive at marine stations, for research purposes as well as for public exhibition. As live material can originate from remote areas, there is always the opportunity for the release of NIS directly into the sea. However, for the cases cited above, it is difficult to determine whether the species were found in the immediate proximity of marine stations due to intense local survey effort, or if they were actually released there. Nevertheless, in some cases, scientists have been deliberately involved in the spread of introduced seaweeds. An example is given by Chemin (1930), who "successfully" introduced the sporophyte phase of *Asparagopsis armata* (Rhodophyta) into the Bay of Morlaix.

Other aspects of scientific research responsible for seaweed introductions include deliberate introductions for aquaculture trials. This has been the main pathway for the establishment of *Undaria pinnatifida* (Ochrophyta, Phaeophyceae) on Atlantic shores of Europe (Pérez et al. 1984, Voisin et al. 2005). Intentional introductions for aquaculture purposes (commercial or scientific trials) have also occurred on many occasions in the Indo-Pacific region, especially of the carrageenophyte species *Kappaphycus alvarezii* and *Eucheuma denticulatum* (Rhodophyta) (Zuccarello et al. 2006, Pickering et al. 2007).

3 Invasion Success of Non-indigenous Species

After introduction, a successful NIS must survive adulthood and then establish a sustainable population. A NIS is considered an invasive species due to its dominance and spread rate in the new habitat. Particularly interesting in invasion ecology is the dramatically greater abundance of some NIS in new ecosystems compared to their points of origin. Overall, three factors are usually cited as determining the fate of invasions: the biology of the introduced species (species invasiveness), the number and frequency of introductions (propagule pressure) and the susceptibility of the native community to invasion (community invasibility) (Lonsdale 1999).

3.1 Species Invasiveness

The question of whether it is possible to determine a set of traits associated with species invasiveness has been a central theme in invasion ecology since it became a discipline (Williamson and Fitter 1996). In the last two decades comparative multispecies studies provided some of the first generalizations and converging results. These studies, largely focused on vascular plants, found that invasive species seem to have a high fecundity and dispersal ability. They are frequently highly efficient using resources and as a result they have a more vigorous and faster growth than native species (Pysek and Richardson 2007). These life traits or fitness differences may help to explain the success of some introduced species, although in many cases, those traits alone are insufficient to explain the outcome of many invasion processes. Recently, MacDougall et al. (2009) proposed that fitness differences interplay with niche differences to drive the fate of invasions. Niche differences between NIS and native species would result from functional traits of invaders quantitatively very different from those occurring in the native assemblages (Mack 2003). The concept of niche differences is derived from coexistence literature (Chesson 2000), and was already suggested by several authors including Elton in his seminal book "The ecology of invasions by animals and plants" (1958). Elton argued that unique traits allowed the invaders to exploit 'empty niches' in species-poor islands.

Multispecies comparison studies on species invasiveness have been rarely performed with seaweeds. Nyberg and Wallentinus (2005) examined 113 seaweeds introduced in Europe looking for species traits that could predict the success of invasions. They used three main categories: dispersal, establishment and ecological impact, later subdivided in more specific categories and scored introduced and randomly selected native species for each category. Later, they compared both group of species. Unfortunately, the lack of reliable information on many traits hindered the results, and the life history traits did not show consistent patterns. The largest differences among native and invasive species occurred in the probability to be transported (higher in the invasive) and in the original range of distribution, which was also wider in the invasive species. No consistent differences were found in other fitness-related features like life span, size, etc.

Molecular techniques used to detect cryptic invasions and sources of introduced marine population (Geller et al. 2010), have been recently used to explore the relationship between ploidy levels and genome size with invasiveness in seaweeds. Varela-Alvarez et al. (2012) examined the DNA contents in three species of *Caulerpa* from the Mediterranean, at individual, population and species levels. Results showed that ploidy levels and genome size vary in these three *Caulerpa* species, with a reduction in genome size for the invasive ones. An association between polyploidy and invasiveness has also been reported for the red invasive seaweed *Asparagopsis taxiformis* (Andreakis et al. 2007, 2009). These results are fairly in agreement with some of the early theories of invasiveness that associated low genome size with invasiveness in vascular plants (Rejmanek 1996).

To date, no further studies have tried to explain species invasiveness by comparing formally different species of seaweeds. However, single species studies are now profuse and some generalities in the traits associated to high invasiveness have emerged. We examine some of these traits later in this chapter by reviewing some of the most conspicuous invasive seaweed in European shores.

3.2 Community Invasibility

Taken as a whole, there are many and diverse theories of invasion success (Sakai et al. 2001, Mitchell et al. 2006), although most of them share the key prediction that successful invaders must be fundamentally different from native species (Titman 1976, Daehler 2003). According to the *Diversity Resistance Hypothesis*, generally credited to Charles Elton (1958), species diversity enhances resistance to biological invasions. This hypothesis suggests that high diversity communities exploit resources more efficiently, leaving less available niches compared to species-poor communities, believed to contain more unoccupied niches. However, an "invasion paradox" has emerged as researchers describe that at regional scales, species-rich communities may be more susceptible to invasions (e.g., Lonsdale 1999, Stohlgren et al. 1999), emphasizing the role of spatial scale in the diversity-invasibility relationship (Levine 2000, Fridley et al. 2007). More recently, different research highlighted the importance of incorporating facilitative relationships to predict ecosystem invasibility (Bruno et al. 2003, Bulleri et al. 2008, Perkins et al. 2011). Theory suggests that a community rich in potential mutualists has less biotic resistance than a community depauperate in mutualists (Shea and Chesson 2002).

Because competition is regarded as a key biotic resistance mechanism acting upon NIS arrival, special attention has been given to the role of natural enemies, i.e., predators and parasites, in the invasion process. Invasion success of NIS has been related to the *Enemy Release Hypothesis* (ERH), which states that the release from natural enemies drives the increase in distribution and abundance of NIS in its new range (Keane and Crawley 2002), by being unrecognized or unpalatable to native enemies. In the marine realm, however, because marine herbivores are often generalists (Hay 1991, Morrison and Hay 2011), the ERH might be of limited use to explain invasion success of marine algae.

The *Fluctuating Resources Theory* (Davis et al. 2000) emerges as an integration of several existing hypotheses regarding community invasibility. This theory suggests that the invasibility of a community changes as the amount of unused resources fluctuates, i.e., a community becomes more susceptible to invasion whenever resource availability increases (Davis et al. 2000). Availability of resources may fluctuate either due to a pulse in resource supply, a decline in resource uptake, or both, which consequently will decrease competition for that resource. Thus, the susceptibility of a community to invasion is not a fixed property and fluctuates in the same way than other community properties fluctuate (e.g., nutrient availability, predation intensity, space availability). Experimental studies reinforce this theory, emphasizing resource fluctuation and competition as the proposed mechanisms affecting invasibility (Davis and Pelsor 2001) and not species richness (Dunstan and Johnson 2007). Most importantly, some studies highlighted the importance of incorporating the different invasion stages over the invasion process. For instance, recent studies with the invasive macroalga *Sargassum muticum* (Ochrophyta, Fucales) suggested that top-down and bottom-up forces, as well as diversity effects comprised specific regulation over different stages of invasion (White and Shurin 2007, Vaz-Pinto et al. 2013a). Moreover, manipulative field experiments found that resource availability was mediated by algal species identity (and not species richness), suggesting that a functional group approach may better describe the different mechanisms of species coexistence acting within a community (e.g., Ceccherelli et al. 2002, Arenas et al. 2006).

3.3 Propagule Pressure

In addition to biotic and abiotic factors, the relative importance of propagule pressure has been largely acknowledged (see Simberloff 2009 for a review). In contrast to species- and community-level particular traits, which remain constant across repeated introduction events separated by a relatively small timeframe, propagule pressure is characteristic of a particular introduction event, differing between introduction events (Lockwood et al. 2005). Thus, propagule pressure could partly explain the idiosyncratic nature of introduction success.

Despite the considerable interest on the problematic widespread of invasive species (Grosholz 2002), relatively few studies have attempted to examine the overall process of invasion. A recent study with the invasive *S. muticum* revealed that the role of propagule pressure in the invasion process showed greater importance over the settlement period, while recruitment and colonisation success of *S. muticum* were mainly affected by resources availability, namely light (Vaz-Pinto et al. 2012).

In conclusion, the invasibility of an ecosystem is the outcome of several factors, from ecological interactions between NIS and native species to the region's climate and related interactions, the disturbance regime, resource availability and propagule supply (Lonsdale 1999, Davis et al. 2005, Perkins et al. 2011).

4 Ecological Impacts of Seaweeds

The introduction and spread of NIS may have dramatic ecological impacts, and it is considered a major threat to biodiversity (Parker et al. 1999, Pejchar and Mooney 2009). Invasion by foundation species, such as many large seaweeds, are considered particularly important and worrying because they may alter both ecosystem structure and function of marine systems (Schaffelke et al. 2006). A recent review described 277 records of introduced seaweeds, which make up a significant proportion of marine introduced species (Schaffelke et al. 2006, Williams and Smith 2007). We will describe below a brief sketch of the range of effects that non-indigenous macroalgal species cause to both biodiversity and ecological processes.

4.1 Genetic-level Effects

In seaweeds, the majority of molecular studies to date sought to identify the source of NIS, where they came from and how they got there, with particular focus on cryptic invasions (e.g., Verlaque et al. 2005, Provan et al. 2008). Tracing the source of introduced species can help to prevent further invasions, as well as to define the ecological characteristics of each species and predict post-invasion population dynamics in the region of introduction (Kolar and Lodge 2001, Grosholz 2002).

A cryptic invasion is any invasion event that stays undetected because the invader is misidentified, leading to the underestimation of the total numbers and impacts of invaders (e.g., multiple species existing within taxa traditionally considered to be single species) (Knowlton 1993, Geller et al. 2010). For instance, a recent study reported the presence of *Ulva pertusa* Kjellman (Chlorophyta) in the NW Iberian Peninsula (Baamonde López et al. 2007). However, revision of herbarium *Ulva* specimens demonstrated that this species has been previously misidentified as *U. rigida*, *U. pseudocurvata*, *U. scandinavica* and, even, *Umbraulva olivascens* (Chlorophyta) (P.J.L. Dangeard) G. Furnari.

Most importantly, the existence of cryptic genetic diversity suggests that corresponding ecological differences could be important for patterns of invasion (Wonham and Carlton 2005). Evidence suggests that the occurrence of multiple introductions may be associated with populations with ecologically relevant physiological differences in the native range. For instance, multiple introductions of the invasive red alga *Polysiphonia harveyi* (Rhodophyta) to the North Atlantic has been suggested to derive from independent native lineages, one associated with populations on the cold-temperate Japanese island of Hokkaido and the other associated with populations from the warm-temperate Honshu (McIvor et al. 2001). Such genetic variation may increase the likelihood of persistence of NIS, and most importantly it can rapidly expand a marine invader range (e.g., Roman 2006).

It becomes necessary to determine the potential genetic or evolutionary changes following seaweed invasions (Booth et al. 2007). As well as cryptic invasions, there is also the possibility of loss of native genotypes through hybridization and introgression of the genes of the invaders into native gene pools. In some cases, hybridization leads to speciation and the new species significantly impacts local communities. With the exception of *Fucus*, there is little evidence of natural interspecific hybridization in seaweeds. Hybridization of the native *F. serratus* and the introduced *F. evanescens* has been reported after identification of fertile hybrids where the two species now occur together (Coyer et al. 2002). Hybridization was described as asymmetric, occurring only between female *F. evanescens* and male *F. serratus* (Ochrophyta, Phaeophyceae) (Coyer et al. 2002, Coyer et al. 2007).

4.2 Population and Community-level Effects

Impacts of non-indigenous macroalgal populations are typically expressed as community dominance through the monopolization of space, and changing competitive relationships in the native assemblage (reviewed by Schaffelke and Hewitt 2007, Williams and Smith 2007). Seaweeds can alter light availability to other species, change nutrient cycling, affect herbivory intensity (Britton-Simmons 2004, Sánchez et al. 2005, Yun and Molis 2012), modify ecosystem properties and ultimately, they may decrease native diversity (e.g., Ambrose and Nelson 1982, Casas et al. 2004). A recent meta-analysis showed that NIS have, on average, small-to-large negative impacts on native species and assemblages (Thomsen et al. 2009). Reported impacts have been related to algal interactions (e.g., Ambrose and Nelson 1982, Britton-Simmons 2004), seaweed-seagrass interactions (Thomsen et al. 2012) as well as on other trophic groups (e.g., Gestoso et al. 2010, Thomsen 2010, Janiak and Whitlatch 2012).

However, some studies described no significant impact of non-indigenous seaweeds in specific areas (e.g., Forrest and Taylor 2002, Cecere et al. 2011), highlighting the unpredictable nature of invasions. For example, a three-year study of sheltered low shore assemblages found little impact from the introduction of *Undaria pinnatifida* (Forrest and Taylor 2002), while large ecological consequences were described after *U. pinnatifida* introduction in sites without large canopy species (Forrest and Taylor 2002, Casas et al. 2004). Other seaweed that has spread dramatically during the last century is *Codium fragile* ssp. *tomentosoides* (Chlorophyta). It has been reported as a fast growing species, growing up to 170 thalli m^{-2}, with high impacts on recipient assemblages by damaging and replacing native kelp (*Laminaria* spp., Phaeophyceae) forests with potential impacts on associated fauna (Trowbridge 1995, Levin et al. 2002). Moreover, the accumulation of masses of *C. fragile* ssp. *tomentosoides* rotting on beaches of the NW Atlantic, Mediterranean, and New Zealand produces a foul odor that drives away visitors. Despite its known capacity for spreading and replacing indigenous species (Carlton and Scanlon 1985, Nyberg and Wallentinus 2005), there are some sites where no impact has been registered for this species, as for example in the Azores (Cardigos et al. 2006) and in the eastern North Atlantic Ocean (Chapman 1998). So far, no link has been found between the ability of an introduced species to spread rapidly (its invasiveness) and the likelihood that it will have a strong impact on the recipient community (Ricciardi and Cohen 2007).

Life history features of invaders may be key factors in determining the fate and the impact of invasions. For instance, invasion by canopy forming macroalgae (e.g., *Sargassum muticum*, *Undaria pinnatifida*) may influence the structure of understory assemblages by modifying levels of light (Jones and Thornber 2010), sedimentation (Airoldi 2003) or water movement (Eckman et al. 1989). Moreover, unique biochemical constituents of invasive species has strong allelopathic effects on its common native competitors, inhibiting its recruitment, e.g., the red alga *Bonnemaisonia hamifera* (Rhodophyta) (Hariot) (Svensson et al. 2013), or productivity, e.g., the green alga *Caulerpa taxifolia* (M. Vahl) C. Agardh (Chlorophyta) (Ferrer et al. 1997). Introduced species often exhibit novel features compared to native species and may have disproportionately high impacts in native ecosystem functioning (Ruesink et al. 2006).

4.3 Ecosystem-level Impacts

Several studies have assessed the issues of biodiversity loss and ecosystem services in general, most of which are reviews of ecological and conservation theory (Tilman 1999, Naeem and Wright 2003, Raffaelli 2006). However, to date, there are virtually no studies focusing on the functional consequences of species additions through biological invasions of NIS in marine habitats (Stachowicz and Byrnes 2006, Vaz-Pinto et al. 2014). Ecosystem level processes are affected by the functional characteristics of the organisms involved (Odum 1969, Díaz and Cabido 2001). However, the consequences of biodiversity change, through changes in species traits, are likely to be idiosyncratic, differing between trophic groups and ecosystems (Emmerson et al. 2001, O'Gorman et al. 2011).

Recently, a manipulative experimental approach with artificial macroalgal assemblages has revealed significant negative impacts linked to the presence of an invader on ecosystem properties and predictability (Vaz-Pinto et al. 2014). The variability of the ecosystem function response found in assemblages invaded by *S. muticum* suggested that invaded assemblages' dynamics are less predictable than native dynamics. Most evidence suggests that the impact of *S. muticum* was related to its high spatial variability (Baer and Stengel 2010), productivity and dominance in the invaded assemblages, which varied drastically between seasons. Further interactions between habitat-modifying species can decrease predictability of community-level effects of an invasion, particularly if invasive species show extremely variable cycles over time (Ward and Ricciardi 2010).

So far, research on the impacts of NIS lacks the ability to disentangle the mechanisms involved in the impacts (Ruesink et al. 2006). In particular, no research has compared whether ecosystem impacts vary with the invader's life-history traits and lack of co-evolution with native species. Nonetheless, the true impact of introduced NIS on ecological and evolutionary processes can still not be predicted, particularly due to the fact that most ecosystem processes are a function of interactions among species, rather than simple presence or absence of species (Chapin III et al. 2000). Native species with a long history of co-evolution are expected to partition resources among them and promote ecosystem functioning throughout resource use complementary effects. In contrast, newly introduced species probably enhance ecosystem functioning by sampling effects where the influence of the invader is well beyond its proportion (Ruesink et al. 2006, Vaz-Pinto et al. 2014).

5 Case Studies

We will now present a series of case studies of well known invasive seaweed species. We will start with *Sargassum muticum*, one of the most well studied invasive macroalgae, then we will describe the patterns of invasion of *Undaria pinnatifida* (Ochrophyta, Phaeophyceae), the only known invasive kelp species, and we will finish with three highly invasive species from the Mediterranean, *Caulerpa taxifolia*, *C. racemosa* (Chlorophyta) and *Lophocladia lallemandii* (Rhodophyta) (see Fig. 1).

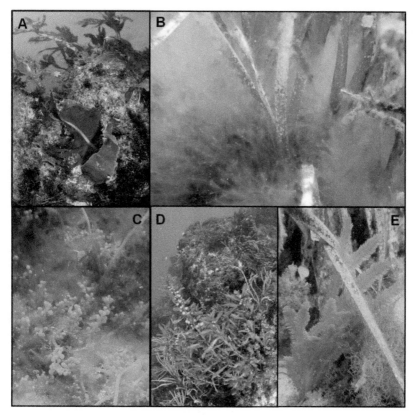

Figure 1. (A) Wakame *Undaria pinnatifida* (Heterokontophyta), native to Japan, [c] David Villegas; (B) Red macroalga *Lophocladia lallemandii*, introduced into the Mediterranean through the Suez Canal [c] Miguel Cabanellas-Reboredo; (C) *Caulerpa racemosa* var. *cylindracea* (Chlorophyta), native to SW Australia, [c] Miguel Cabanellas-Reboredo; (D) Japanese wireweed *Sargassum muticum* (Heterokontophyta), native to SE Asia, [c] David Villegas; (E) *Caulerpa taxifolia* (Chlorophyta), native to the tropical seas, [c] Miguel Cabanellas-Reboredo.

Color image of this figure appears in the color plate section at the end of the book.

5.1 Sargassum muticum *(Yendo) Fensholt (Ochrophyta, Phaeophyceae: Fucales)*

Sargassum muticum, commonly known as the Japanese wireweed, is a large brown seaweed. Thallus is highly differentiated into holdfast, cylindrical main axis, basal leaf-like blades and airbladders. Populations of this species are distributed mainly in sheltered or semi-exposed hard-bottom shores (Strong et al. 2006). It grows attached to rocks by a perennial holdfast up to 5 cm in diameter. This species is often visually dominant, forming beds from sublittoral (Arenas et al. 1995) to the mid-intertidal zone. Individuals also appear in rock pools located on exposed intertidal zones, which gives protection from a low tolerance to desiccation (Norton 1977) and from mechanical stress by wave-action (Viejo et al. 1995).

This species is native to Southeast Asia (Yendo 1907), where it has been described as a relatively minor component of the native macroalgal flora, reaching 1–2 m in length (Norton 1977, Critchley et al. 1983). However, outside of its native range it is found to be highly invasive, and it has been reported to reach 16 m, although most mature plants are 2–3 m in length. It is now widespread throughout the North America Pacific coast, from SE Alaska Mexican (Dawson 1961, Núñez-López and Casas-Valdez 1998) and through most of the Atlantic coast, from Atlantic Morocco to Norway (Rueness 1989, Sabour et al. 2013) and in the Mediterranean (Streftaris et al. 2005).

This species has been introduced to the coastlines of several countries in the northern hemisphere in the past 70 years, mainly associated with the transportation of Japanese oysters (*Crassostrea gigas*) for aquaculture (Scagel 1956). The first register of an established population of *S. muticum* outside its native range was in British Columbia, Canada in 1945, and by the 1970s had made its way south to California and Mexico (Dawson 1961, Abbott and Hollenberg 1976). Within European waters, attached individuals of *S. muticum* were first recorded in the British Isles in 1973 (Critchley et al. 1983), probably arrived from France. On the Iberian Peninsula coast, *S. muticum* was first reported from Asturias in the 1980s; subsequently it was observed on the Galician coast in 1986 (Péres-Cirera et al. 1989) and was first recorded in Portugal in 1989 (Rull Lluch et al. 1994). The first record in Nordic waters is from Denmark in 1984, spreading to Norway in 1988 (Rueness 1989). It was introduced to France, in the Mediterranean in 1980, and to Venice in 1992 (Curiel et al. 1998). Recently, it has been recorded in Andalusia, Spain (Bermejo et al. 2012), and on the Atlantic coast of Morocco (Sabour et al. 2013).

Intrinsic traits such as being a fast growing species, high fecundity, monoecious, self-fertile and pseudo-perennial life history, among others, have been cited as responsible for the success of *S. muticum* as invader (Norton 1976). Moreover, this species has 4 characteristic growth phases, suggesting a highly specialized adaptation to life in a seasonal environment: 1) initial growth phase with large basal leaves which increase photosynthetic surface area; 2) elongation growth where the presence of gas-bladders maintain the thallus erect and closer to the light; 3) reproductive growth (April/May), where energy allocation is now shifted to the production of reproductive tissue (receptacles); and 4) a senescence period (from August onwards) where primary laterals degenerate and float away (Deysher 1984,

Arenas et al. 1995). Multiple-range dispersal mechanisms have been cited for this species, including germling settlement and drifting of fertile thalli (Norton 1976). The characteristic air bladders provide buoyancy, a good mode of rapid dispersal for dislodged individuals (Rueness 1989).

Replacement of native species, increase of filamentous epiphytic algae, changes in composition of flora and fauna, increased sedimentation, interference with coastal fisheries large accumulations of drift algae, blocking of narrow sounds and harbours, and interference with recreational activities are some of the known impacts of *S. muticum* (Critchley et al. 1986, Mack et al. 2000, Pedersen et al. 2005). For example, *S. muticum* may outcompete the giant kelp *Macrocystis pyrifera* (Ochrophyta, Phaeophyceae) by inhibiting its recruitment (Ambrose and Nelson 1982). Moreover, indirect effects affecting abiotic conditions, such as shading or alteration of the current by the dense canopy during the summer months, may affect settlement and recruitment of benthic organisms (Critchley et al. 1990, Staehr et al. 2000, Britton-Simmons 2004). A study showed that *S. muticum* caused strong temperature stratification. Water registered up to 11°C warming above ambient in the canopy and a significant cooling to 2°C below ambient (Strong et al. 2006). Because *S. muticum* undergoes a faster and more complete decomposition than that of the native macroalgae, it has been suggested to alter the energy flow in soft-sediments, by increasing the turnover rate and regeneration of nutrients (Pedersen et al. 2005, Rossi et al. 2011) and rate of primary production (Cebrián and Duarte 1995). In sandy beaches, for example, *S. muticum* can support populations of the beach consumer *Talitrus saltator* (Arthropoda), especially when other seaweeds were absent (Rossi et al. 2011).

Although NIS are often linked to negative impacts on the indigenous biota, some positive effects have also been described. In the North Sea, *S. muticum* provides habitats for epibiota otherwise absent in sediments, with consequent strong effects on diversity of soft sediments (Buschbaum et al. 2006). On rocky shores, recent studies also revealed that *S. muticum* may offer a suitable habitat for many invertebrates (Thomsen 2010, Gestoso et al. 2012).

Concluding remarks

S. muticum is a highly invasive species (Fletcher and Fletcher 1975, Paula and Eston 1987) because it possesses characteristics of both r- and K-selected species (Arenas et al. 1995, Engelen and Santos 2009). It is a species that rapidly colonizes open habitats, and is often associated with the availability of bare rock (e.g., Britton-Simmons 2006, Britton-Simmons and Abbott 2008), rather than a species that displaces native species. As invasive species spread through a new environment, they encounter novel selection pressures and challenges. Evidence suggest that assemblages vary continuously in their susceptibility to invasion over the different stages of invasion (Davis et al. 2000, D'Antonio et al. 2001), and recent research suggests that interaction with environmental factors that co-vary with species diversity may better explain the invasion success of *S. muticum* (Vaz-Pinto et al. 2012).

5.2 Undaria pinnatifida *(Harvey) Suringar (Ochrophyta, Phaeophyceae: Laminariales)*

Undaria pinnatifida, commonly known as wakame, is a large brown kelp commercially grown throughout Asia for human consumption, and it is valuable to the pharmaceutical industry. The stipe has very wavy edges or 'ruffles' at the base, giving it a corrugated appearance. The generic name is derived from *unda*, the Latin word for wave, referring to the ruffled wing (Silva et al. 2002). The blade is broad and flattened with a distinct midrib. The maximum length of *U. pinnatifida* fronds in nature is typically up to 1.5 m length in the wild, but it has been recorded as large as 3 m in cultivation (Pérez et al. 1984). It can be found from the low tide mark down as far as 15 metres attached on hard surfaces forming dense kelp forests in sheltered waters (Floc'h et al. 1996). In its native habitats, *U. pinnatifida* has an annual life cycle, but in its invaded areas, it exhibits considerable plasticity in its life cycle and morphology (Nanba et al. 2011).

U. pinnatifida is native to Japan, northern China and Korea (Akiyama and Kurogi 1982), but it has successfully invaded several areas of the world as far apart as the Atlantic Coast of Europe, the Mediterranean Sea, New Zealand, Tasmania, Argentina, SE Australia and the west coast of the USA and Mexico (Silva et al. 2002, Aguilar-Rosas et al. 2004). *U. pinnatifida* can thrive in a wide range of physical and environmental conditions. A substrate generalist, it is common, but not limited, to sheltered waters, especially harbours.

Undaria's first appearance outside of the NW Pacific was on the Mediterranean coast of France, in 1971, having apparently been imported accidentally with Pacific oysters (*Crassostrea gigas*), and has since spread throughout the Mediterranean coasts (Hay 1990, Fletcher and Manfredi 1995). This species was deliberately introduced into the North Atlantic, to Brittany, in 1983 for commercial reasons (Péres-Cirera et al. 1991). Currently, *U. pinnatifida* can be found in several temperate coasts worldwide. The wide variety of human vectors by which this kelp could be introduced has been largely discussed, including shipping (Hay 1990, Floc'h et al. 1996), transfer of contaminated mariculture stock and equipment (Boudouresque et al. 1985), and other mechanisms such as fishing nets and boat anchors (Sanderson 1997).

The Asian kelp *U. pinnatifida* is ranked among the most invasive seaweeds, based on its wide range of non-native properties (Nyberg and Wallentinus 2005). These include the ability to rapidly colonize new or disturbed substrata and a wide range of artificial structures, a fast growth rate, a large reproductive spore output, a wide seasonal distribution period, a wide physiological tolerance with respect to temperature, light and salinity, and a wide subtidal distribution. This kelp exhibits multiple dispersal strategies. Spore dispersal in *U. pinnatifida* is probably a key mechanism for short-range (metres to hundreds of metres) spread from fixed stands. Dispersal via whole sporophytes or fragments, and spore clumping, is likely to be particularly important in range extensions over scales of hundreds of metres to kilometres, with episodic or chance events potentially leading to spread at even greater scales (Reed et al. 1988).

Concern has been expressed over the spread of this seaweed because of its potential effects on important natural ecosystems and fisheries and the possibility that it could become a fouling pest (Wotton et al. 2004, Irigoyen et al. 2011). *U. pinnatifida* can compete with the native kelp species occupying the shallow sublittoral and infralittoral zones, provoking a reduction in biodiversity (Battershill et al. 1998, Casas et al. 2004). In addition, *Undaria* can tolerate a lowered salinity and grows well in some estuarine regimes, where many native marine species cannot (Farrell and Fletcher 2006). *U. pinnatifida* seems to have a low competitive ability, as shown by a low abundance of this kelp among large seaweeds or sessile macrofauna (Brown and Lamare 1994). However, it is able to rapidly colonize new or recently modified substrata, and it also seems to settle better on artificial substrata such as floats, marinas or piers than on natural rocky shores among the local kelps, which if present are restricted to the most seaward areas (Hay 1990, Floc'h et al. 1996).

Concluding remarks

The mechanisms behind the spread and dominance of *U. pinnatifida* are unknown, although competitive exclusion involving light, nutrient, substrate limitation and the lack of native herbivores preferentially feeding on *U. pinnatifida* are potential explanations (Shea and Chesson 2002). The multiple dispersal strategies of *U. pinnatifida* play a significant role in facilitating rapid spread within regions. The control of *U. pinnatifida* populations has been ineffective worldwide, and restoration efforts based on algal removal have proved to be unsuccessful (Hewitt et al. 2005). Moreover, the eradication of an established population through the use of biocides seems impossible, except perhaps in an enclosed basin (Curiel et al. 1998).

5.3 Caulerpa taxifolia *(M. Vahl) C. Agardh and* C. racemosa *(Forsskål)* J. Agardh *(Chlorophyta: Bryopsidales)*

Caulerpa species are fast-growing marine green algae, showing different growth forms which make them difficult to identify. *Caulerpa taxifolia* shows flattened, feather-like fronds, 5–65 cm long extended upward from horizontal stolons, anchored to underwater surfaces such as rocks, mud, or sand via root-like rhizoids. Branchlets are oppositely attached to midrib, flattened, slightly curved upwards, tapered at both base and tip (Jongma et al. 2013). *Caulerpa racemosa* has a uniaxial thallus mostly divided into a creeping stolon with thin rhizoids and erect fronds, up to approximately 11 cm high from the stolon, either nude, leaf-like or with grape- or feather-like ramuli (Klein and Verlaque 2008).

The genus *Caulerpa* is widespread in the intertidal and subtidal zones of tropical and subtropical regions (Luning 1990). In the last decades, this genus has been attracting considerable research attention in the Mediterranean, where two species, *C. taxifolia* and *C. racemosa* have spread into areas formerly occupied by seagrasses. *C. taxifolia* is a native species to the tropical seas where it grows in small patches and does not present problems. However, this common green alga has gained wide notoriety and the nickname "killer algae" because of its great success in the coastal Mediterranean area (Meinesz et al. 1993). *C. racemosa* has been considered an introduction in the Mediterranean from the Red Sea via the Suez Channel (Ribera and Boudouresque 1995), but on the basis of new morphological and molecular studies, a different variety (identified as *Caulerpa racemosa* (Forsskål) J. Agardh var. *cylindracea* (Sonder) Verlaque, Huisman and Boudouresque) has been reported as the "invasive variety". It is endemic to SW Australia, and is undergoing a continuous expansion throughout the Mediterranean (Verlaque et al. 2000).

In 1984, *C. taxifolia* was accidentally released into the Mediterranean Sea from the Monaco Aquarium, and quickly colonized wide coastal areas of the Mediterranean, possibly disseminated by boats and fishing nets (Meinesz et al. 2001, Williams and Smith 2007). This species also reached the southern California coast in the USA (Jousson et al. 2000) and Sydney, Australia, approximately 600 km south of the known range of autochthonous populations (Meinesz et al. 2001). Genetic studies revealed the extensive homogeneity of the invasive strain of *C. taxifolia* worldwide (Jousson et al. 2000). The sources of introduction and propagation of *C. racemosa* var. *cylindracea* in the Mediterranean are complex, partly because this species includes several distinct strains, which may be distinct species (Famá et al. 2000). This species has been reported as a recent introduction from Australia, probably via ship traffic and aquaria (Klein and Verlaque 2008). By late 2002, it occurred throughout the whole Mediterranean Sea. Lately, it has been detected in the Canary Islands (Verlaque et al. 2004).

These two species are expected to exhibit similar invasiveness, having similar morphologies, growth rates, dispersal strategies and seasonality in productivity (Ceccherelli and Piazzi 2001). They propagate clonally by fragmentation and often show rapid proliferation when growing outside their native ranges, competing with seagrasses (Williams and Smith 2007). The maintenance and spread of *Caulerpa* populations may also take place by sexual reproduction. It has been suggested that *C. taxifolia* might spread mainly clonally in the Mediterranean. Moreover, *C. taxifolia* exhibits larger size, higher growth and resistance to low temperatures, and toxicity to predators than the known tropical populations (Meinesz et al. 1995). *C. racemosa* is capable of reproducing sexually and vegetatively (Klein and Verlaque 2008), achieving even higher propagation speed with respect to *C. taxifolia* (Piazzi et al. 2005).

Both *Caulerpa* species are extremely invasive in the Mediterranean and cause major modifications to benthic communities, mostly in the form of decreased species diversity (Ceccherelli and Cinelli 1997, Piazzi et al. 2001). The potential impact of *C. taxifolia* on biodiversity includes loss of seagrass beds, effects on local fisheries, and general negative effects on the coastal ecosystem (Bouduresque et al. 1995). Some studies have demonstrated that *C. racemosa* have negative effects on native algal community structure, species diversity and abundance (Piazzi et al. 2005). For instance, a recent study showed a negative relationship between this invasive species and native fauna, such as the red gorgonian *Paramuricea clavata* (Cebrián et al. 2012).

Concluding remarks

Despite the fact that *C. racemosa* is comparable to *C. taxifolia* in terms of capacity to colonize and alter native assemblages, there is a wide disparity in the effort and means mobilized to attempt to cope with these two invasive species. It is illegal to import or be in possession of *C. taxifolia* in France, Australia, the USA and certain regions of Spain (Meinesz et al. 2001). To limit the impact of these two invasive species in the Mediterranean, management strategies need to be put into action encompassing all countries affected by the problem. The consequences of the introduction of *Caulerpa* invasive species into the Mediterranean highlight the urgent necessity to inform the public and to prohibit this species from the international aquarium trade.

5.4 Lophocladia lallemandii *(Montagne) F. Schmitz (Rhodophyta: Ceramiales)*

Lophocladia lallemandii is a marine filamentous red alga up to 15 cm in height and pseudo-dichotomously branched, which usually appears as a mat of red filaments intertwined with themselves or with other algae. Erect filaments, 0.5 cm in diameter, arise from a basal disc. This alga displays pronounced seasonal production, with maximum development in summer and autumn and a drastic decline in winter (Ballesteros et al. 2007).

This species shows a native Indo-Pacific distribution, and it was probably introduced into the Mediterranean from the Red Sea via the Suez Canal (Boudouresque and Verlaque 2002). *L. lallemandii* grows well on different types of substrates (bare bedrocks, rocky macroalgal bottoms, seagrass meadows, and over coralligenous communities; Ballesteros et al. 2007), and it has recently become widespread in the Mediterranean Sea, with the exception of Moroccan waters and the north-western Mediterranean (Benhissoune et al. 2003). Attempts to determine the vector of this invasive species' introduction can be only speculative, but dispersal via shipping and other maritime activities is suspected. It is quite common in the eastern part of the Mediterranean, where it has been reported since the early 1900s (Patzner 1998).

L. lallemandii possess very successful strategies for dispersal. It reproduces sexually only during summer and autumn, while its vegetative reproductive activity occurs throughout the year, with minimal growth during late autumn and winter. Moreover, besides reproducing vegetatively through spore dispersal, it can spread by fragmentation. This species is easily broken and free-floating filaments produce small, disc-like holdfasts that are able to attach to a large variety of floating substrates (Cebrián and Ballesteros 2010).

Due to its high invasive potential, *L. lallemandii* is able to cover most kinds of substrate causing homogenization of the benthic landscapes (Boudouresque and Verlaque 2002). The behaviour of this species is very aggressive, especially when colonizing *P. oceanica* meadows, causing a major decrease in seagrass density and growth that can lead to the death of the plants (Ballesteros et al. 2007). This species completely overgrows macroalgal assemblages and also affects the benthic invertebrate community (Cebrián and Ballesteros 2010). Thus, it has been suggested that the habitat alteration caused by *L. lallemandii* at shallow *P. oceanica* meadows from Mallorca Island induced a sharp decline in bryozoans' densities (Deudero et al. 2010).

Concluding remarks

This species has a very pronounced seasonality with disappearance of all thalli in winter, absence in spring, and high growth rates in summer and autumn. Thus, although the effects may be large at the end of the algal growth period, they do not persist a whole year (Ballesteros et al. 2007). However, because it can reproduce and spread so fast, it is complicated to eradicate *L. lallemandii* populations, at least manually (Cebrián and Ballesteros 2010). Currently, little is known about the biology of this invasive species and its invasion mechanisms. However, recent results point to negative effects of *L. lallemandii* colonization, and stress the need to address interaction effects across natural communities and invaded systems before associated and irreversible effects are caused (Ballesteros et al. 2007, Deudero et al. 2010).

6 Interactions with Climate Change and Other Global Stressors

Climate change and global warming are now widely recognized phenomena. Specifically for the marine environment, global changes include increased carbon dioxide that will drive important physical and chemical changes. These include global-scale trends such as ocean acidification, warming and sea-level rise, together with regional changes to patterns of winds and ocean currents, wave heights, upwelling and the associated nutrient supply, ocean stratification, coastal salinity, and changes to rainfall dynamics (Harley et al. 2012).

By 2100, atmospheric CO_2 is expected to exceed 500 ppm decreasing pH by about 0.4–0.5 units, and global temperatures to rise at least 2°C, exceeding conditions of the past 420,000 years (Hoegh-Guldberg et al. 2007). On top of the gradual shift in mean environmental conditions, changes to the occurrence of extreme events are also being registered (IPCC 2007). Climate change models predict an increase in the frequency and intensity of extreme events such as storms, including cyclones and hurricanes, and increasing wave heights, flooding and heat waves (IPCC 2007). Some of these changes are already taking place. Thus, in a recent analysis derived from global sea surface temperatures for the last 30 years, Lima and Wethey (2012) found that an important percentage of the world's coastlines have experienced an increase in extremely hot days. Increasing temperature and CO_2 concentration may affect the physiological and ecological performance of seaweeds, alter marine biodiversity and lead to changes in species composition and biological interactions among species (Harley et al. 2006, Kroeker et al. 2010, Johnson et al. 2012).

Together with climate change, biological invasions are among the most serious global environmental threats (Stachowicz et al. 2002). Both have wide-ranging community and ecosystem-level impacts resulting in local extinctions and in the decline and even collapse of several marine ecosystems (Ruiz et al. 2000, Stachowicz et al. 2002, Frank et al. 2005). Apart from their direct effects, climate change and biological invasions may interact in multiple ways, both directly and indirectly (Crain et al. 2008, Wernberg et al. 2011, Harley et al. 2012). Knowing how climate change interacts with marine invasions will be critical for understanding and predicting successful invasions, as well as managing their impact.

Several forces can affect the prevalence of NIS. Thus, global climate change can modify species distributions and resources availability, inducing biological invasions (Occhipinti-Ambrogi 2007). Climate change also affects NIS by altering native communities and environmental conditions of the invaded habitat (Thomsen et al. 2011). In addition, climate change may favour NIS (Dukes and Mooney 1999), and changes in human transport activities increase the rate of arrival of non-indigenous propagules (Dukes and Mooney 1999, Hoegh-Guldberg and Bruno 2010). As NIS become more prevalent, they may alter ecosystem processes and functioning, many of which will interact with diverse climate change agents.

Global warming may favour the establishment of invasive seaweeds. Evidences from terrestrial systems suggest that invasive species have larger latitudinal ranges than native species, which may be indicative of their ability to tolerate a wider range of environmental conditions and their potential for greater success at increased temperatures (see Dukes and Mooney 1999). For example, the increase of sea surface temperature in the Mediterranean coincides with the establishment of the tropical green seaweed *Caulerpa taxifolia*, which causes *Posidonia oceanica* (Magnoliophyta) die-back (de Villèle and Verlaque 1995). Increasing temperature is also adding pressure on native species, reducing potential competitive advantage over invaders or facilitating their establishment (Harley et al. 2012). Temperature extremes may be even more important than increase of mean temperatures. Such extremes may remove dominant competitors, facilitating the establishment of NIS (Miller et al. 2011). In Australia, summer heat waves episodes have caused large scale die-backs in kelp beds (Smale and Wernberg 2013). This mass mortality facilitated the establishment of the invasive brown alga *Undaria pinnatifida* (Valentine and Johnson 2004). Although seaweeds often show very variable, species-specific response to increasing CO_2 level (Beardall et al. 1998, Johnson et al. 2012), the establishment and spread of some invaders might be favoured by higher level of CO_2 (Hall-Spencer et al. 2008, Porzio et al. 2011). Because recent studies suggested that brown seaweeds might proliferate in a high-CO_2 world (Hall-Spencer et al. 2008, Connell and Russell 2010, Diaz-Pulido et al. 2011, Johnson et al. 2012), it is reasonable to think that some of the most invasive seaweeds in the world, such as *U. pinnatifida* and *S. muticum* (Williams and Smith 2007), could be amongst the ecological winners under an ocean acidification scenario.

Acting simultaneously, ocean warming and acidification might also favour seaweed invasions indirectly by promoting shifts in marine ecosystems through changes in species survival and assemblage composition (Thomsen et al. 2011). Increasing CO_2 and temperature may, for example, cause the decline of coralline algae (Martin and Gattuso 2009) and thus promote the success of existing invaders or the colonization of opportunistic and new invaders, i.e., turf forming and fleshy algae (Connell and Russell 2010). Furthermore, acidification might alter feeding behaviour of certain grazers, limiting their ability to control the climate-enhanced growth of opportunistic turfs (Russell et al. 2009). Increased CO_2 levels may also influence algal-grazer dynamics and competitive interactions, causing changes in structure and function of rocky shore benthic communities (Diaz-Pulido et al. 2011, Johnson et al. 2012). In turn, invaders may be directly affected by the interaction of these two climate change agents. In a recent experiment in the laboratory, the early survivorship of settled germlings of the brown canopy-forming seaweed *S. muticum* responded to an interaction of temperature and CO_2, with survivorship enhanced at high CO_2 and ambient temperature after 3 days, and reduced at ambient CO_2 and high temperature after 10 days (Vaz-Pinto et al. 2013b).

Because invasive seaweeds are often considered ecosystem engineers that can replace native species and their functional role in ecosystems (Britton-Simmons 2004), well-established non-indigenous seaweeds might alter the community response to climate change (Hall-Spencer et al. 2008, Olabarria et al. 2013). Indeed, some evidences indicated that assemblages from rockpools invaded by *S. muticum* might be more resilient to climate change than those dominated by the native canopy-forming species *Cystoseira tamariscifolia* (Olabarria et al. 2013).

Not only climate change, but also other anthropogenic pressures like overfishing, eutrophication, habitat loss, or pollution are affecting marine ecosystems (Halpern et al. 2008). In such contexts, NIS may take advantage. For example, in the Venice lagoon, warmer temperatures in combination with high levels of organic and inorganic pollution have facilitated the introduction and establishment of the non-indigenous algae *U. pinnatifida*, *S. muticum* and *Antithamnion pectinatum*, which have become dominant (Occhipinti-Ambrogi and Savini 2003, Sfriso and Facca 2013). Climate change and changes in biotic interactions resulting from overfishing may also favour invasions. This is the case of the Gulf of Maine, where kelp beds have been replaced by opportunistic and NIS (*Codium fragile tomentosoides*, *Polisyphonia harveyii* and *Bonnemaisonia hamifera*) due to an increase of summer temperatures and a dramatic increase of sea urchins due to overfishing (Harris and Tyrrell 2001).

In the marine realm, studies on potential feedbacks and interactions between seaweed invasions and global change stressors are still very limited. In order to fully understand such relationships, it will be important to understand interactive effects and possible synergies (Crain et al. 2008, Wernberg et al. 2012). Until recently most marine climate change research involved single-stressor experiments focused on organisms in isolation, highlighting an urgent need to conduct more studies at community level, and in particular using multiple-stressors approaches (Johnson et al. 2012, Olabarria et al. 2013).

7 Management of Seaweed Invasions

Recognized as a major component of human-mediated impacts on natural systems (Pimentel et al. 2005), the prevention of biological invasions has become a high priority for governments worldwide (Hewitt et al. 2009), in particular for those species that have significant impacts on the environment, quality of life, economy and/or human health. In general, management of NIS consists of four major steps: 1) prevention, 2) early detection and rapid response, 3) eradication, and 4) control (assessment and management). We review each of these topics separately.

7.1 Prevention

Without any doubt, prevention is the best and most cost-effective management policy (Williams and Grosholz 2008, Olenin et al. 2011). Strategies to prevent new introductions and invasions should be directed at key vectors of introduction, in order to interrupt transfer of a particular target species (Williams and Grosholz 2008). Furthermore, successful management of marine invasive species is only possible if handled at a global and regional level. For instance, for intentional introduction of marine organisms, there are recommended protocols (ICES 2004) and international legislation (IMO 2004) with prevention actions, which need to be applicable beyond national jurisdiction (Bax et al. 2003, Olenin et al. 2011). More recently, risk assessment has been considered one of the most useful management tool in marine biological invasions (see Hewitt and Campbell 2007).

7.2 Early Detection and Rapid Response

Early detection, rapid assessment and rapid response are key management actions against the establishment of NIS, once prevention has failed (Williams and Grosholz 2008). Public awareness and education, together with ongoing monitoring would help to detect a new invasion quickly (Campbell et al. 2007, Olenin et al. 2011). Nevertheless, one of the main weaknesses with early detection is the difficulty to accurately identify any intercepted specimen to the species-level. Recently, molecular diagnostic tools are being developed to detect early stages of NIS not yet identified by morphology (e.g., larvae, spores) (e.g., Armstrong and Ball 2005, Williams and Grosholz 2008). Although the availability of sequence data in GenBank is a great limitation, Australia has already been using genetic probes to detect invasive marine and estuarine species (Hayes et al. 2005).

A rapid response should aim at the rapid eradication of the NIS (Locke and Hanson 2009). Seaweeds, however, often disperse rapidly, increasing the difficulty to control and eradicate introduced species.

7.3 Eradication and Control

Invasive seaweeds are able to establish in a variety of marine environments and there are several species considered high-impact pests that are impossible or costly to eradicate (Piazzi and Ceccherelli 2006, Anderson 2007, Thomsen et al. 2009). Evidence suggests that a successful eradication following a rapid response should be aimed at a small and restricted population of the introduced species, with available human and financial resources (Myers et al. 2000, Wotton et al. 2004).

Several examples of eradication actions are available, although not all of them have been successful. In Britain, an eradication program has been implemented to stop the introduction of the brown macroalga *Sargassum muticum* back in 1973. Hand-gathering was the eradication method employed (time-consuming and labor intensive but highly selective) for three consequent years. After three years, the attempt to eradicate *S. muticum* had failed (Critchley et al. 1986). More recently, an attempt to manually eradicate the invasive green alga *Caulerpa racemosa* var. *cylindracea* in the Mediterranean has also failed (Ceccherelli and Piazzi 2005, Piazzi and Ceccherelli 2006). An example of a successful eradication was reported from the Chatham Islands, New Zealand, where heat-treatment methods were applied to a sunken trawler to kill the invasive kelp *Undaria pinnatifida*. Three years after, no *U. pinnatifida* was found in the sunken vessel (Wotton et al. 2004). When employing chemical agents, the open ocean is a special case, as rapid dilution occurs in flowing waters. Small lagoons are, however, good areas to apply chemical methods, e.g., eradication of the invasive green alga *Caulerpa taxifolia* in California (Anderson 2005).

Once a species is already spreading, eradication will no longer be an efficient management option. Other management options, such as containment (restricting the NIS within a geographic area), control (long-term reduction in abundance by mechanical, chemical, biological and other methods) and mitigation (adaptation and bearing the costs) should then be used to reduce the impacts of NIS, by reducing the invader spread, density and biomass to an acceptable threshold (Anderson 2007, Olenin et al. 2011). Examples of control measures are being applied in Australia with the invasive green alga *Caulerpa taxifolia* (Glasby et al. 2005).

8 Overview

* Despite many hypotheses have been proposed to explain why some ecosystems are more susceptible to invasion than others, it is unlikely that any single hypothesis will apply to all different environments. It has been suggested that no species can maximize growth, reproduction and competitive ability across all environments, so the success of invasive species should be habitat-dependent;
* The establishment of NIS is a complex, multi-factorial process, likely to increase as the global climate changes, the anthropogenic effect increases and the vectors that distribute them proliferate;
* Urgent need to conduct more studies at community level, and in particular using multiple-stressors approaches;
* The management emphasis in most countries must shift from costly eradication and control programs to proactive prevention.

Keywords: Seaweeds, process of invasion, Non-Indigenous Species, vectors of introduction, fouling assemblages, *Sargassum muticum*, *Caulerpa taxifolia*, *Caulerpa racemosa*, *Undaria pinnatifida*, *Lophocladia lallemandii*, invasion success, multiple stressors, management

References Cited

Abbott, I.A. and G.J. Hollenberg. 1976. Marine Algae of California. Stanford University Press, California.

Aguilar-Rosas, R., L.E. Aguilar-Rosas, G. Ávila-Serrano and R. Marcos-Ramírez. 2004. First record of *Undaria pinnatifida* (Harvey) Suringar (Laminariales, Phaeophyta) on the Pacific coast of Mexico. Botanica Marina 47: 255–258.

Airoldi, L. 2003. The effects of sedimentation on rocky coast assemblages. Oceanography and Marine Biology: An Annual Review 41: 161–236.

Akiyama, K. and M. Kurogi. 1982. Cultivation of *Undaria pinnatifida* (Harvey) Suringar, the decrease of crops from the natural plants following crop increase from cultivation. Bulletin of Tohoku Regional Fisheries Research Laboratory 44: 91–100.

Ambrose, R.F. and B.V. Nelson. 1982. Inhibition of giant kelp recruitment by an introduced brown alga. Botanica Marina 25: 265–268.

Anderson, L.J. 2005. California's reaction to *Caulerpa taxifolia*: a model for invasive species rapid response. Biological Invasions 7: 1003–1016.

Anderson, L.W.J. 2007. Control of invasive seaweeds. Botanica Marina 50: 418–437.

Andreakis, N., W.H. Kooistra and G. Procaccini. 2007. Microsatellite markers in an invasive strain of *Asparagopsis taxiformis* (Bonnemaisoniales, Rhodophyta): Insights in ploidy level and sexual reproduction. Gene 406: 144–151.

Andreakis, N., W.H. Kooistra and G. Procaccini. 2009. High genetic diversity and connectivity in the polyploid invasive seaweed *Asparagopsis taxiformis* (Bonnemaisoniales) in the Mediterranean, explored with microsatellite alleles and multilocus genotypes. Molecular Ecology 18: 212–226.

Arenas, F., C. Fernández, J.M. Rico, E. Fernández and D. Haya. 1995. Growth and reproductive strategies of *Sargassum muticum* (Yendo) Fensholt and *Cystoseira nodicaulis* (Whit.) Roberts. Scientia Marina 59: 1–8.

Arenas, F., I. Sánchez, S.J. Hawkins and S.R. Jenkins. 2006. The invasibility of marine algal assemblages: role of functional diversity and identity. Ecology 87: 2851–2861.

Armstrong, K.F. and S.L. Ball. 2005. DNA barcodes for biosecurity: invasive species identification. Philosophical Transactions of the Royal Society B 360: 1813–1823.

Baamonde López, S., I. Baspino Fernández, R. Barreiro Lozano and J. Cremades Ugarte. 2007. Is the cryptic alien seaweed *Ulva pertusa* (Ulvales, Chlorophyta) widely distributed along European Atlantic coasts? Botanica Marina 50: 267–274.

Baer, J. and D.B. Stengel. 2010. Variability in growth, development and reproduction of the non-native seaweed *Sargassum muticum* (Phaeophyceae) on the Irish west coast. Estuarine, Coastal and Shelf Science 90: 185–194.

Ballesteros, E., E. Cebrián and T. Alcoverro. 2007. Mortality of shoots of Posidonia oceanica following meadow invasion by the red alga *Lophocladia lallemandii*. Botanica Marina 50: 8–13.

Battershill, C., K. Miller and R. Cole. 1998. The understorey of marine invasions. Seafood New Zealand 6: 31–33.

Bax, N., A. Williamson, M. Aguero, E. Gonzalez and W. Geeves. 2003. Marine invasive alien species: a threat to global biodiversity. Marine Policy 27: 313–323.

Beardall, J., S. Beer and J.A. Raven. 1998. Biodiversity of marine plants in an era of climate change: some predictions based on physiological performance. Botanica Marina 41: 113–123.

Benhissoune, S., C.F. Boudouresque, M. Perret-Boudouresque and M. Verlaque. 2003. A checklist of the seaweeds of the Mediterranean and Atlantic coasts of Morocco. IV. Rhodophyceae—Ceramiales. Botanica Marina 46: 55–68.

Bermejo, R., J.L. Pérez-Lloréns, J.J. Vergara and I. Hernández. 2012. New records for the seaweeds of Andalusia (Spain). Acta Botanica Malacitana 37: 163–165.

Bolton, J.J., I. Germann and K. Lüning. 1983. Hybridization between Atlantic and Pacific representatives of the Simplices section of Laminaria (Phaeophyta). Phycologia 22: 133–140.

Booth, D., J. Provan and C.A. Maggs. 2007. Molecular approaches to the study of invasive seaweeds. Botanica Marina 50: 385–396.

Boudouresque, C.F. and M. Verlaque. 2002a. Assessing scale and impact of ship-transported alien macrophytes in the Mediterranean Sea. pp. 53–62. *In*: Briand, F. (ed.). Alien marine organisms introduced by ships in the Mediterranean and Black seas. Workshop Monographs n°20, CIESM, Monaco.

Boudouresque, C.F. and M. Verlaque. 2002b. Biological pollution in the Mediterranean Sea: invasive versus introduced macrophytes. Marine Pollution Bulletin 44: 32–38.

Boudouresque, C.F., M. Gerbal and K. Knoeppfler-Peguy. 1985. L'algue Japonaise *Undaria pinnatifida* (Phaeophyceae, Laminariales) en Méditerranée. Phycologia 24: 364–366.

Boudouresque, C.F., A. Meinesz, M.A. Ribera and E. Ballesteros. 1995. Spread of the green alga *Caulerpa taxifolia* (Caulerpales, Chlorophyta) in the Mediterranean: possible consequences of a major ecological event. Scientia Marina 59: 21–29.

Britton-Simmons, K.H. 2004. Direct and indirect effects of the introduced alga *Sargassum muticum* on benthic, subtidal communities of Washington State, USA. Marine Ecology Progress Series 277: 61–78.

Britton-Simmons, K.H. 2006. Functional group diversity, resource preemption and the genesis of invasion resistance in a community of marine algae. Oikos 113: 395–401.

Britton-Simmons, K.H. and K.C. Abbott. 2008. Short- and long-term effects of disturbance and propagule pressure on a biological invasion. Journal of Ecology 96: 68–77.

Brown, M.T. and M.D. Lamare. 1994. The distribution of *Undaria pinnatifida* (Harvey) Suringar within Timaru Harbour, New Zealand. Japanese Journal of Phycology 42: 63–70.

Bruno, J.F., J.J. Stachowicz and M.D. Bertness. 2003. Inclusion of facilitation into ecological theory. Trends in Ecology and Evolution 18: 119–125.

Bulleri, F., J.F. Bruno and L. Benedetti-Cecchi. 2008. Beyond competition: incorporating positive interactions between species to predict ecosystem invasibility. PLoS Biology 6: 1136–1140.

Buschbaum, C., A.S. Chapman and B. Saier. 2006. How an introduced seaweed can affect epibiota diversity in different coastal systems. Marine Biology 148: 743–754.

Cabioc'h, J. and F. Magne. 1987. Première observation du *Lomentaria hakodatensis* (Lomentariaceae, Rhodophyta) sur les côtes françaises de la Manche (Bretagne Occidentale). Cryptogamie, Algologie 8: 41–48.

Callow, M.E. and J.A. Callow. 2002. Marine biofouling: a sticky problem. Biologist 49: 1–5.

Campbell, M.L., B. Gould and C.L. Hewitt. 2007. Survey evaluations to assess marine bioinvasions. Marine Pollution Bulletin 55: 360–378.

Cardigos, F., F. Tempera, S. Ávila, J. Gonçalves, A. Colaço and R.S. Santos. 2006. Non-indigenous marine species of the Azores. Helgoland Marine Research 60: 160–169.

Carlton, J.T. 1985. Transoceanic and interoceanic dispersal of coastal marine organisms: the biology of ballast water. Oceanography and Marine Biology: An Annual Review 23: 313–371.

Carlton, J.T. and J.A. Scanlon. 1985. Progression and dispersal of an introduced alga: *Codium fragile* ssp. *tomentosoides* (Chlorophyta) on the Atlantic Coast of North America. Botanica Marina 28: 155–165.

Carlton, J.T. and R. Geller. 1993. Ecological roulette: the global transport of nonindigenous marine organisms. Science 261: 78–82.

Casas, G., R. Scrosati and M.L. Piriz. 2004. The invasive kelp *Undaria pinnatifida* (Phaeophyceae, Laminariales) reduces native seaweed diversity in Nuevo Gulf (Patagonia, Argentina). Biological Invasions 6: 411–416.

Cebrián, E. and E. Ballesteros. 2010. Invasion of Mediterranean benthic assemblages by red alga *Lophocladia lallemandii* (Montagne) F. Schmitz: Depth-related temporal variability in biomass and phenology. Aquatic Botany 92: 81–85.

Cebrián, E., C. Linares, C. Marschal and J. Garrabou. 2012. Exploring the effects of invasive algae on the persistence of gorgonian populations. Biological Invasions 14: 2647–2656.

Cebrián, J. and C.M. Duarte. 1995. Plant growth-rate dependence of detrital carbon storage in ecosystems. Science 268: 1606–1608.

Ceccherelli, G. and F. Cinelli. 1997. Short-term effects of nutrient enrichment of the sediment and interactions between the seagrass *Cymodocea nodosa* and the introduced green alga *Caulerpa taxifolia* in a Mediterranean bay. Journal of Experimental Marine Biology and Ecology 217: 165–177.

Ceccherelli, G. and L. Piazzi. 2001. Dispersal of *Caulerpa racemosa* fragments in the Mediterranean: lack of detachment time effect on establishment. Botanica Marina 44: 209–213.

Ceccherelli, G. and L. Piazzi. 2005. Exploring the success of manual eradication of *Caulerpa racemosa* var. *cylindracea* (Caulerpales, Chlorophyta): the effect of habitat. Cryptogamie 26: 319–328.

Ceccherelli, G., L. Piazzi and D. Balata. 2002. Spread of introduced *Caulerpa* species in macroalgal habitats. Journal of Experimental Marine Biology and Ecology 280: 1–11.

Cecere, E., I. Moro, M.A. Wolf, A. Petrocelli, M. Verlaque and A. Sfriso. 2011. The introduced seaweed *Grateloupia turuturu* (Rhodophyta, Halymeniales) in two Mediterranean transitional water systems. Botanica Marina 54: 23–33.

Chapin III, F.S., E.S. Zavaleta, V.T. Eviner, R.L. Naylor, P.M. Vitousek, H.L. Reynolds, D.U. Hooper, S. Lavorel, O.E. Sala, S.E. Hobbie, M.C. Mack and S. Díaz. 2000. Consequences of changing biodiversity. Nature 405: 234–242.

Chapman, A. 1998. From introduced species to invader: what determines variation in the success *Codium fragile* ssp. *tomentosoides* (Chlorophyta) in the North Atlantic Ocean? Helgoland Marine Research 52: 277–289.

Chemin, E. 1930. Quelques algues marines nouvelles pour la région de Roscoff. Bulletin de l'Association française pour l'avancement des sciences 20: 200–203.

Chesson, P. 2000. Mechanisms of maintenance of species diversity. Annual Review of Ecology and Systematics 31: 343–366.

Colautti, R.I. and H.J. MacIsaac. 2004. A neutral terminology to define 'invasive' species. Diversity and Distributions 10: 135–141.

Connell, S.D. and B.D. Russell. 2010. The direct effects of increasing CO_2 and temperature on non-calcifying organisms: increasing the potential for phase shifts in kelp forests. Proceedings of the Royal Society B 277: 1409–1415.

Coyer, J.A., A.F. Peters, W.T. Stam and J.L. Olsen. 2002. Hybridization of the marine seaweeds, *Fucus serratus* and *Fucus evanescens* (Heterokontophyta: Phaeophyceae) in a 100-year-old zone of secondary contact. Proceedings of the Royal Society B 269: 1829–1834.

Coyer, J.A., G. Hoarau, W.T. Stam and J.L. Olsen. 2007. Hybridization and introgression in a mixed population of the intertidal seaweeds *Fucus evanescens* and *F. serratus*. Journal of Evolutionary Biology 20: 2322–2333.

Crain, C.M., K. Kroeker and B.S. Halpern. 2008. Interactive and cumulative effects of multiple human stressors in marine systems. Ecology Letters 11: 1304–1315.

Critchley, A.T., W.F. Farnham and S.L. Morrell. 1983. A chronology of new European sites of attachment for the invasive brown alga, *Sargassum muticum*, 1973–1981. Journal of the Marine Biological Association of the United Kingdom 63: 799–811.

Critchley, A.T., W.F. Farnham and S.L. Morrell. 1986. An account of the attempted control of an introduced marine alga, *Sargassum muticum*, in Southern England. Biological Conservation 35: 313–332.

Critchley, A.T., P.R.M. Visscher and P.H. Nienhuis. 1990. Canopy characteristics of the brown alga *Sargassum muticum* (Fucales, Phaeophyta) in Lake Grevelingen, southwest Netherlands. Hydrobiologia 204-205: 211–217.

Curiel, D., G. Bellemo, M. Marzocchi, M. Scattolin and G. Parisi. 1998. Distribution of introduced Japanese macroalgae *Undaria pinnatifida*, *Sargassum muticum* (Phaeophyta) and *Antithamnion pectinatum* (Rhodophyta) in the Lagoon of Venice. Hydrobiologia 385: 17–22.

Daehler, C.C. 2003. Performance comparisons of co-occurring native and alien invasive plants: Implications for conservation and restoration. Annual Review of Ecology, Evolution, and Systematics 34: 183–211.

D'Antonio, C., J.M. Levine and M. Thomsen. 2001. Ecosystem resistance to invasion and the role of propagule supply: A California perspective. Journal of Mediterranean Ecology 2: 233–245.

David, M., S. Gollasch, M. Cabrini, M. Perkovič, D. Bošnjak and D. Virgilio. 2007. Results from the first ballast water sampling study in the Mediterranean Sea—the Port of Koper study. Marine Pollution Bulletin 54: 53–65.

Davis, M.A. and M. Pelsor. 2001. Experimental support for a resource-based mechanistic model of invasibility. Ecology Letters 4: 421–428.

Davis, M.A., J.P. Grime and K. Thompson. 2000. Fluctuating resources in plant communities: a general theory of invasibility. Journal of Ecology 88: 528–534.

Davis, M.A., K. Thompson and J.P. Grime. 2005. Invasibility: the local mechanism driving community assembly and species diversity. Ecography 28: 696–704.

Dawson, E.Y. 1961. A guide to the literature and distributions of Pacific benthic algae from Alaska to the Galapagos Islands. Pacific Science 15: 370–461.

de Villèle, X. and M. Verlaque. 1995. Changes and degradation in a *Posidonia oceanica* bed invaded by the introduced tropical alga *Caulerpa taxifolia* in the north western Mediterranean. Botanica Marina 38: 79–87.

Deudero, S., A. Blanco, A. Box, G. Mateu-Vicens, M. Cabanellas-Reboredo and A. Sureda. 2010. Interaction between the invasive macroalga *Lophocladia lallemandii* and the bryozoan *Reteporella grimaldii* at seagrass meadows: density and physiological responses. Biological Invasions 12: 41–52.

Deysher, L.E. 1984. Reproductive phenology of newly introduced populations of the brown alga, *Sargassum muticum* (Yendo) Fensholt. Hydrobiologia 116-117: 403–407.

Díaz, S. and M. Cabido. 2001. Vive la différence: plant functional diversity matters to ecosystem processes. Trends in Ecology and Evolution 16: 646–655.

Diaz-Pulido, G., M. Gouezo, B. Tilbrook, S. Dove and K.R.N. Anthony. 2011. High CO_2 enhances the competitive strength of seaweeds over corals. Ecology Letters 14: 156–162.

Dukes, J.S. and H.A. Mooney. 1999. Does global change increase the success of biological invaders? Trends in Ecology and Evolution 14: 135–139.

Dunstan, P.K. and C.R. Johnson. 2007. Mechanisms of invasions: can the recipient community influence invasion rates? Botanica Marina 50: 361–372.

Eckman, J.E., D.O. Duggins and A.T. Sewell. 1989. Ecology of under story kelp environments. I. Effects of kelps on flow and particle transport near the bottom. Journal of Experimental Marine Biology and Ecology 129: 173–187.

Elton, C.S. 1958. The Ecology of Invasions by Animals and Plants. University of Chicago Press, Chicago, Illinois, USA.

Emmerson, M.C., M. Solan, C. Emes, D.M. Paterson and D. Raffaelli. 2001. Consistent patterns and the idiosyncratic effects of biodiversity in marine ecosystems. Nature 411: 73–77.

Engelen, A. and R. Santos. 2009. Which demographic traits determine population growth in the invasive brown seaweed *Sargassum muticum*? Journal of Ecology 97: 675–684.

Famà, P., J.L. Olsen, W.T. Stam and G. Procaccini. 2000. High levels of intra and inter-individual polymorphism in the rDNS ITS1 of *Caulerpa racemosa* (Chlorophyta). European Journal of Phycology 35: 349–356.

Farrell, P. and R.L. Fletcher. 2006. An investigation of dispersal of the introduced brown alga *Undaria pinnatifida* (Harvey) Suringar and its competition with some species on the man-made structures of Torquay Marina (Devon, UK). Journal of Experimental Marine Biology and Ecology 334: 236–243.

Feldmann, J. 1942. Les algues marines de la côte des Albères. Fascicule III : Céramiales. Travaux algologiques 1: 23–113.

Feldmann, J. 1956. Sur la parthénogenèse du *Codium fragile* (Sur.) Hariot dans la Méditerranée. Comptes rendus de l'Académie des Sciences 243: 305–307.

Ferrer, E.A., A. Gómez Garreta and M.A. Ribera. 1997. Effect of *Caulerpa taxifolia* on the productivity of two Mediterranean macrophytes. Marine Ecology Progress Series 149: 279–287.

Flagella, M.M., M. Verlaque, A. Soria and M.C. Buia. 2007. Macroalgal survival in ballast water tanks. Marine Pollution Bulletin 54: 1395–1401.

Fletcher, R.L. and S.M. Fletcher. 1975. Studies on the recently introduced brown alga *Sargassum muticum* (Yendo) Fensholt I. Ecology and reproduction. Botanica Marina 18: 149–156.

Fletcher, R.L. and C. Manfredi. 1995. The occurrence of *Undaria pinnatifida* (Phaeophyceae, Laminariales) on the South Coast of England. Botanica Marina 38: 1–4.

Floc'h, J.Y., R. Pajot and I. Wallentinus. 1991. The Japanese brown alga *Undaria pinnatifida* on the coasts of France and its possible establishment in European waters. Journal du Conseil Permanent International pour l'Exploration de la Mer 47: 379–390.

Forrest, B.M. and M.D. Taylor. 2002. Assessing invasion impact: survey design considerations and implications for management of an invasive marine plant. Biological Invasions 4: 375–386.

Frank, K.T., B. Petrie, J.S. Choi and W.C. Leggett. 2005. Trophic cascades in a formerly cod-dominated ecosystem. Science 308: 1621–1622.

Fridley, J.D., J. Stachowicz, S. Naeem, D.F. Sax, E.W. Seabloom, M.D. Smith, T.J. Stohlgren, D. Tilman and B. Von Holle. 2007. The invasion paradox: reconciling pattern and process in species invasions. Ecology 88: 3–17.

Funk, G. 1923. Über einige Ceramiaceen aus dem Golf von Neapel. Beihefte zum botanischen Centralblatt 34: 223–247.

Funk, J.L. and P.M. Vitousek. 2007. Resource-use efficiency and plant invasion in low-resource systems. Nature 446: 1079–1081.

Geller, J.B., J.A. Darling and J.T. Carlton. 2010. Genetic perspectives on marine biological invasions. Annual Review of Marine Science 2: 367–393.

Gestoso, I., C. Olabarria and J.S. Troncoso. 2010. Variability of epifaunal assemblages associated with native and invasive macroalgae. Marine and Freshwater Research 61: 724–731.

Gestoso, I., C. Olabarria and J.S. Troncoso. 2012. Effects of macroalgal identity on epifaunal assemblages: native species versus the invasive species *Sargassum muticum*. Helgoland Marine Research 66: 159–166.

Glasby, T.M., R.G. Creese and P.T. Gibson. 2005. Experimental use of salt to control the invasive marine alga *Caulerpa taxifolia* in New South Wales, Australia. Biological Conservation 122: 573–580.

Godwin, L.S. 2003. Hull fouling of maritime vessels as a pathway for marine species invasions to the Hawaiian Islands. Biofouling 19: 123–131.

Gollasch, S., E. Mac Donald, S. Belson, H. Botnen, J.T. Christensen, P.J. Hamer, G. Houvenaghel, A. Jelmert, I. Lucas, D. Masson, T. Mccollin, S. Olenin, A. Persson, I. Wallentinus, L.P.M.J. Wetseyn and T. Wittling. 2002. Life in ballast tanks. pp. 217–231. *In*: Leppäkoski, E., S. Gollasch and S. Olenin (eds.). Invasive Aquatic Species of Europe, Distribution, Impact and Management. Kluwer Academic Publishers, Dordrecht, Netherlands.

Grosholz, E. 2002. Ecological and evolutionary consequences of coastal invasions. Trends in Ecology and Evolution 17: 22–27.

Hall-Spencer, J.M., R. Rodolfo-Metalpa, S. Martin, E. Ransome, M. Fine, S.M. Turner, S.J. Rowley, D. Tedesco and M.C. Buia. 2008. Volcanic carbon dioxide vents show ecosystem effects of ocean acidification. Nature 454: 96–99.

Halpern, B.S., S. Walbridge, K.A. Selkoe, C.V. Kappel, F. Micheli, C. D'Agrosa, J.F. Bruno, K.S. Casey, C. Ebert, H.E. Fox, R. Fujita, D. Heinemann, H.S. Lenihan, E.M.P. Madin, M.T. Perry, E.R. Selig, M. Spalding, R. Steneck and R. Watson. 2008. A global map of human impact on marine ecosystems. Science 319: 948–952.

Harley, C.D.G., A.R. Hughes, K.M. Hultgren, B.G. Miner, C.J.B. Sorte, C.S. Thornber, L.F. Rodriguez, L. Tomanek and S.L. Williams. 2006. The impacts of climate change in coastal marine systems. Ecology Letters 9: 228–241.

Harley, C.D.G., K.M. Anderson, K.W. Demes, J.P. Jorve, R.L. Kordas, T.A. Coyle and M.H. Graham. 2012. Effects of climate change on global seaweed communities. Journal of Phycology 48: 1064–1078.

Harris, L.G. and M.C. Tyrrell. 2001. Changing community states in the Gulf of Maine: synergism between invaders, overfishing and climate change. Biological Invasions 3: 9–21.

Hay, C.H. 1990. The dispersal of sporophytes of *Undaria pinnatifida* by coastal shipping in New Zealand, and implications for further dispersal of *Undaria* in France. British Phycological Journal 25: 304–313.

Hay, M.E. 1991. Marine-terrestrial contrasts in the ecology of plant chemical defenses against herbivores. Trends in Ecology and Evolution 6: 362–365.

Hayes, K.L., C. Sliwa, S. Migus, F. McEnnulty and P.K. Dunstan. 2005. National priority pests: Part II. Ranking of Australian marine pests. CSIRO Division of Marine Research, Hobart.

Hewitt, C.L. and M.L. Campbell. 2007. Mechanisms for the prevention of marine bioinvasions for better biosecurity. Marine Pollution Bulletin 55: 395–401.

Hewitt, C.L., M.L. Campbell and B. Schaffelke. 2007. Introductions of seaweeds: accidental transfer pathways and mechanisms. Botanica Marina 50: 326–337.

Hewitt, C., R. Everett, N. Parker and M. Campbell. 2009. Marine Bioinvasion Management: Structural Framework. pp. 327–334. *In*: Rilov, G. and J. Crooks (eds.). Biological Invasions in Marine Ecosystems. Springer, Berlin, Heidelberg.

Hewitt, L.C., M.L. Campbell, F. McEnnulty, M.M. Kirrily, N.B. Murfet, B. Robertson and B. Schaffelke. 2005. Efficacy of physical removal of a marine pest: the introduced kelp *Undaria pinnatifida* in a Tasmanian Marine Reserve. Biological Invasions 7: 251–263.

Hoegh-Guldberg, O. and J.F. Bruno. 2010. The impact of climate change on the world's marine ecosystems. Science 328: 1523–1528.

Hoegh-Guldberg, O., P.J. Mumby, A.J. Hooten, R.S. Steneck, P. Greenfield, E. Gomez, C.D. Harvell, P.F. Sale, A.J. Edwards, K. Caldeira, N. Knowlton, C.M. Eakin, R. Iglesias-Prieto, N. Muthiga, R.H. Bradbury, A. Dubi and M.E. Hatziolos. 2007. Coral Reefs under rapid climate change and ocean acidification. Science 318: 1737–1742.

ICES. 2004. ICES Code of Practice on the Introductions and Transfers of Marine Organisms. 2003. International Council for the Exploration of the Sea, Copenhagen. Denmark.

IMO. 2004. International Maritime Organization. International Convention for the Control and Management of Ships' Ballast Water and Sediments. Available from http://www.imo.org.

IPCC. 2007. Climate Change 2007, The Physical Science Basis. Summary for Policymakers. Intergovernmental Panel on Climate Change. Cambridge University Press, Cambridge, UK.

Irigoyen, A.J., G. Trobbiani, M.P. Sgarlatta and M.P. Raffo. 2011. Effects of the alien algae *Undaria pinnatifida* (Phaeophyceae, Laminariales) on the diversity and abundance of benthic macrofauna in Golfo Nuevo (Patagonia, Argentina): potential implications for local food webs. Biological Invasion 13: 1521–1532.

Janiak, D.S. and R.B. Whitlatch. 2012. Epifaunal and algal assemblages associated with the native *Chondrus crispus* (Stackhouse) and the non-native *Grateloupia turuturu* (Yamada) in eastern Long Island Sound. Journal of Experimental Marine Biology and Ecology 413: 38–44.

Johnson, V.R., B.D. Russell, K. Fabricius, C. Brownlee and J.M. Hall-Spencer. 2012. Temperate and tropical brown macroalgae thrive, despite decalcification, along natural CO_2 gradients. Global Change Biology 18: 2792–2803.

Jones, E. and C.S. Thornber. 2010. Effects of habitat-modifying invasive macroalgae on epiphytic algal communities. Marine Ecology Progress Series 400: 87–100.

Jongma, D.N., D. Campo, E. Dattolo, D. D'Esposito, A. Duchi, P. Grewe, J. Huisman, M. Verlaque, M.B. Yokes and G. Procaccini. 2013. Identity and origin of a slender *Caulerpa taxifolia* strain introduced into the Mediterranean Sea. Botanica Marina 56: 27–39.

Jousson, O., J. Pawlowski, L. Zaninetti, F.W. Zechman, F. Dini, G. di Giuseppe, R. Woodfields, A. Millar and A. Meinesz. 2000. Invasive alga reaches California. Nature 408: 157–158.

Keane, R.M. and M.J. Crawley. 2002. Exotic plant invasions and the enemy release hypothesis. Trends in Ecology and Evolution 17: 164–170.

Klein, J. and M. Verlaque. 2008. The *Caulerpa racemosa* invasion: A critical review. Marine Pollution Bulletin 56: 205–225.

Knowlton, N. 1993. Sibling species in the sea. Annual Review of Ecology and Systematics 24: 189–216.

Kolar, C.S. and D.M. Lodge. 2001. Progress in invasion biology: predicting invaders. Trends in Ecology and Evolution 16: 199–204.

Kroeker, K.J., R.L. Kordas, R.N. Crim and G.G. Singh. 2010. Meta-analysis reveals negative yet variable effects of ocean acidification on marine organisms. Ecology Letters 13: 1419–1434.

Levin, P.S., J.A. Coyer, R. Petrik and T.P. Good. 2002. Community-wide effects of nonindigenous species on temperate rocky reefs. Ecology 83: 3182–3193.

Levine, J.M. 2000. Species diversity and biological invasions: relating local process to community pattern. Science 288: 852–854.

Lima, F. and D. Wethey. 2012. Three decades of high-resolution coastal sea surface temperatures reveal more than warming. Nature Communications 3: 704.

Locke, A. and J.M. Hanson. 2009. Rapid response to non-indigenous species. 1. Goals and history of rapid response in the marine environment. Aquatic Invasions 4: 237–247.

Lockwood, J.L., P. Cassey and T. Blackburn. 2005. The role of propagule pressure in explaining species invasions. Trends in Ecology and Evolution 20: 223–228.

Lonsdale, W.M. 1999. Global patterns of plant invasions and the concept of invasibility. Ecology 80: 1522–1536.

Luning, K. 1990. Seaweeds. Their Environment, Biogeography and Ecophysiology. John Wiley, New York.

MacDougall, A.S., B. Gilbert and J.M. Levine. 2009. Plant invasions and the niche. Journal of Ecology 97: 609–615.

Mack, R.N. 2003. Global plant dispersal, naturalization, and invasion: pathways, modes, and circumstances. pp. 3–30. *In*: Ruiz, G.M. and J.T. Carlton (eds.). Invasive Species: Vectors and Management Strategies. Island Press, Washington.

Mack, R.N., D. Simberloff, W.M. Lonsdale, H. Evans, M. Clout and F. Bazzaz. 2000. Biotic invasions: causes, epidemiology, global consequences and control. Ecological Applications 10: 689–710.

Magne, F. 1992. *Goniotrichiopsis* (Rhodophycées, Porphyridiales) en Europe. Cryptogamie, Algologie 13: 109–112.

Martin, S. and P. Gattuso. 2009. Response of Mediterranean coralline algae to ocean acidification and elevated temperature. Global Change Biology 15: 2089–2100.

McIvor, L., C.A. Maggs, J. Provan and M.J. Stanhope. 2001. *rbc*L sequences reveal multiple cryptic introductions of the Japanese red alga *Polysiphonia harveyi*. Molecular Ecology 10: 911–919.

Meinesz, A., J. de Vaugelas, B. Hesse and X. Mari. 1993. Spread of the introduced tropical green alga *Caulerpa taxifolia* in northern Mediterranean waters. Journal of Applied Phycology 5: 141–147.

Meinesz, A., L. Benichou, J. Blachier, T. Komatsu, R. Lemée, H. Molenaar and X. Mari. 1995. Variations in the structure, morphology and biomass of *Caulerpa taxifolia* in the Mediterranean. Botanica Marina 38: 499–508.

Meinesz, A., T. Belsher, T. Thibaut, B. Antolic, K.B. Mustapha, C.F. Boudouresque, D. Chiaverini, F. Cinelli, J.M. Cottalorda, A. Djellouli, A. El Abed, C. Orestano, A.M. Grau, L. Ivesa, A. Jaklin, H. Langar, E. Massuti-Pascual, A. Peirano, L. Tunesi, J. De Vaugelas, N. Zavodnik and A. Zuljevic. 2001. The introduced green alga *Caulerpa taxifolia* continues to spread in the Mediterranean. Biological Invasions 3: 201–210.

Mendez Domingo, C. 1957. Sur l'existence du *Colpomenia peregrina* (Sauv.) Hamel dans la Méditerranée. Vie et Milieu 8: 92–98.

Miller, K.A., L.E. Aguilar-Rosas and F.F. Pedroche. 2011. A review of non-native seaweeds from California, USA and Baja California, Mexico. Hidrobiologica 21: 365–379.

Mineur, F., T. Belsher, M.P. Johnson, C.A. Maggs and M. Verlaque. 2007a. Experimental assessment of oyster transfers as a vector for macroalgal introductions. Biological Conservation 137: 237–247.

Mineur, F., M.P. Johnson, C.A. Maggs and H. Stegenga. 2007b. Hull fouling on commercial ships as a vector of macroalgal introductions. Marine Biology 151: 1299–1307.

Mineur, F., M.P. Johnson and C.A. Maggs. 2008. Macroalgal introductions by hull fouling on recreational vessels: seaweeds and sailors. Environmental Management 42: 667–676.

Mineur, F., A.J. Davies, C.A. Maggs, M. Verlaque and M.P. Johnson. 2010a. Fronts, jumps and secondary introductions suggested as different invasion patterns in marine species, with an increase in spread rates over time. Proceedings of the Royal Society B 277: 2693–2701.

Mineur, F., E.J. Cook, D. Minchin, K. Bohn, A. MacLeod and C.A. Maggs. 2012a. Changing coasts: marine aliens and artificial structures. Oceanography and Marine Biology: An Annual Review 50: 189–234.

Mineur, F., A. Le Roux, H. Stegenga, M. Verlaque and C.A. Maggs. 2012b. Four new exotic red seaweeds on European shores. Biological Invasions 14: 1635–1641.

Mitchell, C.E., A.A. Agrawal, J.D. Bever, G.S. Gilbert, R.A. Hufbauer, J.N. Klironomos, J.L. Maron, W.F. Morris, I.M. Parker, A.G. Power, E.W. Seabloom, M.E. Torchin and D.P. Vásquez. 2006. Biotic interactions and plant invasions. Ecology Letters 9: 726–740.

Morrison, W.E. and M.E. Hay. 2011. Herbivore preference for native vs. exotic plants: generalist herbivores from multiple continents prefer exotic plants that are evolutionarily naïve. PLoS ONE 6: e17227.

Munda, I. 1977. A note on the growth of *Halosaccion ramentaceum* (L.) J. Ag. under different conditions. Botanica Marina 20: 493–498.

Munro, A.L.S., S.D. Utting and I. Wallentinus. 1999. Status of introductions of non-indigenous marine species to North Atlantic waters. ICES Cooperative Research Report 231, 91 pp.

Myers, J.H., D. Simberloff, A.M. Kuris and J.R. Carey. 2000. Eradication revisited: dealing with exotic species. Trends in Ecology and Evolution 15: 316–320.

Naeem, S. and J.P. Wright. 2003. Disentangling biodiversity effects on ecosystem functioning: deriving solutions to a seemingly insurmountable problem. Ecology Letters 6: 567–579.

Nanba, N., T. Fujiwara, K. Kuwano, Y. Ishikawa, H. Ogawa and R. Kado. 2011. Effect of water flow velocity on growth and morphology of cultured *Undaria pinnatifida* sporophytes (Laminariales, Phaeophyceae) in Okirai Bay on the Sanriku coast, Northeast Japan. Journal of Applied Phycology 23: 1023–1030.

Norton, T.A. 1976. Why is *Sargassum muticum* so invasive? British Phycological Journal 11: 197–198.

Norton, T.A. 1977. The growth and development of *Sargassum muticum* (Yendo) Fensholt. Journal of Experimental Marine Biology and Ecology 26: 41–53.

Núñez-López, R.A. and M.M. Casas-Valdez. 1998. Flora ficológica de la laguna de San Ignacio, B.C.S., México. Hidrobiologica 8: 50–57.

Nyberg, C.D. and I. Wallentinus. 2005. Can species traits be used to predict marine macroalgal introductions? Biological invasions 7: 265–279.

Occhipinti-Ambrogi, A. 2007. Global change and marine communities: alien species and climate change. Marine Pollution Bulletin 55: 342–352.

Occhipinti-Ambrogi, A. and D. Savini. 2003. Biological invasions as a component of global change in stressed marine ecosystems. Marine Pollution Bulletin 46: 542–551.

Odum, E.P. 1969. The strategy of ecosystem development. Science 164: 262–270.

O'Gorman, E.J., J.M. Yearsley, T.P. Crowe, M.C. Emmerson, U. Jacob and O.L. Petchey. 2011. Loss of functionally unique species may gradually undermine ecosystems. Proceedings of the Royal Society B: Biological Sciences 278: 1886–1893.

Olabarria, C., F. Arenas, R.M. Viejo, I. Gestoso, F. Vaz-Pinto, M. Incera, M. Rubal, E. Cacabelos, P. Veiga and C. Sobrino. 2013. Response of macroalgal assemblages from rockpools to climate change: effects of persistent increase in temperature and CO_2. Oikos 122: 1065–1079.

Olenin, S., M. Elliott, I. Bysveen, P.F. Culverhouse, D. Daunys, G.B.J. Dubelaar, S. Gollasch, P. Goulletquer, A. Jelmert, Y. Kantor, K.B. Mézeth, D. Minchin, A. Occhipinti-Ambrogi, I. Olenina and J. Vandekerkhove. 2011. Recommendations on methods for the detection and control of biological pollution in marine coastal waters. Marine Pollution Bulletin 62: 2598–2604.

Padilla, D.K. and S.L. Williams. 2004. Beyond ballast water: aquarium and ornamental trades as sources of invasive species in aquatic ecosystems. Frontiers in Ecology and Environment 2: 131–138.

Parker, I.M., D. Simberloff, W.M. Lonsdale, K. Goodell, M. Wonham, P.M. Kareiva, M.H. Williamson, B. Von Holle, P.B. Moyle, J.E. Byers and L. Goldwasser. 1999. Impact: toward a framework for understanding the ecological effects of invaders. Biological Invasions 1: 3–19.

Patzner, R.A. 1998. The invasion of *Lophocladia* (Rhodomelaceae, Lophotalieae) at the northern coast of Ibiza (western Mediterranean Sea). Bolletí de la Societat d'Història Natural de les Balears 41: 75–80.

Paula, E.J. and V.R. Eston. 1987. Are there other *Sargassum* species potentially as invasive as *S. muticum*? Botanica Marina 30: 405–410.

Pedersen, M.F., P.A. Stæhr, T. Wernberg and M.S. Thomsen. 2005. Biomass dynamics of exotic *Sargassum muticum* and native *Halidrys siliquosa* in Limfjorden, Denmark—Implications of species replacements on turnover rates. Aquatic Botany 83: 31–47.

Pejchar, L. and H.A. Mooney. 2009. Invasive species, ecosystem services and human well-being. Trends in Ecology and Evolution 24: 497–504.

Péres-Cirera, J.L., J. Cremades and I. Bárbara. 1989. *Grateloupia lanceola* (Cryptonemiales, Rhodophyta) en las costas de la Península Ibérica: estudio morfológico y anatómico. Lazaroa 11: 123–134.

Péres-Cirera, J.L., J. Cremades, I. Barbara and M.C. López. 1991. Contribución al conocimiento del genero *Phyllariopsis* (Phyllariaceae, Phaeophyta) en el Atlantico europeo. Nova Acta Científica Compostelana (Bioloxía) 2: 3–11.

Pérez, R., R. Kaas and O. Barbaroux. 1984. Culture expérimentale de l'algue *Undaria pinnatifida* sur les côtes de France. Science et Pêche, Bulletin de l'Institut des Pêches maritimes 343: 3–15.

Perkins, L.B., E.A. Leger and R.S. Nowak. 2011. Invasion triangle: an organizational framework for species invasion. Ecology and Evolution 1: 610–625.

Piazzi, L. and G. Ceccherelli. 2006. Persistence of biological invasion effects: Recovery of macroalgal assemblages after removal of *Caulerpa racemosa* var. *cylindracea*. Estuarine, Coastal and Shelf Science 68: 455–461.

Piazzi, L., G. Ceccherelli and F. Cinelli. 2001. Threat to macroalgal diversity: effects of the introduced green alga *Caulerpa racemosa* in the Mediterranean. Marine Ecology Progress Series 210: 161–165.

Piazzi, L., A. Meinesz, M. Verlaque, B. Akçali, B. Antolić, M. Argyrou, D. Balata, E. Ballesteros, S. Calvo, F. Cinelli, S. Cirik, A. Cossu, R. D'Archino, A.S. Djellouli, F. Javel, E. Lanfranco, C. Mifsud, D. Pala, P. Panayotidis, A. Peirano, G. Pergent, A. Petrocelli, S. Ruitton, A. Žuljević and G. Seccherelli. 2005. Invasion of *Caulerpa racemosa* var. *cylindracea* (Caulerpales, Chlorophyta) in the Mediterranean Sea: an assessment of the spread. Cryptogamie Algologie 26: 189–202.

Pickering, T.D., P. Skelton and R.J. Sulu. 2007. Intentional introductions of commercially harvested alien seaweeds. Botanica Marina 50: 338–350.

Pimentel, D., R. Zuniga and D. Morrison. 2005. Update on the environmental and economic costs associated with alien-invasive species in the United States. Ecological Economics 52: 273–288.

Por, F.D. 1978. Lessepsian Migration: The Influx of Red Sea Biota into the Mediterranean by way of the Suez Canal. Ecological Studies 23, 228 pp. Springer-Verlag, Berlin.

Porzio, L., M.C. Buia and J.M. Hall-Spencer. 2011. Effects of ocean acidification on macroalgal communities. Journal of Experimental Marine Biology and Ecology 400: 278–287.

Provan, J., D. Booth, N.P. Todd, G.E. Beatty and C.A. Maggs. 2008. Tracking biological invasions in space and time: elucidating the invasive history of the green alga *Codium fragile* using old DNA. Diversity and Distributions 14: 343–354.

Pysek, P. and D.M. Richardson. 2007. Traits associated with invasiveness in alien plants: Where do we stand? pp. 97–125. *In*: Nentwig, W. (ed.). Biological Invasions. Springer, Berlin.

Raffaelli, D.G. 2006. Biodiversity and ecosystem functioning: issues of scale and trophic complexity. Marine Ecology Progress Series 311: 285–294.

Reed, D.C., D.R. Laur and A.W. Ebeling. 1988. Variation in algal dispersal and recruitment: the importance of episodic events. Ecological Monographs 58: 321–335.

Rejmanek, M. 1996. A theory of seed plant invasiveness: the first sketch. Biological Conservation 78: 171–181.

Ribera, M.A. and C.F. Boudouresque. 1995. Introduced marine plants, with special reference to macroalgae: mechanisms and impact. pp. 187–268. *In*: Round, F.E. and D.J. Chapman (eds.). Progress in Phycological Research, Volume 11. Biopress Ltd., Bristol.

Ribera Siguan, M.A. 2003. Pathways of biological invasions of marine plants. pp. 183–226. *In*: Ruiz, G.M. and J.T. Carlton (eds.). Invasive Species: Vectors and Management Strategies. Island Press, Washington.

Ricciardi, A. and J. Cohen. 2007. The invasiveness of an introduced species does not predict its impact. Biological Invasions 9: 309–315.

Roman, J. 2006. Diluting the founder effect: cryptic invasions expand a marine invader's range. Proceedings of the Royal Society B: Biological Sciences 273: 2453–2459.

Rossi, F., M. Incera, M. Callier and C. Olabarria. 2011. Effects of detrital non-native and native macroalgae on the nitrogen and carbon cycling in intertidal sediments. Marine Biology 158: 2705–2715.

Rueness, J. 1989. *Sargassum muticum* and other introduced Japanese macroalgae: biological pollution of European coasts. Marine Pollution Bulletin 20: 173–176.

Ruesink, J.L., B.E. Feist, C.J. Harvey, J.S. Hong, A.C. Trimble and L.M. Wisehart. 2006. Changes in productivity associated with four introduced species: ecosystem transformation of a 'pristine' estuary. Marine Ecology Progress Series 311: 203–215.

Ruiz, G.M., P.W. Fofonoff, J.T. Carlton, M.J. Wonham and A.H. Hines. 2000. Invasion of coastal marine communities in North America: apparent patterns, processes, and biases. Annual Review of Ecology and Systematics 31: 481–531.

Rull Lluch, J., A. Gómez Garreta, M.C. Barceló and M.A. Ribera. 1994. Mapas de distribución de algas marinas de la Península Ibérica e Islas Baleares. VII. *Cystoseira* C. Agardh (Grupo *C. baccata*) y *Sargassum* C. Agardh (*S. muticum* y *S. vulgare*). Botanica Complutensis 19: 131–138.

Russell, B.D., J.-A.I. Thompson, L.J. Falkenberg and S.D. Connell. 2009. Synergistic effects of climate change and local stressors: CO_2 and nutrient-driven change in subtidal rocky habitats. Global Change Biology 15: 2153–2162.

Sabour, B., A. Reani, H. El Magouri and R. Haroun. 2013. *Sargassum muticum* (Yendo) Fensholt (Fucales, Phaeophyta) in Morocco, an invasive marine species new to the Atlantic coast of Africa. Aquatic Invasions 8: 97–102.

Sakai, A.K., F.W. Allendorf, J.F. Holt, D.M. Lodge, J. Molofsky, K.A. With, S. Baughman, R.J. Cabin, J.E. Cohen, N.C. Ellstrand, D.E. McCauley, P. O'Neil, I.M. Parker, J.N. Thompson and S.G. Weller. 2001. The population biology of invasive species. Annual Review of Ecology and Systematics 32: 305–332.

Sánchez, I., C. Fernández and J. Arrontes. 2005. Long-term changes in the structure of intertidal assemblages after invasion by *Sargassum muticum* (Phaeophyta). Journal of Phycology 41: 942–949.

Sancholle, M. 1988. Présence de *Fucus spiralis* (Phaeophyceae) en Méditerranée occidentale. Cryptogamie, Algologie 9: 157–161.

Sanderson, J.C. 1997. Survey of *Undaria pinnatifida* in Tasmanian coastal waters, January-February 1997. Internal report to Tasmanian Department of Marine Resources.

Scagel, R.F. 1956. Introduction of a Japanese alga, *Sargassum muticum* into the northeast Pacific. Fisheries Research papers, Washington Department of Fisheries 1: 49–59.

Schaffelke, B. and C.L. Hewitt. 2007. Impacts of introduced seaweeds. Botanica Marina 50: 397–417.

Schaffelke, B., J.E. Smith and C.L. Hewitt. 2006. Introduced macroalgae—a growing concern. Journal of Applied Phycology 18: 529–541.

Sfriso, A. and C. Facca. 2013. Annual growth and environmental relationships of the invasive species *Sargassum muticum* and *Undaria pinnatifida* in the lagoon of Venice. Estuarine, Coastal and Shelf Science 129: 162–172.

Shea, K. and P. Chesson. 2002. Community ecology theory as a framework for biological invasions. Trends in Ecology and Evolution 17: 170–176.

Silva, P.C., R.A. Woodfield, A.N. Cohen, L.H. Harris and J.H.R. Goddard. 2002. First report of the Asian kelp *Undaria pinnatifida* in the northeastern Pacific Ocean. Biological Invasions 4: 333–338.

Simberloff, D. 2009. The role of propagule pressure in biological invasions. Annual Review of Ecology, Evolution, and Systematics 40: 81–102.

Smale, D. and T. Wernberg. 2013. Extreme climatic event drives range contraction of a habitat-forming species. Proceedings of the Royal Society Biological Sciences 280: 2012–2829.

Smith, D.L., M.J. Wonham, L. McCann, G.M. Ruiz, H.A. Hines and J.T. Carlton. 1999. Invasion pressure to a ballast-flooded estuary and an assessment of inoculant survival. Biological Invasions 1: 67–87.

Sorte, C.J.B., S.L. Williams and J.T. Carlton. 2010. Marine range shifts and species introductions: comparative spread rates and community impacts. Global Ecology and Biogeography 19: 303–316.

Stachowicz, J.J. and J.E. Byrnes. 2006. Species diversity, invasion success, and ecosystem functioning: disentangling the influence of resource competition, facilitation, and extrinsic factors. Marine Ecology Progress Series 311: 251–262.

Stachowicz, J.J., J.R. Terwin, R.B. Whitlatch and R.W. Osman. 2002. Linking climate change and biological invasions: ocean warming facilitates non-indigenous species invasions. Proceedings of the National Academy of Sciences 99: 15497–15500.

Staehr, P.A., M.F. Pedersen, M.S. Thomsen, T. Wernberg and D. Krause-Jensen. 2000. Invasion of *Sargassum muticum* in Limfjorden (Denmark) and its possible impact on the indigenous macroalgal community. Marine Ecology Progress Series 207: 79–88.

Stam, W.T., J.L. Olsen, S.F. Zaleski, S.N. Murray, K.R. Brown and L.J. Walters. 2006. A forensic and phylogenetic survey of *Caulerpa* species (Caulerpales, Chlorophyta) from the Florida coast, local aquarium shops, and e-commerce: Establishing a proactive baseline for early detection. Journal of Phycology 42: 1113–1124.

Stohlgren, T.J., D. Binkley, G.W. Chong, M.A. Kalkhan, L.D. Schell, K.A. Bull, Y. Otsuki, G. Newman, M. Bashkin and Y. Son. 1999. Exotic plant species invade hot spots of native plant diversity. Ecological Monographs 69: 25–46.

Streftaris, N., A. Zenetos and E. Papathanassiou. 2005. Globalisation in marine ecosystems: the story of non-indigenous marine species across European seas. Oceanography and Marine Biology: An Annual Review 43: 419–453.

Strong, J.A., M.J. Dring and C.A. Maggs. 2006. Colonisation and modification of soft substratum habitats by the invasive macroalga *Sargassum muticum*. Marine Ecology Progress Series 321: 87–97.

Svensson, J.R., G.M. Nylund, G. Cervin, G.B. Toth and H. Pavia. 2013. Novel chemical weapon of an exotic macroalga inhibits recruitment of native competitors in the invaded range. Journal of Ecology 101: 140–148.

Thomsen, M.A. 2010. Experimental evidence for positive effects of invasive seaweed on native invertebrates via habitat-formation in a seagrass bed. Aquatic Invasions 5: 341–346.

Thomsen, M.S., T. Wernberg, F. Tuya and B.R. Silliman. 2009. Evidence for impacts of nonindigenous macroalgae: a meta-analysis of experimental field studies. Journal of Phycology 45: 812–819.

Thomsen, M.S., J.D. Olden, T. Wernberg, J.N. Griffin and B.R. Silliman. 2011. A broad framework to organize and compare ecological invasion impacts. Environmental Research 111: 899–908.

Thomsen, M.S., T. Wernberg, A.H. Engelen, F. Tuya, M.A. Vanderklift, M. Holmer, K.J. McGlathery, F. Arenas, J. Kotta and B.R. Silliman. 2012. A meta-analysis of seaweed impacts on seagrasses: generalities and knowledge gaps. Plos ONE 7: e28595.

Titman, D. 1976. Ecological competition between algae: experimental confirmation of resource-based competition theory. Science 192: 463–465.

Tilman, D. 1999. The ecological consequences of changes in biodiversity: a search for general principles. Ecology 80: 1455–1474.

Trowbridge, C.D. 1995. Establishment of the green alga *Codium fragile* ssp. *tomentosoides* on New Zealand rocky shores: current distribution and invertebrate grazers. Journal of Ecology 83: 949–965.

Valentine, J.P. and C.R. Johnson. 2004. Establishment of the introduced kelp *Undaria pinnatifida* following dieback of the native macroalga *Phyllospora comosa* in Tasmania, Australia. Marine and Freshwater Research 55: 223–230.

Varela-Álvarez, E., A. Gómez Garreta, J. Rull Lluch, N. Salvador Soler, E.A. Serrão and M.A. Ribera Siguán. 2012. Mediterranean species of *Caulerpa* are polyploid with smaller genomes in the invasive ones. Plos ONE 7: e47728.

Vaz-Pinto, F., C. Olabarria and F. Arenas. 2012. Propagule pressure and functional diversity: interactive effects on a macroalgal invasion process. Marine Ecology Progress Series 471: 51–60.

Vaz-Pinto, F., C. Olabarria and F. Arenas. 2013a. Role of top-down and bottom-up forces on the invasibility of intertidal macroalgal assemblages. Journal of Sea Research 76: 178–186.

Vaz-Pinto, F., C. Olabarria, I. Gestoso, E. Cacabelos, M. Incera and F. Arenas. 2013b. Functional diversity and climate change: effects on the invasibility of macroalgal assemblages. Biological Invasions 15: 1833–1846.

Vaz-Pinto, F., C. Olabarria and F. Arenas. 2014. Ecosystem functioning impacts of the invasive seaweed *Sargassum muticum* (Ochrophyta). Journal of Phycology 50: 108–116.

Verlaque, M. 2001. Checklist of the macroalgae of Thau Lagoon (Hérault, France), a hot spot of marine species introduction in Europe. Oceanologica Acta 24: 29–49.

Verlaque, M. and R. Riouall. 1989. Introduction de *Polysiphonia nigrescens* et d'*Antithamnion nipponicum* (Rhodophyta, Ceramiales) sur le littoral méditerranéen français. Cryptogamie, Algologie 10: 313–323.

Verlaque, M., C.F. Boudouresque, A. Meinesz and V. Gravez. 2000. The *Caulerpa racemosa* complex (Caulerpales, Ulvophyceae) in the Mediterranean Sea. Botanica Marina 43: 49–68.

Verlaque, M., J. Afonso-Carrillo, M.C. Gil-Rodriguez, C. Durand, C.F. Boudouresque and Y. Le Parco. 2004. Blitzkrieg in a marine invasion: *Caulerpa racemosa* var. *cylindracea* (Bryopsidales, Chlorophyta) reaches the Canary Islands (north-east Atlantic). Biological Invasions 6: 269–281.

Verlaque, M., P.M. Brannock, T. Konmatsu, M. Villalard-Bohnsack and M. Marston. 2005. The genus *Grateloupia* C. Agardh (Halymeniaceae, Rhodophyta) in the Thau Lagoon (France, Mediterranean): a case study of marine plurispecific introductions. Phycologia 44: 477–496.

Viejo, R.M., J. Arrontes and N.L. Andrew. 1995. An experimental evaluation of the effect of wave action on the distribution of *Sargassum muticum* in Northern Spain. Botanica Marina 38: 437–442.

Voisin, M., C.R. Engel and F. Viard. 2005. Differential shuffling of native genetic diversity across introduced regions in a brown alga: aquaculture vs. maritime traffic effects. Proceedings of the National Academy of Sciences of the United States of America 102: 5432–5437.

Walters, L.J., K.R. Brown, W.T. Stam and J.L. Olsen. 2006. E-commerce and *Caulerpa*: unregulated dispersal of invasive species. Frontiers in Ecology and the Environment 4: 75–79.

Ward, J.M. and A. Ricciardi. 2010. Community-level effects of co-occurring native and exotic ecosystem engineers. Freshwater Biology 55: 1803–1817.

Wernberg, T., B.D. Russell, P.J. Moore, S.D. Ling, D.A. Smale, A. Campbell, M.A. Coleman, P.D. Steinberg, G.A. Kendrick and S.D. Connell. 2011. Impacts of climate change in a global hotspot for temperate marine biodiversity and ocean warming. Journal of Experimental Marine Biology and Ecology 400: 7–16.

Wernberg, T., D.A. Smale and M.S. Thomsen. 2012. A decade of climate change experiments on marine organisms: procedures, patterns and problems. Global Change Biology 18: 1491–1498.

White, L.F. and J.B. Shurin. 2007. Diversity effects on invasion vary with life history stage in marine macroalgae. Oikos 116: 1193–1203.

Williams, S. and E. Grosholz. 2008. The invasive species challenge in estuarine and coastal environments: marrying management and science. Estuaries and Coasts 31: 3–20.

Williams, S.L. and J.E. Smith. 2007. A global review of the distribution, taxonomy, and impacts of introduced seaweeds. Annual Review of Ecology, Evolution, and Systematics 38: 327–359.

Williamson, M. 2006. Explaining and predicting the success of invading species at different stages of invasion. Biological Invasions 8: 1561–1568.

Williamson, M.H. and A.A. Fitter. 1996. The characters of successful invaders. Biological Conservation 78: 163–170.

Wonham, M.J. and J.T. Carlton. 2005. Trends in marine biological invasions at local and regional scales: the Northeast Pacific Ocean as a model system. Biological Invasions 7: 369–392.

Wotton, D.M., C. O'Brien, M.D. Stuart and D.J. Fergus. 2004. Eradication success down under: heat treatment of a sunken trawler to kill the invasive seaweed *Undaria pinnatifida*. Marine Pollution Bulletin 49: 844–849.

Yendo, K. 1907. The fucaceae of Japan. Journal of the College of Science, Imperial University of Tokyo 21: 1–174.

Yun, H.Y. and M. Molis. 2012. Comparing the ability of a non-indigenous and a native seaweed to induce anti-herbivory defenses. Marine Biology 159: 1475–1484.

Zuccarello, G.C., A.T. Critchley, J. Smith, V. Sieber, G. Bleicher Lhonneur and J.A. West. 2006. Systematics and genetic variation in commercial *Kappaphycus* and *Eucheuma* (Solieriaceae, Rhodophyta). Journal of Applied Phycology 18: 643–651.

CHAPTER 5

Marine Algae as Carbon Sinks and Allies to Combat Global Warming

Francisco Arenas[a],* and Fátima Vaz-Pinto[b]

1 Global Carbon Cycle and Marine Algae

Human activities, particularly the utilization of fossil fuels for energy are quickly changing the trace gas composition of Earth's atmosphere. The current concentration of carbon dioxide (CO_2) in the atmosphere has increased 40% since the 280 parts per million (ppm) estimated around 1750, at the beginning of the Industrial Era (Le Quéré et al. 2012). In May 2013, the concentration measured at Mauna Loa Observatory, Hawaii, reached 400 ppm (see www.CO2Now.org), which means that for the first time in the history of humankind, on average for every one million molecules in the atmosphere, 400 of them were CO_2 molecules. In fact, for the past ten years the average annual rate of increase has been 2.07 parts per million (ppm), which is more than double the increase in the 1960s (0.9 ppm year^{-1}). Considering the overall mass of the atmosphere 5.148×10^{21} grams (Trenberth and Smith 2005), and correcting by the molecular mass, every single ppm of CO_2 in our atmosphere corresponds approximately to 7.8 petagrams (Pg) of CO_2 (7.822×10^{15} grams of CO_2) or 2.134 petagrams of C (carbon).[1] This means that on average and at the current increase rates,

Laboratory of Coastal Biodiversity, CIIMAR—University of Porto, Rua dos Bragas 289, 4050-123 Porto, Portugal.

[a] Email: farenas@ciimar.up.pt
[b] Email: f_vazpinto@yahoo.com
* Corresponding author

---

[1] To avoid confusion with units, it is important to realize that 1 petagram (1 Pg = 10^{15} gram) equals to 1 gigatonne (10^9 tonne). All units are presented in petagrams of carbon or petagrams of CO_2. Units of gigatonnes of CO_2 (or billion tonnes of CO_2) are frequently used in policy circles and are equal to 3.67 multiplied by the value in units of Pg C.

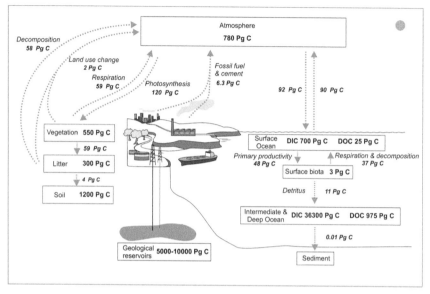

Figure 1. Current global carbon cycle. Boxes are carbon reservoirs with their size. Italics correspond to carbon fluxes among reservoirs. Units are Pg C for reservoirs and Pg C year[-1] for carbon fluxes (Adapted from Houghton 2007).

15.6 petagrams of CO_2 or 4.2 petagrams of carbon are added yearly to our atmosphere and are not being absorbed by other CO_2 reservoirs.

Including only those reservoirs relevant in a time frame of years to centuries, there are four major reservoirs of carbon on Earth: reserves of fossil fuels, terrestrial ecosystems with vegetation and soils, atmosphere and oceans (Houghton 2007). Although diminishing as a result of human activities, it is estimated that between 5000–10000 Pg C are still stored as coal, gas and oil fuels in the first reservoir. Land ecosystems store around 550±100 Pg C, while the organic matter in soils is around 1500–2000 Pg C in the first meter and as much as 2300 Pg in the top 3-meter layer (Jobbagy and Jackson 2000). In the case of the atmosphere and given the current concentration of CO_2 (400 ppm), it contains almost 850 Pg C. Finally oceans are the largest reservoir of carbon; the total amount of carbon in the world's oceans is approximately 38,000 Pg C, nearly 50 times more carbon than in the atmosphere. The majority of this oceanic carbon is in intermediate and deep waters; only 700–1000 Pg C are in the surface ocean in direct contact with the atmosphere (Houghton 2007).

Thus, most of the carbon dioxide in our planet is in the oceans (approximately 90%). Each year the oceans takes up about 40% of anthropogenic CO_2 produced from fossil fuel burning and cement production, buffering the effects of climate change. Unlike other gases, solubility of CO_2 in water is relatively high. This is because carbon dioxide is not simply dissolved in water as other gases, but it reacts with seawater and forms carbonic acid (H_2CO_3) and its dissociation products, bicarbonate (HCO_3^-) and carbonate ion (CO_3^{2-}) (Zeebe and Wolf-Gladrow 2001).

$$CO_2(aq) + H_2O \leftrightarrow H_2CO^3 \leftrightarrow HCO_3^- + H^+ \leftrightarrow CO_3^{2-} + 2 H^+$$

The equilibrium between these three species is not always straightforward. At the current ocean pH (~8.07), 90% of the total carbon dioxide is bicarbonate HCO_3^- (i.e., 2,200 µM), around 9% is carbonate CO_3^{2-} and finally only 1% remains as un-dissociated CO_2 and H_2CO_3 (Sabine and Tanhua 2010). All these products from the former CO_2 are known as DIC, dissolved inorganic carbon.

While in the pre-industrial Holocene era there was an overall time-space equilibrium between the concentration of CO_2 in the atmosphere and that dissolved in the surface ocean (Raven and Falkowski 1999), the massive release of CO_2 to the atmosphere in the last century has resulted in a net flux of CO_2 from the atmosphere to the ocean surface (Reid et al. 2009). The ocean has a vast capacity to store CO_2 so it will continue to take up CO_2 as long as it is increasing in the atmosphere. However, despite being the largest pool of CO_2, the current rate of ocean carbon storage does not seem to be keeping pace with the rate of growth in CO_2 emissions. One of the reasons is because the transport of CO_2 toward deeper layers (where CO_2 it is temporarily stored) is two to three orders of magnitude slower than the uptake of CO_2 from the atmosphere and this imbalance is the primary process controlling large-scale CO_2 uptake (Sabine and Tanhua 2010). In fact about 30% of the CO_2 released from human activities is found at depths shallower than 200 m and nearly 50% at depths above 400 m (Sabine et al. 2004).

Several mechanisms called carbon pumps are responsible for the removal of CO_2 from ocean surface to deeper waters. The first is the solubility pump. The solubility pump occurs because of the down-welling of superficial ocean water containing dissolved carbon. It is estimated that the solubility pump is responsible for about 25–40% of the difference on dissolved inorganic carbon (DIC) concentration between surface and the deep ocean. This pump is driven by the ocean thermohaline circulation patterns, i.e., the formation of deep water at high latitudes, and the fact that at low seawater temperatures and salinities the uptake of CO_2 as DIC is much higher due to increased solubility. Thus, ice forming areas like sub-polar regions of North Atlantic and Southern Oceans are major CO_2 sinking areas through the solubility pump. In other areas, wintertime convective mixing may provide transitory (up to decades) carbon sinks by keeping CO_2 at intermediate depths. These convective processes are very relevant in regions where subtropical mode and intermediate waters are formed (Sabine et al. 2004).

In coastal areas at temperate and sub-polar latitudes, there is a special case of the solubility pump named as the continental shelf pump (Tsunogai et al. 1999) which also involves the biological activity of phytoplankton. In these areas and during winter, cold and dense water with high CO_2 solubility is formed at the surface; this input of carbon is enhanced by the increased biological production characteristic of coastal areas. This seawater enriched in dissolved and particulate carbon, is cold and dense and sinks to the shelf bottom from where it is transported by isopycnal mixing (advection and diffusion) into the subsurface layer of the open ocean.

The second carbon pump is the carbonate counter pump, also known as the alkalinity pump. The carbonate pump is a consequence of the formation, sedimentation and dissolution of carbonate salts mostly from shell material of calcifying plankton like coccolithphorides, cysts of dynopflagellates, forminifera and pteropods. Even small bacteria seem to be involved in the formation of carbonates (Heldal et al. 2012). Carbonate pump runs in parallel and counteracts partially the biological pump in terms of its effects on the air-sea CO_2 exchange. During the biogenic formation of carbonates (calcite, aragonite and magnesium calcites), the removal of CO_3^{2-} by precipitation of calcium carbonate ($CaCO_3$) increases dissociation of HCO_3^- to restore the ion CO_3^{2-} lost, releasing additional H^+ ions. Thus during the formation of carbonates, the pH is lowered, there is a release of CO_2 but the overall DIC is reduced. Finally, carbonate minerals act as effective ballast for organic carbon, either by increasing the density of sinking particles or by providing some protection against degradation, enhancing the efficiency of the biological pump.

The third CO_2 sinking mechanism is the biological pump. Solar energy, water and nutrients are required to support photosynthesis, whereby oxidized inorganic carbon is converted to energy-rich organic carbon molecules, a process known as primary production (PP) (Chavez et al. 2011). The biological pump is the mechanism that transfers this CO_2 fixed by photosynthesis to the deep ocean, primarily as dead organisms including the organic skeletal and faecal material (particulate organic carbon, POC), and carbonate skeletons (particulate inorganic carbon, PIC) (Reid et al. 2009). This results in the sequestration (storage) of carbon for periods of decades to centuries, depending on the depth of re-mineralization, or even more permanently in the sediments. It is estimated that phytoplankton and other organisms living in the sunlit ocean surface pump about 15% of the organic material produced each year to the deep sea and 0.1% of it gets buried in the seafloor (Falkowski 2012). Overall, the biological pump is responsible for up to two-thirds of the surface-to-depth DIC gradient (Riebesell et al. 2009). In fact, if the biological pump was "turned off" atmospheric CO_2 would rise about 200 ppm. Most of the nutrients and carbon in phytoplankton are not lost from the surface layer (as mentioned above, only 15% of the organic material sinks). Instead, they are recycled and used to support the growth and photosynthesis of new phytoplankton. This recycling system involves a high diversity of planktonic organisms including bacteria, viruses and zooplankton and is referred to as the 'microbial loop'.

Marine free living algae account for almost 50% of the world carbon biogenic fixation, even though accounting for less than 1% of the photosynthetic biomass on Earth (Falkowski 2012). This is possible because on average, the entire global population of phytoplankton is replaced on average every two to six days (Falkowski et al. 1998, Behrenfeld 2011), whereas in land turnover, average time is 19 years. Overall, the world's phytoplankton incorporate yearly around 45–50 Pg C into their cells (Falkowski 2012) making them as important in modifying the planet's cycle of carbon and carbon dioxide as all the world's land plants combined. Per unit area, the productivity of the oceans is around 140 g C m^{-2} y^{-1}, approximately one third of that found on land plants (Field 1998). These differences seem to be related with light use efficiency (7% of incident PAR absorbed at the ocean by phytoplankton *versus* 31% of land plants).

Besides light, net primary production in the ocean is highly controlled by nutrients. In low nutrient waters, i.e., oligotrophic oceanic waters, phytoplankton biomass is dominated by small plankton cells, known as picophytoplankton and constituted mostly by cyanobacteria and nanoflagellates with sizes ranging from 1 to 5 μm (Falkowski et al. 1998). These small phytoplankton cells, due to a higher surface area-to-volume ratio and a smaller thickness of the diffusion boundary layer, have competitive advantage over larger cells in nutrient-impoverished environments (Raven 1998). Picophytoplankton cells have a ubiquitous distribution and may contribute to significant portions of the total phytoplankton biomass and production. For example in the North Atlantic small cells (<2 μm) account for 44% of primary production (Jardillier et al. 2010), with integrated primary production rates ranging from 14 to 800 mg C m^{-2} d^{-1} (Teira et al. 2005). Despite being dominant primary producers in oligotrophic oceans, they may also become important in coastal seas (Morán et al. 2010). Oligotrophic zones occupy large areas in the ocean and overall, their NPP was estimated as 11 Pg C year^{-1} (Field 1998).

In nutrient-rich waters, when solar energy is enough and the thermocline prevent cells sinking below the euphotic zone, large phytoplankton cells like diatoms are stimulated (Falkowski et al. 1998). These larger cells may grow quick and escape grazing through temporal decoupling with large zooplankton life-cycles, which need to go through larval stages before reaching adult status. These decouplings promote the burst of blooms that will increase vertical fluxes of organic matter, contributing significantly to carbon sequestration. Planktonic net productivity of the eutrophic areas of the oceans is estimated as 9.1 Pg C year^{-1}. More than half of the annual net primary productivity in the ocean seems to occur in mesotrophic areas with intermediate nutrient and chlorophyll levels and an overall annual productivity values of around 27.4 Pg C year^{-1} (Field 1998).

As we have seen, on Earth half of the carbon captured by photosynthetic organisms is linked to phytoplankton activity. This carbon captured at the sea is known as "blue carbon". However blue carbon is not exclusive from phytoplankton. Coastal benthic macrophytes also have a relevant role as carbon sinks and a very high value because of their contribution to marine biodiversity and coastal functioning, including the supply of important goods and services. Marine coastal macrophytes include mangroves, salt marshes, seagrasses and seaweed assemblages. In particular, marine macroalgae make up a small proportion of ocean primary production, but they are the dominant primary producers of rocky shore coastal habitats (Mann 1973), providing an essential ecological function for these ecosystems. Marine macrophytes (seaweeds and seagrasses) occupied about 2 x 10^6 Km^2 and account for a carbon uptake for around 2.55 Pg C year^{-1} (Duarte and Cebrian 1996)—not that much when compared with phytoplankton which cover an area of approx. 3.6 x 10^8 Km^2 (Gattuso et al. 1998) and have an overall C uptake almost 20 times bigger (45–50 Pg C year^{-1}) (Falkowski 2012).

Nonetheless, marine macrophytes dominate shallow coastal areas where most of the seafloor lies within the euphotic zone, supporting highly productive benthic communities (Mann 1973). Thus several species of macrocalgae such as *Ascophyllum nodosum, Macrocystis integrifolia, Sargassum horneri, Postelsia capillaceae* and *Ecklonia radiata* are capable of productivities per unit area substrate in excess of 1,000 g C m^{-2} year^{-1} (Chung et al. 2010).

Kelp forests are especially relevant macroalgae assemblages. Kelp forests are dominated by large brown algae from the order Laminariales (although some seaweeds regarded as kelps are not members of this order). Kelps dominate seaweed assemblages on shallow rocky shores in temperate and arctic regions. The overall standing crop of kelp forest is estimated to be between 0.015 Pg C to 0.02 Pg C (de Vouys 1979), but it could be 2 times higher (Reed and Brzezinski 2009). Average net primary productivity of kelp forest is around 1000 g C m^{-2} y^{-1} (Reed and Brzezinski 2009), but could reach productivity values up 3,000 g C m^2 year^{-1} like those recorded for natural beds of kelps from the genera *Macrocystis* and *Laminaria* (Gao and Mckinley 1994). Evidences from a field study highlighted the potential of natural dense *Macrosystis* kelp beds as CO$_2$ sinks, with a decrease of dissolved inorganic carbon (DIC) in the water column between 30 and 60 µmol kg^{-1} day^{-1} (Delille et al. 2000).

Natural seaweed assemblages may act as valuable carbon sinks compared to phytoplankton due to their higher biomass and larger turnover time (ca. 1 year compared to days). These turnover times are much more rapid than terrestrial plants (years to decades) and clearly limit the efficiency of natural seaweed systems as carbon sinks, because most of the carbon sequestered outflows within one year. Little is known regarding the carbon cycling in these assemblages, but estimates show that seaweed assemblages export most of the produced biomass out of the system (±40% biomass), the rest either is quickly consumed by grazers (±30%) or decomposed *in situ* (±30%) (Duarte and Cebrian 1996).

2 Farming Marine Algae: Promoting both Carbon Sequestration and Sustainability

Current atmospheric CO_2 levels (400 ppm) are well above those considered safe for the preservation of the planet similar to that on which civilization developed and to which life on Earth is adapted. This barrier was established by researchers as 350 ppm (Rockstrom et al. 2009), see www.350.org for further information. Proactive policies that combine actions to remove CO_2 from the atmosphere with actions to reduce emissions could decrease CO_2 concentrations faster than possible via natural processes (Lemoine et al. 2012). Among these carbon negative actions is the bio-energy with carbon capture (BEBCCS) which combines biological carbon capture and sequestration with biomass production (Mathews 2008). Because of their very high productivity, marine algae offer great potential for carbon negative bioremediation approaches.

2.1 Macroalgae Carbon Sequestration

For centuries, humans have harvested natural seaweed populations for their food and chemical constituents. Nowadays, as food and energy demand continues to rise, marine macroalgae are receiving increasing attention as an attractive renewable sources of food, fuels and other chemicals. Around 15.7 million tonnes wet weight seaweeds are harvested annually both from wild and cultivated sources, with an estimated total annual value of US$ 6 billion (Roesijadi et al. 2008). Macroalgae contains on average 30% carbon, so this figure represents 4.71×10^6 tonnes of carbon harvested from macroalgae annually. High levels of exploitation of natural stocks are difficult to achieve and need to be well controlled due to the exceptional importance of macroalgae in supporting marine biodiversity. In fact, seaweed exploitation in Europe is currently restricted to manual and mechanised harvesting of natural stocks, whereas the majority of Asian seaweed resources are cultivated (Bruton et al. 2009). Thus, cultivation is the most likely way to generate significant volumes of seaweed biomass. Cultivated seaweeds from the genera *Laminaria*, *Porphyra*, *Undaria*, *Kappaphycus*, and *Gracilaria* comprise 90% of the annual production (Roesijadi et al. 2010) and are now the predominant seaweed source for human consumption. Seaweed cultivation industry is mostly based in Asia and Chile, with China as the largest producer of edible seaweeds, harvesting more than 7 million wet tonnes. Macroalgal farms may be integrated as off-shore farms, near-shore farms, and land-based ponds.

There is evidence that current culture production is enough to meet the needs of the food and commodity markets (Roesijadi et al. 2010); however recent advances in the technology to convert carbohydrates from seaweeds into biofuels have burst the interest in improving seaweed cultivation methods. Worldwide energy demand is increasing and there is a growing dilemma about finding a feedstock capable of keeping up with demand. Currently, one fifth of the global CO_2 emissions are from the transport sector (Rawat et al. 2013). The continued use of fossil fuels is not sustainable and biofuel production, i.e., derived from biological carbon sequestration, has emerged as an effective alternative (Demirbas 2007, Singh et al. 2011). The use of renewable biomass as an energy resource has good benefits to the environment (lower toxicity and lower emission profiles). Biodiesel and bioethanol synthesis from terrestrial food crops are attractive and widely used because of well-established farming practices and simple and cheap extraction processes (Wei et al. 2013). However, conventional biofuels from classic food crops require high-quality agricultural land and there are large concerns on future competition with food agriculture for arable land use (Goh and Lee 2010). One plausible alternative to food crops is the use of non-food ligno-cellulosic biomass such as agricultural residues or wood waste. However, current technologies have yet to solve the problems related with sugar extraction from these products (Wei et al. 2013).

In this context, marine macroalgae seems to offer an attractive alternative to terrestrial crops. No competition for arable land would occur as algae production can be placed anywhere from brown field land to the open ocean. In addition, algal farms can be implemented in waters close to CO_2-emitting industrial centres and power plants resulting in further advantage. Also, estimates of seaweed carbon capture during photosynthesis are ~1600 g C m^{-2} y^{-1} (Duarte et al. 2005) whereas the global net primary productivity of crop land is ~470 g C m^{-2} y^{-1} (Field et al. 2008). In particular, microalgae are being widely investigated as a biodiesel feedstock while macroalgae, with low lipid content, can either produce biogas or alcohol-based fuels.

Among macroalgae, brown seaweed are the most likely feedstock candidates for conversion to liquid fuels due to the presence of polysaccharides and sugar alcohols, e.g., laminarin and mannitol (Bruton et al. 2009, Hughes et al. 2012). Furthermore, an advantage for macroalgae compared to terrestrial biomass is the scarcity of lignin type materials, which is resistant to conversion during biofuels production and increases the cost of the process. Natural sugars and other carbohydrates present in seaweeds can be processed through fermentation, either anaerobic digestion to create biogas, or ethanol fermentation (Bruton et al. 2009). Biogas (~60% methane) can be used to produce heat and electricity or compressed for use as a transport fuel. Some seaweed genera such as *Macrosystis, Sargassum, Laminaria, Ascophyllum, Gracilaria, Ulva* and *Cladophora* have been tested as potential methane sources (Gunaseelan 1997). A realistic average production of 22 m^3 of methane per tonne wet weight was obtained from *Laminaria* sp., with a conservative yielding of 171 GJ ha^{-1} (Hughes et al. 2012). Nonetheless, although the existing seaweed farming is considered a significant commercial market for macroalgae, it is still far from reaching the biomass needed to achieve energy demands for developing countries. For example, it has been estimated that using all of the current brown alga produced in culture (6.8 million tonnes $year^{-1}$) will reach only ~0.06% of the UK total energy demand for 2010 (9518 PJ). Still, to meet 1% of UK's total energy demand would require a marine cultivation area of approximately 5440 km^2 whereas for terrestrial biofuel production, a land area of 7700 km^2 would be necessary (Hughes et al. 2012).

Recent and very ambitious initiatives propose to use large scale macroalgal ocean afforestation ecosystems to use seaweeds and seaweed-digesting microbes to concentrate carbon from the 0.04% in air into 40 bio-CO_2 and 60% bio-methane (N'Yeurt et al. 2012).

Ocean afforestation aims to grow seaweeds in euthropic oceans; once harvested, seaweeds will be digested anaerobically to separate energy-rich content (biogas) from nutrients. N'Yeurt et al. (2012) estimated that to reduce 100 CO_2 ppm in the atmosphere (i.e., moving from the predicted 450 CO_2 ppm next decade to the 350 CO_2 ppm considered as the safe limit) in 50 years, it is necessary to afforest 6% of ocean surface.

Although the biogas production from seaweed process has been demonstrated to be technically viable, the associated cost of the process is still high and not competitive in the current market (Gunaseelan 1997, Bruton et al. 2009). Nonetheless, the resource potential is high, and the ability of the world's oceans to produce marine biomass as a biofuel feedstock supply is largely untapped (Roesijadi et al. 2008). Also, the current world economic situation suggests that fossil fuel prices will keep increasing whereas macroalgal production costs will inevitably fall as production is expanded and intensified.

2.2 Microalgae Carbon Sequestration

Microalgae have also received much attention as a source for biodiesel. The concept of using microalgae for CO_2 utilization and conversion to fuels was first developed in 1960 (Oswald and Golueke 1960). Later on, in the 1970s and early 1980s, research in this field was pushed forward in response to the energy crisis (Benemann 1997). Allied to its high photosynthetic efficiency, microalgae are capable of producing 30 times the amount of oil per unit area of land, compared to terrestrial oilseed crops (Rawat et al. 2013). Microalgal oil and spent biomass have good potential sources as biodiesel feedstocks. Microalgae are particularly interesting because of their high growth rates, sometimes doubling their biomass every 4–6 hours, and tolerance to varying environmental conditions (Rawat et al. 2013). Also, their oil contents can approach 80%, depending on the algal strain selected and the nutrient conditions observed in the culture. Nonetheless, its production costs are still too expensive (Hughes et al. 2012) and one of the current challenges in research is to reduce costs, namely of those high energy requirement processes such as harvesting and dewatering (Singh and Ahluwalia 2013).

Microalgae are particularly suitable for enhanced biological carbon fixation or CO_2 mitigation. Using microalgae cultures as an energy production system combined with waste-water treatment or petroleum-based power stations could be the key for an economic cost-effective process to be achieved (Kumar et al. 2010). Industrial exhaust gases can contain 10–20% CO_2 (Ho et al. 2011) and are considered the primary contributor to excess carbon dioxide (Brown and Zeiler 1993). Thus, the benefits are two-fold, reduction in greenhouse gas emissions from the power plant stack, and a potentially more efficient route for CO_2 bio-fixation. Microalgal growth actively utilizes 1.83 kg of CO_2 for every 1 kg dry biomass produced (Chisti 2007). Traditionally, microalgae are cultivated in enclosed photo-bioreactors (PBR) systems or open raceway ponds (open system), which are aerated or exposed to air to allow microalgal growth. Outdoor open systems are the closest to the natural environment but closed PBRs offer more regulated and well-controlled cultivation conditions with a low contamination risk, but are also more costly (Rawat et al. 2013). Thus, building effective PBRs represents the most important step towards a successful microalgal-CO_2 fixation process. Several closed PBR designs have been developed and tested in order to optimize biomass productivity. These include: tubular, flat panel, vertical-column, and internally illuminated systems (Chisti 2007, Pulz 2001, Tredici and Materassi 1992, Ugwu et al. 2008). High-quality microalgal-CO_2 fixation PBR system should include good mixing, gas transfer, and light distribution (Eriksen 2008). Tubular and bubble column-type PBRs are the most commonly applied for use in algal PBRs. In addition, hybrid systems using both open ponds and PBR systems in combination have been developed as an attempt to achieve favourable economic conditions. In these hybrid systems, the first stage of growth is undertaken within an enclosed PBR to prevent contaminations and then the second stage is achieved in an open raceway pond. Hybrid systems can produce as much as 20–30 tonnes ha^{-1} of lipid annually depending on climate favourability (Rodolfi et al. 2009).

Also, the selection of optimal microalgae species is vital. Microalgae can differ in shape, size, chemical structure and composition. To develop the most effective CO_2 emission mitigation process, it is necessary to select a fast-growing microalgal species with high CO_2 fixation efficiency and a large amount of valuable by-products (Ono and Cuello 2006). For instance, microalgal genera such as *Chlorella, Dunaliella, Chlamydomonas, Scenedesmus* and *Spirulina* are known to contain a large amount (>50% of the dry weight) of starch and glycogen, useful as raw materials for ethanol production. It has been found that *Scenedesmus* sp. could convert approximately 15–25% atmospheric CO_2 into biodiesel for transportation fuel (Ho et al. 2010). In particular, for open pond production systems, alga should be able to dominate wild strains and for photobioreactors cultivation it should be robust enough to survive the shear stresses common in photobioreactors. Also, the ideal algal strain for power plant flue gas remediation should have high CO_2 tolerance and sinking capacity (Singh and Ahluwalia 2013). The major components in the flue gas are CO_2, SOx and NOx and thus growth inhibitory effects may arise from the presence of high concentrations of NOx and SOx (Ho et al. 2011). The microalgae *Chlorella vulgaris's* greatest growth was observed in the culture grown on 15% CO_2-in-Air, suggesting that this particular alga may be appropriate for a microalgae-based system for CO_2 capture from flue gas.

Algal productivity is estimated in terms of biomass produced per day per unit of available surface area (e.g., g dry biomass m^{-2} day^{-1}). Open raceway ponds can facilitate multi-strain cultivation while PBRs are well suitable for single strain cultivation. Productivity levels in the order of 27–62 g DCW m^{-2} day^{-1} may be achieved in PBRs, whereas algal productivity in open raceway systems range from 5 to 50 g DCW m^{-2} day^{-1} (Rawat et al. 2013). As algae grows, excess culture overflows and is harvested. However, microalgae have low densities and are found in suspension, which makes its harvesting the major challenge in biodiesel production (Mata et al. 2010). Biomass harvesting can contribute to 20–30% of the total biomass production costs, and thus further research is needed to lessen these challenges. Under open pond system conditions, it is particularly important to control microalgal population dynamics, as seasonal growth variations have been observed. Suitable species for open pond cultivation include *Chlorella* sp., *Duniella salina* and *Spirulina* sp. (Borowitzka 1999).

Concerning CO_2 mitigation using microalgae, open raceway ponds are more favourable than PBRs. Carbon fixation rates of 6 to 15 g C m^{-2} day^{-1} have been obtained at high-rate open pond systems. A recent study proposed a model for CO_2 capture using an algal biomass production system from flue gas CO_2 from a power plant (Brune et al. 2009). The study was based on a 50 MW, 50% base-load, natural gas fired electrical generation plant, with an average production of 450 tons of CO_2 day^{-1}. Using an estimated algal productivity of 20 g dry algae m^{-2} day^{-1} and a CO_2 capture rate of ~37 g m^{-2} day^{-1}, they estimated that an area of 8.8 km^2 would be required to capture 70% of the CO_2 generated. Over a 240-day season, a capture of 70% of the flue-gas CO_2 would produce 42.4 million Kg algal dry weight season^{-1}. The efficiency of flue-gas CO_2 capture would, however, depend, in particular, on the amount of sunlight hours available and in the efficiency of CO_2 transfer into the pond and out-gassing from the ponds. In addition, total parasitic energy required to produce, harvest, and concentrate algal biomass is estimated at 0.35–0.50 kWh kg^{-1} of dry algal biomass. By extracting 20% of the algal oil content, it would yield a potential replacement transportation fuel equivalent to 20% of plant gross GHG emissions (Brune et al. 2009). This study did not, however, test the economics of the processes considered.

In conclusion, there are two main ways in which biomass can reduce atmospheric CO_2 levels, as carbon store and as an energy source. Marine primary producers such as phytoplankton, seaweeds and seagrasses are more efficient carbon sequestering agents than their terrestrial counterparts. Additionally, the utilization of anthropogenic CO_2 as an industrial by-product for algal biomass production can contribute to some extent in meeting the global food, fodder, fuel and pharmaceuticals requirements. Although production costs are still high, the use of marine algae for CO_2 mitigation and biofuel production are considered realistic approaches for the long-term (Rawat et al. 2013).

3 Links between Marine Algae and Climate Change

Oceans absorb both heat and carbon from the atmosphere, alleviating the impacts of global warming in the environment. However in the last few decades, human activities are modifying fundamental physical and chemical properties of the ocean. These changes are expected to impact marine ecosystem structure and functioning and have the potential to alter the cycling of carbon and nutrients on the ocean surface with likely feedbacks on the climate system (Riebesell et al. 2009).

Increases in seawater temperature will reduce CO_2 solubility, decreasing anthropogenic carbon uptake by 9–15%. Global warming will also intensify the hydrological cycle, increasing precipitation in some areas and evaporation in others; both processes will probably balance each other with minor changes on CO_2 uptake rates. Higher seawater temperatures, reduced sea ice formation and freshening of polar oceans may reduce the formation of deep water both in the North Atlantic and the Southern Ocean around Antarctica, reducing the intensity of the thermohaline circulation and thus affecting CO_2 sinking rates (Schmittner et al. 2008). The impact of these processes on the efficiency of the solubility are estimated between 3–20% reduction on the CO_2 uptakes (Riebesell et al. 2009).

Marine primary producers will be also directly affected by ocean warming. Thermal stratification will reduce mixing and nutrient availability but will increase light availability which could be a limiting factor at higher latitudes. Temperature will affect metabolic rates altering species phenology, phytoplankton productivity and ultimately the assemblage's structure. Predicting the effect of ocean warming is challenging not only because of the large direct and indirect effects involved, but also because of the ecological interactions among different species and functional groups (Behrenfeld 2011). For example, it is expected that higher temperatures will promote photosynthetic activity enhancing primary productivity in areas with no nutrient limitation, but simultaneously higher temperature will promote heterotrophic metabolism intensifying top-down control of phytoplankton assemblages and reducing primary productivity (Winder and Sommer 2012).

Similarly, the increasing extent of anthropogenic CO_2 in the ocean will have both physical and biological consequences. Increasing aqueous CO_2 concentration reduces ocean pH and modifies the balance among the different carbon speciation in the water. The reduction of CO_3^{2-} leads to a progressive reduction of atmospheric CO_2 uptake, and thus of the ocean's buffering capacity. Biologically, effects of increasing CO_2 will differ among organisms. Ocean acidification will have obvious adverse effects in calcifying organisms. On the other hand, the ocean carbonation may bring benefits to some groups of photosynthetic organisms, particularly those that operate a relatively inefficient CO_2 acquisition pathway. However, unlike terrestrial plants, many marine algae have biochemical tools to concentrate CO_2 inside their cells; thus higher carbon availability may not have large effects on productivity (Kroeker et al. 2010). Experiments in natural phytoplankton assemblages show that increased CO_2 level would aid the growth of some groups of phytoplankton such as cyanobacteria, and decrease the competitiveness of others (e.g., some calcifiers). As in the case of temperature, interactions between CO_2 with other physical driving forces could reduce our abilities to predict future impacts. For example, the responses of both phytoplankton and macroalgae to elevated CO_2 seems to be highly modified by light intensity (Hepburn et al. 2011, Gao et al. 2012).

Understanding how stressors affect natural communities and ecosystems and the prevalence of synergistic or non-synergistic interactions remains a challenge for researchers (Darling et al. 2010). Thus, many of our current predictions from studies of impacts of single climate stressors on the productivity of algal assemblages under future environmental conditions will probably need re-examination after the wide recognition of interactions and feedbacks among physical and biological factors. An additional difficulty is to up-scale results from experiments to relevant global effects. The opposite results on global trends of the ocean's primary productivity described in recent literature, negative in some studies (Behrenfeld et al. 2006, Boyce et al. 2010) and positive in others (Chavez et al. 2011), is a clear example of our great uncertainty regarding climate change and its consequences.

Acknowledgement

FA was partially supported by COMPETE - Operational Competitiveness Program and National Funds through FCT under the project "PEst-C/MAR/LA0015/2011" and the FCT funded project CLEF (PTDC-AAC-AMB-102866-2008). FVP was supported by a PhD grant from the Portuguese Foundation for Science and Technology –FCT (SFRH/BD/33393/2008).

Keywords: Marine algae, CO_2, Climate change, Carbon sequestration, Blue carbon

References Cited

Behrenfeld, M.J. 2011. Biology: uncertain future for ocean algae. Nature Climate Change 1(1): 33–34.

Behrenfeld, M.J., R.T. O'Malley, D.A. Siegel, C.R. McClain, J.L. Sarmiento, G.C. Feldman, A.J. Milligan, P.G. Falkowski, R.M. Letelier and E.S. Boss. 2006. Climate-driven trends in contemporary ocean productivity. Nature 444(7120): 752–755.

Benemann, J.R. 1997. CO_2 mitigation with microalgae systems. Energy Conversion and Management 38: S475–S479.

Borowitzka, M.A. 1999. Commercial production of microalgae: ponds, tanks, tubes and fermenters. Journal of Biotechnology 70: 313–321.

Boyce, D.G., M.R. Lewis and B. Worm. 2010. Global phytoplankton decline over the past century. Nature 466(7306): 591–596.

Brown, L.M. and K.G. Zeiler. 1993. Aquatic biomass and carbon dioxide trapping. Energy Conversion and Management 34: 1005–1013.

Brune, D.E., T.J. Lundquist and J.R. Benemann. 2009. Microalgal biomass for greenhouse gas reductions: potential for replacement of fossil fuels and animal feeds. J. Environ. Eng-ASCE 135(11): 1136–1144.

Bruton, T., H. Lyons, Y. Lerat, M. Stanley and M.B. Rasmussen. 2009. A review of the potential of marine algae as a source of biofuel in Ireland. Dublin, Ireland Sustainable Energy Ireland.

Chavez, F.P., M. Messié and J.T. Pennington. 2011. Marine Primary Production in Relation to Climate Variability and Change. Annual Review of Marine Science 3(1): 227–260.

Chisti, Y. 2007. Biodiesel from microalgae. Biotechnology Advances 25(3): 294–306.

Chung, I.K, J. Beardall, S. Mehta, D. Sahoo and S. Stojkovic. 2010. Using marine macroalgae for carbon sequestration: a critical appraisal. Journal of Applied Phycology 23(5): 877–886.

Darling, E.S., T.R. McClanahan and I.M. Cote. 2010. Combined effects of two stressors on Kenyan coral reefs are additive or antagonistic, not synergistic. Conservation Letters 3(2): 122–130.

Delille, B., D. Delille, M. Fiala, C. Prevost and M. Frankignoulle. 2000. Seasonal changes of pCO(2) over a subantarctic Macrocystis kelp bed. Polar Biology 23(10): 706–716.

Demirbas, A. 2007. Alternatives to petroleum diesel fuel. Energy Sources Part B-Economics Planning and Policy 2(4): 343–351.

De Vouys, C.G.N. 1979. Primary production in aquatic environments. *In*: B. Bolin, E.T. Degens, S. Kempe and P. Ketner (eds.). SCOPE 13: The global carbon cycle. Wiley. U.K. 491 pp.

Duarte, C.M. and J. Cebrian. 1996. The fate of marine autotrophic production. Limnology and Oceanography 41(8): 1758–1766.

Duarte, C.M., J.J. Middelburg and N. Caraco. 2005. Major role of marine vegetation on the oceanic carbon cycle. Biogeosciences 2(1): 1–8.

Eriksen, N.T. 2008. The technology of microalgal culturing. Biotechnology Letters 30(9): 1525–1536.

Falkowski, P. 2012. Ocean science: The power of plankton. Nature 483(7387): S17–S20.

Falkowski, P.G., R.T. Barber and V. Smetacek. 1998. Biogeochemical controls and feedbacks on ocean primary production. Science 281(5374): 200–206.

Field, C.B. 1998. Primary production of the biosphere: integrating terrestrial and oceanic components. Science 281(5374): 237–240.

Field, C.B., J.E. Campbell and D.B. Lobell. 2008. Biomass energy: the scale of the potential resource. Trends in Ecology and Evolution 23(2): 65–72.

Gao, K. and K.R. Mckinley. 1994. Use of macroalgae for marine biomass production and CO_2 remediation—a review. Journal of Applied Phycology 6(1): 45–60.

Gao, K., J. Xu, G. Gao, Y. Li, D.A. Hutchins, B. Huang, L. Wang, Y. Zheng, P. Jin, X. Cai, D. Häder, W. Li, K. Xu, N. Liu and U. Riebesell. 2012. Rising CO_2 and increased light exposure synergistically reduce marine primary productivity. Nature Climate Change 2: 519–523.

Gattuso, J.P., M. Frankignoulle and R. Wollast. 1998. Carbon and carbonate metabolism in coastal aquatic ecosystems. Annual Review of Ecology and Systematics 29: 405–434.

Goh, C.S. and K.T. Lee. 2010. A visionary and conceptual macroalgae-based third-generation bioethanol (TGB) biorefinery in Sabah, Malaysia as an underlay for renewable and sustainable development. Renewable & Sustainable Energy Reviews 14(2): 842–848.

Gunaseelan, V.N. 1997. Anaerobic digestion of biomass for methane production: A review. Biomass & Bioenergy 13(1-2): 83–114.

Heldal, M., S. Norland, E.S. Erichsen, T.F. Thingstad and G. Bratbak. 2012. An unaccounted fraction of marine biogenic $CaCO_3$ particles. PLoS ONE 7(10): e47887.

Hepburn, C.D., D.W. Pritchard, C.E. Cornwall, R.J. McLeod, J. Beardall, J.A. Raven and C.L. Hurd. 2011. Diversity of carbon use strategies in a kelp forest community: implications for a high CO_2 ocean. Global Change Biology 17(7): 2488–2497.

Ho, S.H., W.M. Chen and J.S. Chang. 2010. Scenedesmus obliquus CNW-N as a potential candidate for CO_2 mitigation and biodiesel production. Bioresource Technology 101(22): 8725–8730.

Ho, S.H., C.Y. Chen, D.J. Lee and J.S. Chang. 2011. Perspectives on microalgal CO2-emission mitigation systems—a review. Biotechnology Advances 29(2): 189–198.

Houghton, R.A. 2007. Balancing the global carbon budget. Annual Review of Earth and Planetary Sciences 35: 313–347.

Hughes, A.D., M.S. Kelly, K.D. Black and M.S. Stanley. 2012. Biogas from macroalgae: is it time to revisit the idea? Biotechnology for Biofuels 5: 86.

Jardillier, L., M.V. Zubkov, J. Pearman and D.J. Scanlan. 2010. Significant CO_2 fixation by small prymnesiophytes in the subtropical and tropical northeast Atlantic Ocean. The ISME Journal 4: 1180–1192.

Jobbagy, E.G. and R.B. Jackson. 2000. The vertical distribution of soil organic carbon and its relation to climate and vegetation. Ecological Applications 10(2): 423–436.

Kroeker, K.J., R.L. Kordas, R.N. Crim and G.G. Singh. 2010. Meta-analysis reveals negative yet variable effects of ocean acidification on marine organisms. Ecology Letters 13(11): 1419–1434.

Kumar, A., S. Ergas, X. Yuan et al. 2010. Enhanced CO_2 fixation and biofuel production via microalgae: recent developments and future directions. Trends in Biotechnology 28(7): 371–380.

Le Quéré, C., R.J. Andres, T. Boden, T. Conway, R.A. Houghton, J.I. House, G. Marland, G.P. Peters, G.R. van der Werf, A. Ahlstrom, R.M. Andrew, L. Bopp, J.G. Canadell, P. Ciais, S.C. Doney, C. Enright, P. Friedlingstein, C. Huntingford, A.K. Jain, C. Jourdain, E. Kato, R.F. Keeling, K. Klein Goldewijk, S. Levis, P. Levy, M. Lomas, B. Poulter, M.R. Raupach, J. Schwinger, S. Sitch, B.D. Stocker, N. Viovy, S. Zaehle and N. Zeng. The global carbon budget 1959–2011. Earth System Science Data Discussions 5(2): 1107–1157.

Lemoine, D.M., S. Fuss, J. Szolgayova, M. Obersteiner and D.M. Kammen. 2012. The influence of negative emission technologies and technology policies on the optimal climate mitigation portfolio. Climatic Change 113(2): 141–162.

Mann, K.H. 1973. Seaweeds—their productivity and strategy for growth. Science 182(4116): 975–981.

Mata, T.M., A.A. Martins and N.S. Caetano. 2010. Microalgae for biodiesel production and other applications: A review. Renewable & Sustainable Energy Reviews 14(1): 217–232.

Mathews, J.A. 2008. Carbon-negative biofuels. Energy Policy 36(3): 940–945.

Morán, X.A., A. López-Urrutia, A. Calvo-Díaz and William K.W. Li. 2010. Increasing importance of small phytoplankton in a warmer ocean. Global Change Biology 16(3): 1137–1144.

N'Yeurt, A.D., D.P. Chynoweth, M.E. Capron, J.R. Stewart and M.A. Hasan. 2012. Negative carbon via ocean afforestation. Process Safety and Environmental Protection 90(6): 467–474.

Ono, E. and J. Cuello. 2006. Feasibility assessment of microalgal carbon dioxide sequestration technology with photobioreactor and solar collector. Biosystems Engineering 95(4): 597–606.

Oswald, W.J. and C.G. Golueke. 1960. The biological conversion of solar energy. Advances in Applied Microbiology 11: 223–242.

Pulz, O. 2001. Photobioreactors: production systems for phototrophic microorganisms. Applied Microbiology and Biotechnology 57(3): 287–293.

Raven, J.A. 1998. The twelfth Tansley Lecture. Small is beautiful: the picophytoplankton. Functional Ecology 12: 503–513.

Raven, J.A. and P.G. Falkowski. 1999. Oceanic sinks for atmospheric CO(2). Plant Cell and Environment 22(6): 741–755.

Rawat, I., R.R. Kumar, T. Mutanda and F. Bux. 2013. Biodiesel from microalgae: A critical evaluation from laboratory to large scale production. Applied Energy 103: 444–467.

Reed, D.C. and M.A. Brzezinski. 2009. Kelp forests. pp. 31–38. In: Laffoley, D. and G. Grimsditch (eds.). The Management of Natural Coastal Carbon Sinks. IUCN, Gland, Switzerland.

Reid, Philip C., Astrid C. Fischer, Emily Lewis-Brown et al. 2009. Impacts of the Oceans on Climate Change. Advances in Marine Biology 56: 1–150.

Riebesell, U., A. Kortzinger and A. Oschlies. 2009. Sensitivities of marine carbon fluxes to ocean change. Proc. Natl. Acad. Sci. USA 106(49): 20602–9.

Rockstrom, J., W. Steffen, K. Noone, Å. Persson, F.S. Chapin, E.F. Lambin, T.M. Lenton, M. Scheffer, C. Folke, H.J. Schellnhuber, B. Nykvist, C.A. de Wit, T. Hughes, S. van der Leeuw, H. Rodhe, S. Sörlin, P. K. Snyder, R. Costanza, U. Svedin, M. Falkenmark, L. Karlberg, R.W. Corell, V.J. Fabry, J. Hansen, B. Walker, D. Liverman, K. Richardson, P. Crutzen and J.A. Foley. 2009. A safe operating space for humanity. Nature 461(7263): 472–475.

Rodolfi, L., G.C. Zittelli, N. Bassi, G. Padovani, N. Biondi, G. Bonini and M.R. Tredici. 2009. Microalgae for oil: strain selection, induction of lipid synthesis and outdoor mass cultivation in a low-cost photobioreactor. Biotechnology and Bioengineering 102(1): 100–112.

Roesijadi, G., A.E. Copping, M.H. Huesemann, J. Forster and J.R. Benemann. 2008. Techno-Economic Feasibility Analysis of Offshore Seaweed Farming for Bioenergy and Biobased Products. Battelle Pacific Northwest Division Report Number PNWD-3931.

Roesijadi, G., S.B. Jones, L.J. Snowden-Swan and Y. Zhu. 2010. Macroalgae as a biomass feedstock: a preliminary analysis. Edited by U.S.D.O. Energy: Pacific Northwest National Laboratory.

Sabine, C.L. and T. Tanhua. 2010. Estimation of anthropogenic CO_2 inventories in the ocean. Annual Review of Marine Science 2(1): 175–198.

Sabine, C.L., R.A. Feely, N. Gruber et al. 2004. The oceanic sink for anthropogenic CO_2. Science 305(5682): 367–371.

Schmittner, A., A. Oschlies, H.D. Matthews and E.D. Galbraith. 2008. Future changes in climate, ocean circulation, ecosystems, and biogeochemical cycling simulated for a business-as-usual CO(2) emission scenario until year 4000 AD. Global Biogeochemical Cycles 22(1): 1–21.

Singh, A., P.S. Nigam and J.D. Murphy. 2011. Renewable fuels from algae: An answer to debatable land based fuels. Bioresource Technology 102(1): 10–16.

Singh, U.B. and A.S. Ahluwalia. 2013. Microalgae: a promising tool for carbon sequestration. Mitigation and Adaptation Strategies for Global Change 18(1): 73–95.

Teira, E., B. Mouriño, E. Marañon, V. Pérez, M.J. Pazó, P. Serret, D. de Armas, J. Escánez, E.M.S. Woodward and E. Fernández. 2005. Variability of chlorophyll and primary production in the Eastern North Atlantic Subtropical Gyre: potential factors affecting phytoplankton activity. Deep Sea Research I 52: 569–588.

Tredici, M.R. and R. Materassi. 1992. From open ponds to vertical alveolar panels—the Italian experience in the development of reactors for the mass cultivation of phototrophic microorganisms. Journal of Applied Phycology 4(3): 221–231.

Trenberth, K.E. and L. Smith. 2005. The mass of the atmosphere: A constraint on global analyses. Journal of Climate 18: 864–875.

Tsunogai, S., S. Watanabe and T. Sato. 1999. Is there a "continental shelf pump" for the absorption of atmospheric CO_2? Tellus Series B-Chemical and Physical Meteorology 51(3): 701–712.

Ugwu, C.U., H. Aoyagi and H. Uchiyama. 2008. Photobioreactors for mass cultivation of algae. Bioresource Technology 99(10): 4021–4028.

Wei, N., J. Quarterman and Y.S. Jin. 2013. Marine macroalgae: an untapped resource for producing fuels and chemicals. Trends in Biotechnology 31(2): 70–77.

Winder, Monika and Ulrich Sommer. 2012. Phytoplankton response to a changing climate. Hydrobiologia 698(1): 5–16.

Zeebe, Richard E. and Dieter A. Wolf-Gladrow. 2001. CO_2 in Seawater: Equilibrium, Kinetics, Isotopes. Elsevier Oceanography Series. Elsevier. Amsterdam, New York.

CHAPTER 6

Review of Marine Algae as Source of Bioactive Metabolites: a Marine Biotechnology Approach

Loïc G. Carvalho* and Leonel Pereira

1 Introduction

Marine algae are the primary food source of a vast ecosystem on our planet. The marine environments are totally dependent of this resource and so are we. It has been estimated that 80% of the oxygen from the atmosphere comes from phytoplankton. Phytoplankton is composed of a considerable variety of photosynthetic microorganisms. In the scientific community there is a concept, which is finding increasing acceptance, that cyanobacteria are not microalgae. Thus, we distinguish two groups: cyanobacteria or blue-green algae (Prokaryota) and microalgae (Eukaryota). According to this concept, we can consider marine microalgae as eukaryotic and photoautotrophic microorganisms.

There are three major groups of macroalgae, Chlorophyta (green algae), Rhodophyta (red algae) and Ochrophyta-Phaeophyceae (brown algae). Macroalgae (seaweed) are multicellular photoautotrophic organisms which besides being primary producers they play an important role in the structuring and maintenance of the marine ecosystems, introducing important nursery spots for a variety of marine species.

IMAR-CMA (Institute of Marine Research) and MARE (Marine and Environmental Sciences Centre), Department of Life Sciences, FCTUC, University of Coimbra, 3001-455 Coimbra, Portugal.
Email: leonel.pereira@uc.pt
* Corresponding author: loicgoncalvesdecarvalho@gmail.com

The first industrial interest in studying algae started with the aquaculture industry for both microalgae and macroalgae (seaweed). The demand for seaweed as a direct food resource and the demand for fish production needed the application of aquaculture techniques. In the case of seaweed, this was the only choice for sustainable production. The production of microalgae was necessary for the feeding phase of fish larvae to ensure the survival of newly born juveniles and also for the feeding of zooplankton.

Microalgae production in fish tanks started to show some other advantages such as better water conditions and even more healthy fish production. Then the algae pigments, and their antioxidant capacity, were shown to be of considerable interest, not only for aquaculture but for the human food industry as well. β-carotene was a big production success from the marine microalgae *Dunnaliela salina* (Chlorophyta). This carotenoid is used as food additive, to give more orange to the food and to conserve it better as an antioxidant. This compound can also be found as a food supplement and there are many published studies that refer to β-carotene as a good antioxidant for human health among other carotenoids (Dufossé et al. 2005, Spolaore et al. 2006).

Besides the antioxidants, algae have other relevant compounds such as fatty acids, proteins, polysaccharides, vitamins and minerals. One of the reasons that make algae so suitable for biofuel production is their richness in fatty acids which can be converted into biodiesel, in the case of microalgae, and their richness in sugars which can be converted into bioethanol, in the case of seaweeds.

Polysaccharides and fatty acids derivatives, among other compounds, are the new potential resources for the pharmaceutical and cosmetic industry. Some of these products have shown antimicrobial, antiviral, antitumor and many other qualities. Algae extracts have considerable hydrating skills and cosmetic effects, leading to healthier treatments for human skin.

Another field where marine algae can be applied is in bioremediation. The problem of water pollution like industrial, urban or mining effluents and even crude oil spills, may have an answer in algae treatment. As referred before, the use of algae in aquaculture tanks results in an improvement in the quality of water and healthier growth of fish. This is accomplished with microalgae or seaweed and results in a new concept of integrated multi-trophic aquaculture (IMTA) (Fig. 1; Pinto and Abreu 2011). These characteristics along with some others, like metal accumulation, nutrients uptake or the capacity of crude oil metabolism, turns marine algae into a potential tool for remediation.

The aim of this chapter is to approach the potential that marine algae biotechnology has for different industrial interests and what can we expect of this science in the near future. For that reason, themes such as

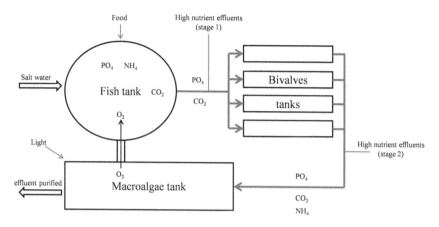

Figure 1. Illustration of an IMTA system. High nutrient effluents (stage 1): dissolved and suspended nutrients. High nutrient effluents (stage 2): dissolved nutrients.

pharmaceutical, food, cosmetics, bioremediation and industrial production will be discussed as potential fields for marine algae biotechnology application.

2 Bioactive Compounds of Marine Algae and their Potential Pharmaceutical Applications

The capacity that algae have in growth inhibition or toxicity against pathogenic microorganisms has been reported a long time ago. Take the example of Chlorellin, the metabolite isolated from *Chlorella vulgaris* (Chlorophyta) (Pratt et al. 1945). It is the presence of pathogenic agents and the need for defense against them that may contain the answer as to how it is possible for these organisms to produce so many interesting antimicrobial compounds.

Extensive work for the screening of antimicrobial effects as a way of research for the new antibiotics discovery, are very common and can usually reveal incredible work. Normally, for seaweeds, their antimicrobial capacity can change among groups of algae (green, red and brown), seasons of the year and area of the globe.

2.1 Antibacterial

Manivannan and collaborators (2011) made a screening for antibacterial effects. A series of bacteria, *Klebsiella pneumoniae*, *Escherichia coli*, *Staphylococcus aureus*, *Enteroccoci* sp., *Proteus* sp., *Streptococcus* sp.,

Pseudomonas aeruginosa, Vibrio parahaemolyticus, Salmonella sp., *Shewanella* sp., *Vibrio fluvialis* and *V. splendidus, Enterococcus faecalis, Vibrio cholerae, Shigella flexneri, Staphylococcus epidermitis, Aeromonas liquefaciens* and *Bacillus subtilus*, were tested through *Turbinaria conoides, Padina gymnospora, Sargassum tenerrimum* (Phaeophyceae) extracts. All the tree algae could inhibit bacteria growth, some extracts better than others.

There is a good quantity of screening work for antimicrobial effects done all over the globe which has always revealed inhibition of bacteria growth (Osman et al. 2010, Rosaline et al. 2012).

The lipophilic phase, the result of an extraction from a *Skeletonema costatum* culture, showed significant antibacterial effect on *Listonella anguillarum* (Naviner et al. 1999). A study indicates that *Isochrysis galabana* (Prymnesiophyceae) extracts had growth inhibition effect on the multiple drug resistant (MDR) *Mycobacterium tuberculosis*. Further analyses suggested unsaturated fatty acids fractionated from those extracts may be the metabolites responsible for the growth inhibition (Prakash et al. 2010).

There is remarkable work on antibacterial effects from supernatant extracts obtained from *Nostoc caeruleum, Limnothrix redekei* (formerly *Oscillatoria redeckei*) (Cyanobacteria), *Thalassiosira profunda* (Bacillariophyceae) and *Leucocryptos marina* (Cryptophyta). The same work revealed antibacterial activity from the biomass extract of *Anabaena oscillatorioides* (formerly *Anabaena circinalis*), *Nostoc punctiforme, Pseudanabaena mucicola* (Cyanobacteria), *Monoraphidium contortum* (Chlorophyta), *Thalassiosira profunda* (Bacillariophyceae), *Leucocryptos marina* (Cryptophyta), *Amphidinium carterae* and *Gyrodinium estuariale* (Dinophyceae) (Scholz and Liebezeit 2012).

Phaeodactylum tricornutum (Bacillariophyceae) produces fatty acids with antibacterial capacity against multidrug resistant specimen of *Staphylococcus aureus* (Gram +) and marine pathogen *Listonella anguillarum* (Gram –) (Desbois et al. 2008).

The constant appearance and evolution of new antibiotic resistant organisms are the strong motivation for the continuation of this research. The revealing of all these potential compounds suggests that the next generation of antibiotics could have a significant origin in marine algae.

2.2 Antifungal

Some studies showed the capacity of certain algae to have antifungal capacity. It is more often encountered in microalgae, probably because of competition for the same resources.

Sometimes it is the strong toxicity of some compounds that make that capacity present, like stored polyketide, as the polycyclic ether macrolides, and the openchain polyketides (Cardozo 2006). The only problem is that they demonstrated such high toxicity that they cannot be applied in human therapy.

Amphidinolide B is a kind of compound that has antifungal capacity with high cytotoxic and antitumor capacity, too (Cardozo 2006).

Notable work was done with screening of the antifungal potential of seaweeds. The aim of the work was antimicrobial screening and while antibacterial effects from algae extracts are very common in other cited bibliography, antifungal properties are not that common although a variety of extracts from a group of algae were tested against a series of fungi and revealed positive results as antifungal reaction. The seaweeds *Turbinaria conoides*, *Padina gymnospora*, *Sargassum tenerrimum* (Phaeophyceae) revealed antifungal properties against all the fungal pathogens of the study, such as *Candida albicans*, *Penicillium* sp., *Aspergillus flavus*, *Aspergillus tetreus*, *Candida glabrata*, *Cryptococcus neoformans*. The exception was for *Aspergillus niger* (Manivannan 2011).

More screenings revealed the potential antifungal activity of some algae extracts of *Codium decorticatum*, *Caulerpa scalpelliformis* (Chlorophyta), *Turbinaria conoides*, *Sargassum wightii* (Phaeophyceae), and *Acanthophora spicifera* (Rhodophyta) (Lavanya and Veerappan 2012).

2.3 Antiviral

Fabregas and collaborators (1999) revealed endocellular extracts with virus inhibiting properties from microalgae *Porphyridium cruentum* (Rhodophyta), *Chlorella autotrophica* (Chlorophyta), *Isochrysis galbana* (Prymnesiophyceae) and *Dunaliella tertiolecta* (Chlorophyta) against rhabdovirus of viral haemorrhagic septicaemia (VHSV). The same extracts obtained from *P. cruentum* and *C. autotrophica* also inhibited African swine fever virus (ASFV). The exocellular extracts of all these microalgae had inhibitory properties too, except with the extracts of *I. galbana* against both VHSV and ASFV and *C. autotrophica* and *D. tertiolecta* against ASFV.

Assembled work revealed that seaweed are a potential tool for future application in antiviral research, with particular interest in HIV-1 infection due to some inhibition results; the same is the case for dengue (Mayer et al. 2010).

Phlorotannins, from *Ecklonia cava* and *Ishige okamurae* (Phaeophyceae), are phloroglucinol-based polyphenols responsible for anti-HIV activity. Lectins and polysaccharides of algae demonstrate some kind of effect against the infection capacity of the virus. Sulfated polysaccharides have been shown to inhibit the replication of enveloped viruses including members of the flavivirus, togavirus, arenavirus, rhabdovirus, orthopoxvirus, herpes virus families and HIV virus too. Sulfated polisaccharides of *Dictyota mertensii, Lobophora variegata, Spatoglossum schroederi, Fucus vesiculosus* (Phaeophyceae) and the genus of red algae *Grateloupia*, demonstrate some anti-HIV activity (Vo and Kim 2010).

Seaweed rich sulphated polysaccharides, such as fucans and alginates, produced from brown algae are good antiviral tools. Assembled works about these compounds consider *Undaria pinnatifida* (Phaeophyceae) fucans antiviral tools (Wijesinghea and Jeon 2012).

2.4 Antitumor

Herbivores have a persistent presence in the environment of marine algae. Among microalgae this behavior is mostly produced by the zooplankton, but adult fish and other adult organisms that feed by filtration play their role in this food chain too. Some microalgae develop silica structures that do not allow a predator to consume them because it just cannot swallow the algae. This way of defense is dominated by diatoms (Bacillariophyceae), but the avoidance of being eaten by size and structural changes can be accomplished through the formation of colonial structures, too (Donk et al. 2010). So, as the colony grows bigger, with higher numbers of individuals, the more difficult it becomes for predators to swallow it, at least for small predators. Other algae develop ways of moving in the water column making them a more difficult target. Yet another way is the development of biochemical weapons.

These biochemical weapons can be toxins, repellents or compounds that can inhibit mitosis provoking some lower hatching rates or abnormal physical development in their predators or even compromising their egg development and metabolism (Donk et al. 2010).

Normally toxins are very harmful to organisms and some case studies have revealed that those produced by some marine microalgae have considerable consequences because of their strong interaction and toxicity.

Microalgae produce compounds that repel their predators. It is not yet known how the produced compounds react towards the predators to induce such behavior, but they work like an alert signal that induces other microalgae to produce mechanisms of predatory defense.

Antitumor compounds may be the ultimate and most enthusiastic discovery in microalgae. This type of compound is the most wanted because of the possibility of discovering a tool against cancer. Briefly, cancer is a direct result of abnormal cell division by mitosis, so, if microalgae produce compounds that act specifically inhibiting mitosis, they might be an answer for future cancer therapies. *Skeletonema costatum, Chaetoceros calcitrans, Guinardia delicatula, Guinardia striata, Odontella regia, Rhizosolenia setigera, Stephanopyxis turris, Thalassiosira pseudonana* (Bacillariophyceae), are some of the microalgae producing these metabolites (Donk et al. 2010).

It is reported that diatoms such as *Thalassiosira, Skeletonema, Odontella, Chaetoceros, Navicula, Nitzschia,* among others, together with some dinoflagellates (Dinophyceae) such as *Prorocentrum micans, Gymnodinium sanguineum,* and *Lingulodinium polyedra* (formerly *Gonyaulax polyedra*) induce low hatching and low fertilization success in copepods, even in high egg production. Those that could hatch were presented with some deformities, expressing marked morphological asymmetry which resulted in their death a few days after (Ianora and Poulet 1993, Ianora et al. 1999, Lee et al. 1999, Caldwell et al. 2002, Poulet et al. 2007). These morphological development alterations were proposed to be the result of compounds with some kind of mitosis inhibition effect that could accumulate in the reproductive organs of the copepods that were fed with a diatom or dinoflagellate rich diet and somehow were transferred to the eggs. Studies revealed that the strongly reduced hatching success of wild copepods feed on diatoms were due to the existence of short chain polyunsaturated aldehydes. These compounds are produced by cell disruption of the microalgae (Caldwell et al. 2002). Somehow, these compounds exist in the living cell at standard concentration, but the disruption of the cell leads to the liberation of these compounds to the surrounding medium which are recognized by other diatoms as a kind of signaling that might induce the overproduction of the compounds.

These events were already named as diatom-derived embryonic inhibition, and some of the agents were identified as two C_{10} short chain polyunsaturated aldehydes (PUAs) from *Thalassiosira rotula, Skeletonema costatum* and *Pseudo-nitzschia delicatissima* (Bacillariophyceae). Continuous research was carried out on this subject leading to the development of new methodologies for the study and detection of these compounds which gave us the idea of the existence of a range of PUAs: 2-trans,4-trans decadienal (DDE); 2-trans,4-trans,7-cis decatrienal and 2-trans,4-cis,7-cis decatrienal (Caldwell et al. 2002).

Some seaweed have demonstrated good effects against cancer cells and tumors. Compounds such as alginates, a type of polysaccharide produced in brown alga *Sargassum vulgare,* revealed antitumor effect by *in vivo* growth-inhibition (Sousa et al. 2006). Reviewed assembled work about different fucans of *Undaria pinnatifida, Ecklonia cava, Fucus evanescens, Saccharina gurjanovae* (formerly *Laminaria gurjanovae*), *Cladosiphon okamuranus* (Phaeophyceae), are considered as antitumor, anti-proliferation or anticancer candidates (Wijesinghea and Jeon 2011).

2.5 Biological Tools

Squalene is a polyunsaturated triterpene and is a natural biochemical precursor of cholesterol and other steroids. It is naturally produced by animals and plants. Known work was done with olive tree, sharks, cereals and other plants and animals (Reddy and Couvreur 2009, Grigoriadou et al. 2006, Nakazawa et al. 2011). Research on squalene is mostly applied in human health studies because of its great capacity for better delivery of vaccines, drugs and other substances through squalene emissions (Reddy and Couvreur 2009).

The stability related to some other individual capacities of this compound, such as antioxidant and even antitumor, make squalene an exceptional tool in medicine application as an adjuvant or other kind of tool. Binding this compound to drugs, or other compounds for therapy purposes, turns the therapy more successful than other emulsions. The reason is related to the protection of the drug due to squalene's stability and antioxidant aspect. This way, the therapeutic compound is protected from the surrounding metabolism influence which leads to its prolonged effect (Reddy and Couvreur 2009).

Because squalene is naturally produced in the human organism, found even in the skin, excreted as constitute of sebum, its use assures biocompatibility, thus making it less suitable for an immune reaction. This also is the key for the success of squalene emulsions for vaccination. Somehow, not only does it protect the drug or antigen but, due to its biocompatibility, the drug transportation and even recognition is higher than in traditional vaccine solutions (Reddy and Couvreur 2009).

There are microalgae that are known for high levels of squalene production which might be a better and more sustainable solution than liver oil from high depth sharks and better than other plant resources because of higher production rates (Yue and Jiang 2009, Kebelmann et al. 2012).

This compound is just an example of a biological tool with high commercial value that can be extracted from algae. Algae can reveal new biological tools that can be applied in research or other fields.

2.6 Toxins—Pharmacological Tools and Potential Drug Leads

It is rare to find eukaryotic microalgae that produce toxins because, in most aquatic environments, blue-green algae (Cyanobacteria) are responsible for toxins production by bloom formation (Table 1), although, there are some eukaryotic microalgae, such as some diatoms (Bacillariophyceae) and dinoflagellates (Dinophyceae), which originate harmful algal blooms (HAB) and produce some dangerous toxins that can affect the nervous system (Vasconcelos et al. 2010).

The subject of HAB is a very important area of research, especially if we consider the risk of human health safety at all water resources and touristic destinations such as beaches, estuaries, lakes and rivers. All this is to say that, the already known toxins produced by these organisms can provoke some serious problems, and depending on the toxins, even simple

Table 1. Marine algae biotoxins and their syndromes and mechanisms of action.

Microalgae	Toxins	Effects and/or syndromes	References
Anabaena spp.; *Oscillatoria* spp.	Anatoxins	Post-synaptic nicotinic agonist and acetylcholinesterase inhibitor	Cardozo et al. 2006
Pseudo-nitszchia spp.	Domoic Acid	Glutamate receptors; Amino acid Limited TNF-a and matrix metalloproteinase-9 release; amnestic shellfish poisoning (ASP)	Cardozo et al. 2006, Mayer and Hamann 2005
Gambierdiscus toxicus	Ciguatoxins	Voltage dependent sodium channel blockers; ciguatera fish poisoning (CFP)	Cardozo et al. 2006
Dinophysis spp., *Prorocentrum* spp.	Dinophysistoxins Okadaic acid	Ser/Thr protein phosphatases inhibitors; diarrhetic shellfish poisoning (DSP)	Cardozo et al. 2006
Nodularia spp.	Nodularins	Protein phosphatase types 1 and 2A inhibition; Liver failure and death by hipovolemic chock	Cardozo et al. 2006, Vasconcelos et al. 2010
Gymnodinium breve	Brevetoxins	Voltage-dependent sodium channel site 5; neurotoxic shellfish poisoning (NSP)	Cardozo et al. 2006
Alexandrium spp., *Gonyaulax* spp., *Gymnodinium* spp.	Saxitoxins	Voltage-dependent sodium channel site 1; paralytic shellfish poisoning (PSP)	Cardozo et al. 2006
Protoperidinium crassipes	Azaspiracid (AZA)	Diarrhetic shellfish poisoning (DSP)	Twiner et al. 2008
Dinophysis spp.	Pectenotoxin	Macrolidee Disruption of F-actin cytoskeletal; Induction of F-actin depolymerization	EFSA 2008, Mayer and Hamann 2005
Protoceratium reticulatum, *Lingulodinium polyedrum* and *Gonyaulax spinifera*	Yessotoxin	Lymphocyte [Ca2+]i homeostasis modulation; Inhibition of calcium channels	Paz et al. 2008, Mayer and Hamann 2005
Ostreopsis spp.	Palytoxin	General malaise and weakness, associated with myalgia, respiratory effects, impairment of the neuromuscular apparatus and abnormalities in cardiac function	Tubaro et al. 2011
Symbiodinium spp.	Zooxanthellatoxins	Vasoconstrictive	Gordon and Leggat 2010
Gambierdiscu sp., *Ostreopsis* sp.	Maitotoxin	Modulation of calcium and sodium influx	Mayer and Hamann 2005
Lyngbya majuscula	Antillatoxin	Potent neurotoxin; depolarization-evoked Na+ load, glutamate release, relief of Mg2+ block of NMDA receptors, and Ca, + influx	Li et al. 2001, Mayer and Hamann 2005

Microalgae: known species that produce toxins; toxins: known marine toxins produced by microalgae; effects and/or syndromes: known consequences or mechanism of action of the toxins.

skin contact or breathing during a swim could let to intoxication (Hinder et al. 2011).

In most environments, toxins bioaccumulate in the food chain and normally, high concentrations can be detected in the top food chain predator. In the aquatic/marine environment, seafood such as shellfish, which feed through a process of filtration, can accumulated high amounts of toxins because of their resistance to them.

The ingestion of contaminated shellfish can let to high intoxication leading to toxicological effects such as hepatotoxicity, dermal toxicity, cytotoxicity, neurotoxicity and, in some extreme cases, death (Hinder et al. 2011).

Marine biotoxins are classified in ten groups: azaspiracid (AZA), brevetoxin, cyclic imine, domoic acid (DA), okadaic acid (OA), pectenotoxin (PTX), saxitoxin (STX), yessotoxin (YTX), palytoxins (PlTX) and ciguatoxins (CTX) (EFSA 2008). Azaspiracid (AZA) is one example of a marine toxin that can be found in shellfish and one of the known microalga responsible for its presence is *Azadinium spinosum*, which allows the assumption that the entire group *Azadinium* spp. (Dinophyceae) might produce AZA.

The known toxicity of marine toxins and their common presence in food resources makes it necessary to study and develop methods of analysis, detection and clinical tests. The problem is the need for some quantities of the toxin for those tests. Once more, biotechnology enters to help beginning with the first step of culturing techniques, followed by extraction processes and purification of toxins (EFSA 2008, Jauffrais et al. 2012).

The most interesting part is the possibility of finding some other potential applications of the toxins in the study because of their known mechanisms of action in the organism (Table 1). For example, during some toxicological essays, the understanding of the toxin's action mechanism could lead to the discovery of unknown pharmacological tools (Mayer and Hamann 2005).

2.7 Other Pharmaceutical Capacities

Other capacities from algae that we can find are anticoagulant and antithrombotic activity from fucans of *Ecklonia cava, Fucus evanescens, Padina gymnospora, Ascophyllum nodosum, Sargassum fulvellum, Hizikia fusiformis, Saccharina cichorioides* (formerly *Laminaria cichorioides*) (Phaeophyceae).

There are anti-inflammatory compounds and tissue protection from the fucans of *A. nodosum, C. okamuranus, Saccharina japonica* (formerly *Laminaria japonica*); and immunomodulation from the fucan of *Fucus vesiculosus* (Wijesinghea and Jeon 2012, Xia et al. 2005). There are other capacities as the tissue protection, like the neuroprotective effect found in microalgae (Pangestuti and Kim 2011).

Algae are a source of new bioactive compounds against protozoan diseases (Felício et al. 2010). The genus *Symbiodinium*, belonging to the Zooxanthellae, accumulates some unique macrolides, an interesting toxin with a 62-membered macrolactone structure and potent vasoconstrictive activity (Cardozo et al. 2006). There is even interest in the antifouling property of algae (Maréchal et al. 2004).

Unfortunately in the pharmaceutical world, it is not enough to discover a new compound. Clinical research is done to study if the compound produces good results in its own function, with good success rates; biospecific behavior, with very low or no toxicity for human or animal treatment and some other subjects have to be considered. For the production to be suitable, some key processes are important, such as the biomass/compound ratio, the extraction and purification rate, the stability of the compound and finally if it is worth the trouble, or in other words, if it is commercially viable.

The problem here is that for every compound found, extraction and purification methods have to be studied. After finding the best method we have to be certain that the quantity of the compound produced by the algae is the highest; so, culturing studies are a necessary way to achieve the best compound/biomass ratio. Resuming, finding new marine drugs takes a lot of qualified research that can take a good many years.

3 Marine Algae and the Food Industry

As stated earlier, algae are the primary producers of most aquatic habitats. They are well known for their rich nutrient composition such as vitamins, lipids, proteins, carbohydrates and minerals. This makes these organisms a good and healthy source of food for those who realize these facts. Some of the most famous names of seaweed as direct source of food are *Ulva* (Chlorophyta), *Porphyra* (Rhodophyta), *Undaria*, *Laminaria*, *Himanthalia* and *Saccharina* (Phaeophyceae) (Pereira 2011). The high demand for this resource from some countries results in their increased production through aquaculture. Seaweed are gathered too but mostly as an unsustainable way. An ecological approach with sustainable gathering of seaweed for food is possible but it will never be enough for the total demand.

The same aspects are true for microalgae. They too are a primary food resource for human populations, but with less diversity of resources. The name *Spirulina* as a food source is not so recent. *Spirulina* spp. was used for many years as direct food supply in different regions of the globe for centuries, such as Central America and Africa. In the case of China, it was *Nostoc* (Milledge 2010). These genera of Cyanobacteria grow naturally with abundance in some salty or mineralized lakes or river channels. Local populations learned how to harvest, store and preserve these resources when nothing else was available to eat (Spolaore et al. 2006).

Besides their application as a human food resource by gathering, microalgae's first industrial production and applications were as feeding stocks for aquaculture. The advances in microalgae culturing techniques made possible their use as a source of food for all stages of some fish and bivalve mollusks or for larval stages of some crustaceans and fish. Thus, the aquaculture industry did not need to capture plankton and young fish from the oceans anymore.

There is a considerable variety of microalgae used for these means; those more often used, such as *Skeletonema* spp., *Thalassiosira* spp., *Chaetoceros* spp., *Phaeodactylum* spp. (Bacillariophyceae), *Isochrysis* spp. (Haptophyceae), *Tetraselmis* spp. (Prasinophyceae), *Chlamydomonas* spp. (Cryptophyceae), *Dunaliella* spp. (Chlorophyta) and *Spirulina* spp. (Cyanobacteria), are those whose nutrient composition is better known.

Today *Arthrospira platensis* (formerly *Spirulina platensis*) and *Spirulina maxima* are produced on a large scale for many different objectives from a food supplement perspective. They are easily found as capsules in natural products stores, as rations or supplement for aquaculture fish or ornamental fish and as rations for cattle and pets. Some farmers let this organism or other Cyanobacteria grow naturally in pounds which enrich the water for cattle (Cohen et al. 1991).

All this application of *Arthrospira* and *Spirulina* spp. is supported by the nutrient composition found, like the protein content of the biomass which is known to reach a high percentage of the total biomass content. This aspect together with the know-how of culturing techniques for *Spirulina* spp., makes this organism an alternative for feeding cattle rather than crops such as maize or soy. Even though we are talking of a fresh water resource, the marine species, *Spirulina subsalsa*, is also available and does not differ that much from the other fresh water genera.

The advantages of algae in diet due to their nutritional value, are well known, although in the future marine algae might be considered as functional food or nutraceutics (Gupta and Abu-Ghannam 2011, Mohamed et al. 2012).

There are products for quality and healthy feeding produced from diatoms and other microalgae for pet, cattle and other farmed animals (Milledge 2010). These are presented as ration foods, supplements and diatom sand, which is said to help and prevent against internal and external parasites (animal skin parasites).

There are many more interesting bioactive compounds that are of interest to the food industry (Pereira 2011, Gupta and Abu-Ghannam 2011), and some will be further discussed.

3.1 PUFAs

The lipid content of marine algae is well appreciated. Microalgae are those which produce such compounds in higher quantities than seaweed. What makes them so interesting are not only the saturated fatty acids, which are the primary reason for the burst of research on microalgae as a fuel resource. What really makes them a target is because they are excellent producers of unsaturated fatty acids and polyunsaturated fatty acids (PUFAs) (Cohen et al. 1991, Spolaore et al. 2006).

These metabolites are considered high value products because of their properties in general health when consumed. PUFAs can be found in fish biomass and are obtained by fish through an algae-rich diet in their environment. Initially, these PUFAs were industrially produced by the use of fish biomass; more properly, the fish waste produced from the fisheries. However, there was too much risk of getting undesired compounds in the final product such as toxins and other pollutants. Now, the best way of producing PUFAs for consumption is with microalgae production (Barclay et al. 1994).

There are studies that establish the importance of these metabolites in the early development of human fetus and during the development of the human baby. The presence of PUFAs in the organism ensures the good development of the nervous system and optic system and the general growth and well-being of the organism by means of homeostasis. The consumption of PUFAs works like we are giving the best tools for our organism to function resulting in an healthier status, for example, lowering the risk of heart diseases and, perhaps, giving better chances against some cancers, improving the immune system or protecting against inflammatory like diseases (Cardozo 2006, Milledge 2010).

That is why pregnant women and mothers who breast feed are advised to take PUFAs as supplements or to adopt a diet rich in PUFAs, in order to ensure the uptake of these supplements by the fetus or baby for its healthier development (Milledge 2010).

Another important role of these molecules, besides development regulation, is in regulatory physiology (Cardozo 2006).

Most marine microalgae produce PUFAs but the record holder in variety and quantity of these metabolites, compared with all biomass content, are the algae from Ochrophyta phylum, especially diatoms (Bacillariophyceae) (Milledge 2010). Work was done to find the most PUFA-rich algae among those that potentially might be used for aquaculture (Patil et al. 2006). The ω-3 kind of PUFA is the most common such as docosahexaenoic acid (DHA, 22:6ω3) and Eicosapentaenoic acid (EPA, 20:5ω3), produced, for example, by *Phaeodactylum tricornutum* (Desbois et al. 2008).

PUFA ω-6 is present in a wide variety of microalgae and is as much desired. Gamma-linolenic acid (GLA, 18:3ω6), which is industrially produced by *Spirulina* spp. for human consumption is also an example (Cyanobacteria). But there are marine algae used for producing the same compounds, though they are more applied for aquaculture feeding. Arachidonic acid (AA, 20:4ω6) can be found on *Porphyridium purpureum* (formerly *Porphyridium cruentum*) (Rhodophyta) (Bigognoa et al. 2002).

3.2 Phycocolloids (Polysaccharides)

Macroalgae are known for their high content in polysaccharides. In the algae group, seaweeds are the richest sources of sulphated polysaccharides. These compounds are very valuable resources for the food industry. They are used as additive agents because of gelling and thickening properties (e.g., alginates, agar and carrageenan) and can be found as the following European codes: alginic acid—E400, sodium alginate—E401, potassium alginate—E402, ammonium alginate—E403, calcium alginate—E404, propylene glycol alginate—E405, agar—E406, carrageenan—E407, semirefined carrageenan or "processed *Eucheuma* seaweed"—E407A (Pereira et al. 2013) (see also the Chapter 8).

These additives can be found in aqueous based jellifying deserts, low calorie jelly, flans, puddings, chocolate milk, milk derivatives, ice-creams, soy milk, cheeses, processed and canned meat, beer, gravy, sauce, jam and other processed foods.

Members of the red algae (Rhodophyta) produce galactans (e.g., carrageenans and agars).

Brown algae (Ochrophyta, Phaeophyceae) produce uronates (alginates) and other sulphated polysaccharides (e.g., fucoidan and laminaran).

3.2.1 Agar

Agar is well known as an inert support medium for microbial culture but it also has applications as a gelling agent for food.

Although agar was found in a variety of red algae, those industrially more important are the genera *Gelidium*, *Pterocladiella*, *Gelidiella*, and *Gracilaria* (Pereira et al. 2003, Pereira 2011, Pereira et al. 2013).

3.2.2 Carrageenan

There are three commercial carrageenans, kappa-, iota-, and lambda-carrageenans but in nature, there is a variety of carrageenan hybrids due to the existence of different precursors and their concentrations in the algae content. Each type of carrageenan has a property, the kappa type forms hard, strong, and brittle gels, iota-carrageenan forms soft and weak gels and lambda-carrageenan acts as a thickening agent.

This compound is industrially obtained by *Kappaphycus alvarezii*, for kappa-carrageenan, *Eucheuma denticulatum*, for iota-carrageenan, and genera *Gigartina* and *Chondrus*, for lambda-carrageenan (Pereira et al. 2009a,b, Pereira 2011, Pereira et al. 2013).

3.2.3 Alginic acid and derivates

Alginic acid is extracted from brown algae (Ochrophyta, Phaeophyceae) as a mixed salt of sodium and/or potassium, calcium and magnesium and its composition varies in each algae.

Alginic acid is the precursor of its derivatives such as the famous alginate. The commercial species selected by industry for the extraction of this compound are *Macrocystis pyrifera* and *Ascophyllum nodosum* but an extended range of other raw material such as *Laminaria, Lessonia, Alaria, Ecklonia, Eisenia, Nereocystis, Sargassum, Cystoseira* and *Fucus* are also used (Pereira 2011, Pereira et al. 2013).

This type of polysaccharide not only shows appropriate use in the food industry, but there are studies revealing some antitumor and other types of pharmaceutical applications (Sousa et al. 2006).

3.2.4 Fucans

Fucans are an example of sulphated polysaccharides extracted from brown algae (Ochrophyta, Phaeophyceae). The most studied is fucoidan and it is already commercially produced using *Fucus vesiculosus* as raw material (Pereira et al. 2013).

It is not a candidate for the food industry as an additive, but it shows great prospects for the pharmaceutical field. Now, the most appropriate use for this compound is its application as a food resource, or better, as a nutraceutical.

Functional food as a health improver is more and more common today. It is easier to put a functional food in the market than a new pharmaceutical drug because of the long processes needed to assure the safety procedures of use for that drug. If in a short term we want to put a functional derivative from seaweed in the market, the best way is to select one from an already considered edible seaweed. This way the safety of the product is already secure, although, we still have to prove the beneficial use of our product and how there is more advantage in the use of this pure derivative product from a seaweed instead of the all seaweed.

3.2.5 Laminaran

Laminaran and its derivatives are other sulphated polysaccharides that could become functional food. Extracted from some brown seaweed (Laminariales), they have a lot of potential in the area of health, for example, it has known antitumor effects (Ibrahim et al. 2005).

3.3 Pigments

3.3.1 Carotenoids

Carotenoids are widely explored compounds by the food industry because of their color and, in the case of β-carotene, astanxantin and fucoxanthin, their known antioxidant capacity (radical scavenging activity). Recent studies reveal a range of properties that some pigments have, such as antitumor effects, apoptosis induction on cancer cells and anti-inflammatory effects (Dufossé et al. 2005, D'Orazio et al. 2012).

3.3.2 β-carotene

Although most of the time, the β-carotene used is the synthetic one, producers and consumers are increasingly demanding the use of the naturally produced one by *Dunaliella salina* (Chlorophyta) (Spolaore et al. 2006, Milledge 2010). Chicken eggs enriched with β-carotene or lutein can be found in the supermarkets, although there is no reference or publicity of these supplements in the eggs sold, they are widely used to enhance the appearance of egg yolks to a deeper shade of orange. To achieve this result the carotenoids are given through the food consumed by chickens which will later be part of the egg yolk.

This kind of strategy is not so recent; moreover, for human health purposes, the efforts are focused on the production of functional foods (nutraceuticals) (Spolaore et al. 2006, Milledge 2010, Gupta and Abu-Ghannam 2011, Mohamed et al. 2012). Although some communities who know the nutritional advantages of an algae rich diet consume it traditionally, it may not be available to all, and neither does it appeal to everyone. However, people who may never have eaten algae in their life might reconsider its inclusion in their diet if studies show that it confers significant health benefits.

The use of algae pigments as an objective for food application might have started with fish aquaculture. Aquaculture industry is the industry with the most focus on this matter (Dufossé et al. 2005). As said before, the food resources for aquaculture fish were and still are, mostly rations obtained and developed from fisheries industries that use fish waste (body parts of fish considered unfit for human consumption). In aquaculture, these rations are suited to carnivorous fish. They have high levels of proteins and oils and are enriched with vitamins and minerals if needed. The problem arose when consumers did not want to buy full grown salmon reared for commercial purposes, because it did not have the typical "salmon color".

Research was done to resolve this problem and it was discovered that the color of the wild salmon, originates from its diet that is rich in krill, an orange/pink crustacean. This crustacean is rich in carotenes and β-carotene. But krill does not produce these pigment, it gets them through its microalgae rich diet. The problem was solved by enrichment of the fish rations with a good percentage of microalgae or directly with the pigments. The salmon started to gain its true color and got more acceptance from consumers.

Carotenoids have proved to be of interest to human health. They have provitamin A activity, and can still remain active where the normal vitamin C cannot in low oxygen rates. These pigments, such as β-carotene, modulate UVA induced gene expression, protect the skin and eyes from photo-oxidation against UV light, and prevent human eye disease such as cataract (Dufossé et al. 2005).

3.3.3 Astaxanthin

Astaxanthin can play a diversity of roles, such as prevention of some human pathology, like skin UV-mediated photo-oxidation, inflammatory processes and even cancer (Spolaore et al. 2006).

When natural astaxanthin started to be industrially produced, some salmon aquacultures preferred the use of this pigment or added it to the salmon diet. This pigment, as with β-carotene, is a large scale production compound extracted from the freshwater microalgae *Haematococcus pluvialis* (Chlorophyta) (Dufossé et al. 2005, Spolaore et al. 2006). It is not only the living red color of this compound that the food industry contemplates, but its antioxidant capacity, higher than β-carotene that considerably increases the value of this product. While it is the "know-how" of *H. pluvialis* culturing techniques that makes this organism the target for astaxanthin, this metabolite is present in a large variety of marine microalgae.

3.3.4 Fucoxanthin

Fucoxanthin is a xanthophyll present in Ochrophyta phylum of algae that, with the presence of the other pigments, gives a brown-yellow or olive-green appearance. As with all carotenoids, the antioxidant effect is remarkable, but in this case, antitumor effects, apoptosis induction on cancer cells, anti-inflammatory effects and radical scavenging activity were reported. Ultimately, we have to focus on the most probable application of this compound. The most promising one is in the food industry or, more precisely, in functional food (nutraceutical). It is a shorter way to reach the consumer and the health benefits are strongly supported in the bibliography which, recently, has been focused on the capacity that fucoxanthin has to modulate the expression of specific genes responsible for cell metabolism (D'Orazio et al. 2012).

The results of those studies are those which can affirm the activities recently mentioned for fucoxanthin, such as apoptotic cell death in primary effusion lymphoma (antitumor effect), through the functional inhibition of Hsp90 chaperon, abdominal white adipose tissue burner as mRNA, protein inducer through modulation of the UCP1 gene, and other good consequences that might be called positive secondary effects of the uptake of this remarkable compound (D'Orazio et al. 2012).

3.3.5 Chlorophylls

Chlorophylls are other metabolites well appreciated by the food industry. Although chlorophyll is easily found in every plant, they are in all algae too. There is no algae production for exclusive extraction of chlorophylls, but its obtention as a secondary product of an algae production is more likely to happen. It would be a loss of value if in the process of producing a determinant product we could not separate the chlorophylls and isolate them. The reason is because these pigments are largely used in food industry and mixed with other pigments to obtain a large variety of colors. Above all of these advantages, it is important to refer that these compounds are increasing in use to put an end to the use of synthetic pigments for the food industry.

3.3.6 Phycobiliproteins

Phycobiliproteins are the pigments that give most organisms of the Cyanobacteria group their blue-green color and Rhodophyta group their red color, from such algae like blue-green algae *Arthrospira* and the red algae *Porphyridium* respectively (Spolaore et al. 2006). Again, these colors offer natural pigments for food processing (Milledge 2010), substituting for synthetic colorants. For the same reasons, cosmetic and pharmaceutical industries too use these pigments for their products as colorants. Beyond these applications, phycobiliproteins possess some other interesting characteristics that make them excellent tools in research. Because of their capacity of fluorescence in known wavelength, studies in immunology are using these pigments as markers in assays for those research studies (Spolaore et al. 2006).

4 Marine Algae as Tools for Bioremediation

Algae are very profitable tools for bioremediation. For instance, high levels of nutrients in water are always the promoters of algae blooms, not only microalgae but seaweeds too, and it normally leads to eutrophication of the water systems. In this situation, algae might look to be the problem, although they are only responding to environmental changes. The most important to retain of this algae behavior is that it can be a helpful tool for eutrophication problems. The strategy is to treat these high nutrient effluents before they reach the water environment. Not all waste water treatment stations (WWTS) have the tertiary treatment required for high nutrient water effluents, and when they do, it is a chemical treatment that is usually adopted. Nowadays the possibility of using microalgae instead of chemical treatment for high nutrients effluents is a well known matter.

Microalgae are the most suitable for this kind of job—they grow fast, are in suspension and there is the potential for biomass recovery for industrial production, such as biofuel and cattle feed depending on their quality. What made this possible is the knowledge of some algae with good growth rates and high lipid production that can really be a solution for the future of pollution issues (Rawat et al. 2010).

Macroalgae are not mentioned because most of the industry and even WWTS's are located near fresh water sources where effluents are more likely to be discharged. There are no known fresh water macroalgae suitable for these kinds of treatments but there are a lot of industries with near-shore activity that could use macroalgae as a bioremediation tool for high nutrient problems.

We can take the example of the fish aquaculture industry. Fish production in aquaculture is a polluting activity because of the big population of fish that needs to be fed and that produces high quantities of excrement in a limited area. Not all the food is consumed which, together with the excrement, produces effluents of high nutrient concentrations. Those effluents could be used by algae before getting to the seawater. This would solve the fish production problem and at the same time there would be profit with the algae biomass recovery.

The other benefits from introducing a controlled algae culture is the reduction of CO_2 and production of O_2. This simple task performed by algae result in a healthier environment for fish, reducing the risk of anoxia and establishing some balance within the bacteria population. If the environment is rich in O_2, only aerobic bacteria can develop, normally, those less problematic. This way we might reduce infections, not only with O_2 but with healthy competition for nutrients between algae and bacteria.

Nitzschia sp., a benthic microalga, is a good example of how algae can serve as a tool. Studies have revealed that biofilm development of this alga in organic enriched sediments led to the recovery of those sediments, not only by direct action of the algae but because of the production of O_2 that promoted aerobic biodegradation among bacteria (Yamamoto et al. 2008).

Bioremediation in the algae world is more related to the microalgae domain. Studies about heavy metal uptake, like cadmium uptake by *Chlorella* sp. and other Chlorophyta and Cyanobacteria are common in the published bibliography for other heavy metals such as iron and aluminium (Matsunaga et al. 1999, Harun et al. 2010, Richards and Mullins 2012). In this case we are talking about the incorporation of the metal into the algae body which makes possible to recover the pollutant source with the biomass harvesting.

Seaweed might seem a difficult tool to apply because of a slower growth rate and the need for substrate fixation. Would that problem be solved with the use of extracts? Research with *Ascophyllum nodosum* demonstrates it as a good tool for heavy metal absorption, as is *Fucus vesiculosus* (Phaeophyceae). This capacity happens to be related with the biochemical composition of the cell wall (polysaccharides) (Harun et al. 2010).

There is record of the capacity that algae have on the degradation of lindane, an organochlorine. An experiment with microalgae, regarding their quick capacity to adapt when environment conditions change reported that *Desmodesmus intermedius* (formerly *Scenedesmus intermedius*) (Chlorophyta) was able to adapt and grow with a high concentration of this contaminant among other species that could not. This involved not only proof of a quick adaptation capacity but also the capability to metabolize organochlorines (González et al. 2012). *Navicula* sp., *Phaeodactylum tricornutum*, *Nitzschia* sp. and *Synedra* sp. (Bacillariophyceae) are examples of algae that can degrade phloroglucinol and naphthalene; other freshwater algae can degrade herbicides including the marine diatom *Skeletonema costatum* (Yang et al. 2012). This makes way for new research about finding microalgae that can solve the problem of degrading other toxic compounds.

Oil spills are some of the most complicated situations. In places with such problems, the toxicity is so high that biodiversity is drastically diminished and it takes years before the local ecosystem gets restored. The petroleum compounds have a strong negative effect on the primary producers, such as algae although, the same organisms are, possibly, an advantage for crude polluted environments. The study of some microalgae has shown their capacity for physiological adaptation and was supposed to result in genetic adaptation too (Romero-Lopez et al. 2012). If that is true, microalgae genetic mutations due to petroleum contaminants exposure could result in the creation of new natural tools for these kinds of bioremediation.

Algae can really surprise us with their versatile and complex behavior. Overall there is enough knowledge for the application of algae in bioremediation, at least in the near future in high nutrients effluents.

5 Cosmetic Potential of the Marine Algae

Algae are used for their therapeutic capacity in cosmetic application or in areas of health and body care, like spa treatments. Thalassotherapy is the use of the therapeutic benefits from the sea. Normally, it is the application of seawater bath treatments with the help of some mixes of salts, sand and seaweed infusions. We might even say that this therapy makes good use of natural and biological tools to help the skin and body of a person through the combination of ingredients from the sea (Pereira 2010).

The sea ingredients all together with some relaxing approaches and techniques are why thalassotherapy is so appealing by those whom had the opportunity to experiment it.

Although, there is particular interest to fully understand what are the agents responsible for such therapeutic effects. Separate these sea ingredients and study them individually seems to be a smart strategy to the development of new products for the cosmetic industry. For example, some sunscreen contain algae extracts or more purified compounds that protect the skin from solar radiation. Some seaweed (Ochrophyta, Phaeophyceae) and microalgae are widely used for sunscreen production.

If we still think about the possibility to use only one compound from algae, and that it is more reliable to isolate only that compound from the algae to achieve our case-study goal we are looking away from a problem. Sometimes research of new natural compounds for health and care applications reveals that it is not the isolated compound that makes the difference, but the synergy from the interaction of a group of compounds which results in therapeutic solution.

Is the thalassotherapy pure result from the algae used in it? Maybe. But why does it results better with seawater and salt? Perhaps because of the behavior and reaction of the algae compounds together with other compounds present in the seawater and salt.

Today it is very common to encounter the words "algae extract" and/or "seaweed extract" on the labels of many products. It might be because the product genuinely contains an extract although, when the industry uses a specific extract or compound from the algae for the formulation, it is very common to use the strategy of only revealing in the label that there is an extract instead of revealing the name of the compound. Nevertheless, what cannot be forgotten is that the product's utility might be due to a complex group of compounds instead of just one.

With that said, it is in our interest to analyze which algae and compounds are used in the cosmetic and body-care industry.

Algae applications are focused on the skin and body treatment market through skin care products. Today we have several products that indicate the inclusion of algae extracts in the composition of anti-aging, regenerating, anti-irritant, slimming and exfoliating creams.

Algae have antioxidants, gene expression inducers, antimicrobials, vitamins, and a diversity of minerals and hydrating compounds such as some proteins, lipids and polysaccharides. If we collate this information and extrapolate it to the skin care and treatment area of research, we could get some interesting results. The simple coloration of the pigments is of interest to the cosmetic industry.

Such work has been done by some cosmetic companies and it has resulted in skin care products like tissue regeneration, wrinkle reduction, prevention of striae formation, cell proliferation, regulation of skin metabolism, etc., with the use of *Chlorella vulgaris* (Chlorophyta) and *Spirulina* sp. (*Arthrospira*) (Cyanophyceae) (Spolaore et al. 2006).

Some sunscreens include microalgae compounds rather than the usual macroalgae compounds for skin and hair. Reference to *Nannochloropsis* sp. (Eustigmatophyceae) as having anti-wrinkling properties and *Dunaliela salina* (Chlorophyta) as having cellular proliferation stimulation with positive effect on skin metabolism are examples. There are anti-aging creams, refreshing, regenerative care products, and emollients as an anti-irritant (Spolaore et al. 2006).

Squalene was already subject for the pharmaceutical field, even though, it finds its application in the cosmetic field too. Substance protection, metabolism precursor, and biocompatibility are among some of the remarkable known functions of this compound (Mayer 2005, Xia et al. 2005, Spolaore 2006, Cardozo 2006, Reddy 2009, Pangestuti 2011, Wijesingher and You-Jin 2011, Mohamed 2012, Scholz 2012, Pereira 2013).

For skin care applications, it not only helps in stabilizing emulsions and protecting the skin tissue through its capacities, but brings more quality for a product because it is a natural alternative among other synthetic compounds used in the cosmetic industry. Thanks to its biocompatibility, squalene helps the product to be better absorbed by the skin tissue, resulting in superior effects.

The high demand for squalene not only for the pharmaceutical industry but for the cosmetic industry too, results in work related with the discovery of algae as a new source for it (Grigoriadou et al. 2006, Cai-Jun and Jiang 2009, Nakazawa et al. 2011, Kebelmann et al. 2012).

6 Marine Algae Biotechnology: Some Industrial Principles

Now with all the information mentioned before we can declare that marine algae are a research field of interest with economic potential.

If we want to start in this field we have to consider all the steps, since algae growth until compound extraction and final product production. This way, the first step would be to choose between macroalgae or microalgae.

6.1 Macroalgae Biotechnology

The advantage in the use of macroalgae is, for sure, the naturally available biomass and the simple method of handling this resource. Seaweed grows naturally in the environment and is easily collected for further application. Today we feel more responsible for the ecosystems and the environment. We know that actions have consequences and that if we develop an industrial project based on the wild harvesting of seaweed for industrial production, we are not making a sustainable choice. Not only for the future of the selected seaweed and the ecosystem where it was taken from, but for the company as well because as soon as the seaweed was gone the company could not continue its activity.

The natural biomass production of seaweed takes time and can be seasonal depending on its location in the globe. With a production through gather, there is the risk of a local decimation of the entire selected specie of algae.

Unless the location, where the seaweed gather takes place, is well handled with sustainable management and coordination, with instructed personnel that respect the seasonal growth of the seaweed, the next generation of biomass growth will not be secured and would result in drastic consequences for the industrial activity.

This is why large scale seaweed production around the globe is done through aquaculture techniques. Off-shore and in-shore, the long-line strategy of growing seaweed is a very effective method. Sometimes it can be more difficult, with the need of a previous embryo production and seeding (Morrondo 2011). Sometimes, we just need to get an adult specimen, fix it in a line and collect only part of the algae body so that vegetative reproduction for the next gathering season is assured.

This might seem fairly simple to replicate, but in fact, it is not so. Not every country has a stable shore, such as in some of the equatorial and tropical regions of the globe, for off-shore production. The alternative could be in-shore production, although, once more, not all shores have such calm waters that guarantee the safety of investing in a long-line strategy. That is why the installation of such aquaculture production is typically made in estuaries, lagoons and other brackish-water environments.

In brief, to grow seaweed, we need water, nutrients and light (Figs. 1 and 2), all available in the environments referred to earlier. Sometimes, nutrient availability is more critical in biomass production; normally, higher levels of nutrients than those found in the natural environment are always more suitable for biomass production.

If the seaweed production is in a closed system, we can control the amount of nutrients available in the water column and if needed add more to it. There are different solutions that may be considered. For example, estuarine waters naturally contain more organic nutrients, not only because of the normal input from river waters but more because of excess fertilizers from agriculture fields that enter the water courses of the rivers. The use of these waters may be an alternative to the use of fertilizers. We can even say that we are practicing bioremediation through biological uptake of excessive nutrients. Even though, the freshwater from river will still be a problem for

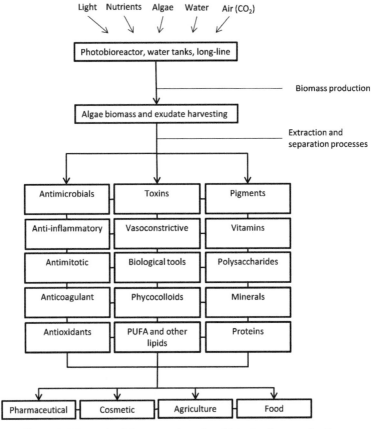

Figure 2. Schematic of the steps of an algae biotechnology production.

the aquaculture due to low salt concentration during raining periods. Once more, a closed aquaculture system may do the work, but the best way of solving this problem is through the algae itself. If our seaweed is already familiar to all these salinity changes, even the stress of low salinity might affect the biomass production, the seaweed will still survive and carry on.

One of the most suitable strategies is the integrated multi-trophic aquaculture system (IMTA) (Fig. 1). This situation resolves the nutrient issue in more than one way. As mentioned before, fish aquaculture causes the problem of high nutrient production due to fish or shellfish feeding and excrement. If these waters get to a seaweed production tank before reaching the environment again, the nutrient uptake by seaweed will solve the high nutrient pollution and stimulate the biomass production of algae (Pinto and Abreu 2011, Abreu et al. 2011).

Now, for the industrial part, the best way to profit from seaweed is to use all its content, or at least attempt to do so. Normally, a company focuses on a single compound of interest; however, it would be wrong to not take into consideration, the remaining contents of the seaweed. We know, for instance, that some seaweed are used as food, that they have potential health and body-care properties, that they are rich in sugars, vitamins, minerals, proteins—in other words, they are a powerful and versatile resource (Fig. 2).

Thus, it is in a company's best interests to consider not only the compound selected for its high market value, but all the other compounds as well. Algae are a very good resource as fertilizers for agriculture, and vegetal solution for animal feeding such as pet and cattle. They originate extracts and/or other isolated compounds for the cosmetic, food and pharmaceutical industries. Their notable sugar content is a profitable way to produce bioethanol through the seaweed biomass residues produced from another industry.

Sometimes the aggressive methods used in the industry do not allow the efficient use of the total content of the biomass, but with a carefully designed method we can improve the efficiency and obtain more than just one compound/product.

6.2 Microalgae Biotechnology

An industry sector for microalgae production is already in existence. It would appear that it all started with freshwater species because freshwater is easier to get then seawater. The truth, however, is that the history of microalgae culture, from an industrial perspective, started with the fish aquaculture industry. The need of these organisms for the purpose of fish feed led to research in culturing techniques for quality biomass production. A diverse group of algae were cultivated for several years which resulted in a considerable update of the knowledge in culturing techniques. Over the years, continued studies of these algae revealed some compounds of interest like vitamins, PUFA, carotenoids, etc. It seems that it is only a matter of adjusting the already known culturing technique for the algae to produce our desired compound.

This made some names of algae famous such as *Chlorella* sp., *Dunaliella* sp., *Spirulina* sp., *Haematococcus pluvialis*, *Phaeodactylum tricornutum*, *Skeletonema costatum*, *Isochysis galabana*, *Oscillatoria* sp., *Thalassiosira* sp., *Nostoc* sp.

Although it might appear easy to buy some of these strains from culture collections of algae and start an industrial production, there is hard work that needs to be carried out which together with the time needed to achieve satisfying results for industrial purposes, may be the reason for poor investment in this field comparing to other industrial fields.

Fortunately, the times are changing. Research on microalgae continues and is waiting to be applied. Today the word "microalgae" is more recognized along with photobioreactors production, industrial scale photobioreactor production.

The development of photobioreactors technology was and still is critical. There are always new developments that determine new advantages in microalgae biomass production. One reason for that constant dynamic in this technology is because we have to consider the demands of our selected microalga and build an adequate reactor for it.

Nowadays, if we want to start a business in the field of microalgae we could produce already existing commercial microalgae because there is high demand for f microalgae biomass. Fish, together with other organisms of aquaculture industry, are an example of consumers for those microalgae because it seems that there is not enough biomass availability to meet the demands of the aquaculture industry.

Does this situation look like a good opportunity? Perhaps it is for plans in the near future. However, it is still is a replication of what has already been done.

Every country and every region has a specific kind of microalgae flora, adapted to the local weather conditions all through the year. This is a powerful tool that is not often exploited. Industrial production of algae will always be dependent on local environmental conditions. We are talking about weather conditions throughout a year, like temperature and light radiation oscillations. For open pounds or open reactors we have also to consider the surrounding biota that can influence microalgae culture.

A very good strategy to take in consideration is the study of local microalgae flora. First, we should carry out the observation of samples, identification of most of the species and the identification of those which are dominant in local weather conditions and those that are easy to culture.

Then, the next step is to access the value of those algae by studying their potential and compounds through quality analyses (Fig. 2).

There will always be compounds of interest in algae for the industrial domain (Fig. 2). The only doubt that remains is if industrial production of the selected algae will be a commercially viable proposition. This not only depends on algae itself and the culture technique used but also on the biorefinery sector and its extraction and separation techniques.

Industrial production of microalgae results in a biomass production that is much more like a microalgae paste obtained from the culture. The remaining culture medium also has compounds, exudates of algae that have commercial interests too.

It is not possible to defragment and isolate all the algae compounds at the same time because there are different techniques for each type or group of compounds that influence the possibility of later isolating other compounds. But there is the possibility of separation, isolation and extraction of some compounds that assures that the rest of the biomass can be used. All will depend on the priority of the company, in what it selects as the final compound.

For example, lipids and living algae can be separated from the aqueous mixture. This leads to three fronts on which we can work: living microalgae, lipids, aqueous mixture. The most complex work is with the living algae because of the abundance of compounds in it. The right strategy might be in developing the best methods of extraction to get the most compounds and use the remaining biomass as valuable extracts for cosmetic, agriculture or other industries.

There are always ways to maximize profit of algae, it all depends on the industrial design of the production unit of the company. Above all, it depends on the time expended in laboratory research before scaling up production levels.

7 Conclusion

Marine algae are important resources in human society. Their use comes far from the past and still reveals extreme importance today, proven by their industrial application and continuous research that reveal new discoveries. Algae can lead to new drug therapeutics against pathogens with antibacterial, antifungal, antiviral and antiprotozoal activity. This research field is so important because not only there is a lack of antiviral and antiprotozoal medicines, but the continuous rise of antibiotic resistant strains of fungi and bacteria demands new antibiotic resources.

Besides, the effective antiviral activity of algae against HIV is a motivation against a disease that is very difficult to treat and affects a considerable number of the human population. Cancer too is a motivation to research tools and mechanisms that will guarantee some effective therapies in the future. Cancer is a very complex pathology and its characteristics differ from one patient to another. Some new discoveries among marine algae could let to the next generation of cancer treatments.

Algae toxins are of major concern because of their presence in food or bathing waters. Continuous work for monitoring and discovery of new biotoxins from these organisms are of high relevance, though, we must always consider the possibility of their having potential applications that need to be studied, especially in the pharmaceutical field.

There are plenty of other fields of research—anticoagulants, antithrombotics, anti-inflammatory, tissues protectors (neuroprotectors), immunomodulators, and vasoconstrictors—where marine algae can be part of future work.

The food industry is a field where algae are widely used. Possibly, the lack of acceptance in western society might change in the coming years thanks to the consistent work and propaganda by the scientific community related to the nutritional value and health benefits of algae consumption. However, nothing should obstruct studies and intention of production for their application as functional food. Studying algae as a known food resource but for pharmaceutical properties only encourages more algae production with a commercial objective, not only for unprocessed algae, but for the production of extracts too.

As pharmaceutical properties are studied among algae, new products for cosmetic application can be produced from new compounds. As long as the algae source of such compounds are reported from those which are used as food and have no toxic effect, their cosmetic application can be considered as biological, natural, and biocompatible, with no harmful effects. Or, perhaps, algae can become the source of already known compounds that were more difficult to obtain or whose acquisition was not ecologically acceptable and sustainable (e.g., squalene).

An interesting field of intervention for marine algae is in the bioremediation department. There is enough work that could lead to *in situ* application specifically in high nutrients concentration effluents that dramatically affect some natural ecosystems. It has been proved possible and even profitable. The biomass produced through this mechanism can be used for a variety of purposes. Using algae in heavy metals remediation is a different matter. Algae have the tendency to incorporate and accumulate such pollutants. This makes the algae unsuitable for use as food; however, there must be other applications for the heavy metal contaminated biomass. There is the possibility of using such biomass as biofuel and particular extracts production.

The culturing technology of algae is a dynamic sphere with continuous updates. There is always the possibility to improve and there is always new developments in technology that can help to make the difference. What may be important to know is that a good strategy is the use of local algae resources and the association of culturing methods together with other trophic groups of aquaculture. This way there is innovation and competition between other producers. The problem is that to start such activity some scientific research as to get done and it takes a lot of time and resources. But, in the end, it could originate new strategies with high value compounds from other algae than the commercial ones.

To use the available know-how of to initiate an industrial activity with a commercial specie sounds tempting. The problem is that this strategy might not be the most adequate because the algae is not indicated for the climate of our region of activity. Besides that, we would steel not be different from other industrial algae producing companies which would make it very hard to get into the market.

In conclusion, marine algae are a source of new bioactive compounds and are tools for biotechnological application with wide industrial uses, not only for commercial interest but as a versatile ecological and sustainable resource.

Acknowledgments

The authors acknowledge financial support from the Portuguese Foundation for Science and Technology—IMAR-CMA (Institute of Marine Research).

Keywords: Macroalgae, microalgae, biotechnology, bioactive compounds, thallassotherapy, pharmaceutical applications, food industry, bioremediation

References Cited

Abreu, M.H., R. Pereira, L. Mata, A. Nobre and I.S. Pinto. 2011. Aquacultura multi-trófica integrada em Portugal. pp. 54–77. *In*: Ferreiro, U.V., M.I. Filgueira, R.F. Otero and J.M. Leal (eds.). Macroalgas en la acuicultura multitrófica integrada peninsular. Valorización de su biomasa. Centro Tecnológico del Mar—Fundación CETMAR, Spain. ISBN: 978-84-615-4974-0.

Barclay, W.R., K.M. Meager and J.R. Abril. 1994. Heterotrophic production of long chain omega-3 fatty acids utilizing algae and algae-like microorganisms. Journal of Applied Phycology 6(2): 123–129.

Bigognoa, C., I. Khozin-Goldberga, S. Boussibaa, A. Vonshaka and Z. Cohena. 2002. Lipid and fatty acid composition of the green oleaginous alga *Parietochloris incisa*, the richest plant source of arachidonic acid. Phytochemistry 60(5): 497–503.

Caldwell, G.S., P.J.W. Olive and M.G. Bentley. 2002. Inhibition of embryonic development and fertilization in broadcast spawning marine invertebrates by water soluble diatom extracts and the diatom toxin 2-trans,4-trans decadienal. Aquatic Toxicology 60(1-2): 123–137.

Cardozo, K.H.M., T. Guaratini, M.P. Barros, V.R. Falcão, A.P. Tonon, N.P. Lopes, S. Campos, M.A. Torres, A.O. Souza, P. Colepicolo and E. Pinto. 2006. Metabolites from algae with economical impact. Comparative Biochemistry and Physiology, Part C: Toxicology & Pharmacology 146: 60–78.

Cohen, Z., S. Didi and Y.M. Heimer. 1992. Overproduction of 'y-linolenic and eicosapentaenoic acids by algae. Plant Physiology 98(2): 569–572.

D'Orazio, N., E. Gemello, M.A. Gammone, M. Girolamo, C. Ficoneri and G. Riccioni. 2012. Fucoxantin: a treasure from the sea. Marine Drugs 10(3): 604–616.

Desbois, A.P., T. Lebl, L. Yan and V.J. Smith. 2008. Isolation and structural characterisation of two antibacterial free fatty acids from the marine diatom, *Phaeodactylum tricornutum*. Applied Microbiology and Biotechnology 81: 755–764.

Donk, E.V., A. Ianora and M. Vos. 2010. Induced defences in marine and freshwater phytoplankton: a review. Hydrobiologia 668(1): 3–19.

Dufossé, L., P. Galaupa, A. Yaronb, S.M. Aradb, P. Blancc, K.N.C. Murthyd and G.A. Ravishankar. 2005. Microorganisms and microalgae as sources of pigments for food use: a scientific oddity or an industrial reality? Trends in Food Science & Technology 16(9): 389–406.

European Food Safety Authority [EFSA]. 2008. Marine biotoxins in shellfish—azaspiracid group. Scientific Opinion of the Panel on Contaminants in the Food Chain. The EFSA Journal 723: 1–52.

Fabregas, J., D. García, M. Fernandez-Alonso, A.I. Rocha, P. Gómez-Puertas, J.M. Escribano, A. Otero and J.M. Coll. 1999. *In vitro* inhibition of the replication of haemorrhagic septicaemia virus (VHSV) and african swine fever virus (ASFV) by extracts from marine microalgae. Antiviral Research 44(1): 67–73.

Felício, R., S. Albuquerque, M.C.M. Young, N.S. Yokoya and H.M. Debonsi. 2010. Trypanocidal, leishmanicidal and antifungal potential from marine red alga *Bostrychia tenella* J. Agardh (Rhodomelaceae, Ceramiales). Journal of Pharmaceutical and Biomedical Analysis 52(5): 763–769.

González, R., C. García-Balboa, M. Rouco, V. Lopez-Rodas and E. Costas. 2012. Adaptation of microalgae to lindane: a new approach for bioremediation. Aquatic Toxicology 109: 25–32.

Gordon, B.R. and W. Leggat. 2010. Symbiodinium—invertebrate symbioses and the role of metabolomics. Marine Drugs 8(10): 2546–2568.

Grigoriadou, D., A. Androulaki, E. Psomiadou and M.Z. Tsimidou. 2006. Solid phase extraction in the analysis of squalene and tocopherols in olive oil. Food Chemistry 105(2): 675–680.

Gupta, S. and N. Abu-Ghannam. 2011. Recent developments in the application of seaweeds or seaweed extracts as a means for enhancing the safety and quality attributes of foods. Innovative Food Science and Emerging Technologies 12: 600–609.

Harun, R., M. Singh, G.M. Forde and M.K. Danquah. 2010. Bioprocess engineering of microalgae to produce a variety of consumer products. Renewable and Sustainable Energy Reviews 14(3): 1037–1047.

Hinder, S.L., G.C. Hays, C.J. Brooks, A.P. Davies, M. Edwards, A.W. Walne and M.B. Gravenor. 2011. Toxic marine microalgae and shellfish poisoning in the British Isles: history, review of epidemiology, and future implications. Environmental Health 10: 54.

Ianora, A. and S.A. Poulet. 1993. Egg viability in the copepod *Temora stylifera*. Limnology and Oceanography 38(8): 1615–1626.

Ianora, A., A. Miralto, I. Buttino, G. Romano and S.A. Poulet. 1999. First evidence of some dinoflagellates reducing male copepod fertilization capacity. Limnology and Oceanography 44(1): 147–153.

Ibrahim, A.M.M., M.H. Mostafa, M.H. El-Masry and M.M.A. El-Naggar. 2005. Active biological materials inhibiting tumor initiation extracted from marine algae. Egyptian Journal of Aquatic Research 31(1): 146–155.

Jauffrais, T., J. Kilcoyne, V. Sechet, C. Herrenknecht, P. Truquet, F. Herve, J.B. Berard, C. Nulty, S. Taylor, U. Tillmann, C.O. Miles and P. Hess. 2012. Production and isolation of azaspiracid-1 and -2 from *Azadinium spinosum* culture in pilot scale photobioreactors. Marine Drugs 10(6): 1360–1382.

Kebelmann, K., A. Hornung, U. Karsten and G. Griffiths. 2012. Intermediate pyrolysis and product identification by TGA and Py-GC/MS of green microalgae and their extracted protein and lipid components. Biomass and Bioenergy 49: 38–48.

Lavanya, R. and N. Veerappan. 2012. Pharmaceutical properties of marine macroalgal communities from Gulf of Mannar against human fungal pathogens. Asian Pacific Journal of Tropical Disease 2(1): 320–323.

Lee, H.-W., S. Ban, Y. Ando, T. Ota and T. Ikeda. 1999. Deleterious effect of diatom diets on egg production and hatching success in the marine copepod *Seudocalanus newmani*. Plankton Biology and Ecology 46(2): 104–112.

Li, W.I., F.W. Berman, T. Okino, F. Yokokawa, T. Shioiri, W.H. Gerwick and T.F. Murray. 2001. Antillatoxin is a marine cyanobacterial toxin that potently activates voltage-gated sodium channels. PNAS 98(13): 7599–7604.

Manivannan, K., G. Karthikai devi, P. Anantharaman and T. Balasubramanian. 2011. Antimicrobial potential of selected brown seaweeds from Vedalai coastal waters, Gulf of Mannar. Asian Pacific Journal of Tropical Biomedicine 1(2): 114–120.

Maréchal, J.-P., G. Culioli, C. Hellio, H. Thomas-Guyon, M.E. Callow, A.S. Clare and A. Ortalo-Magné. 2004. Seasonal variation in antifouling activity of crude extracts of the brown alga *Bifurcaria bifurcata* (Cystoseiraceae) against cyprids of *Balanus amphitrite* and the marine bacteria *Cobetia marina* and *Pseudoalteromonas haloplanktis*. Journal of Experimental Marine Biology and Ecology 313(1): 47–62.

Matsunaga, T., H. Takeyama, T. Nakao and A. Yamazawa. 1999. Screening of marine microalgae for bioremediation of cadmium-polluted seawater. Journal of Biotechnology 70(1-3): 33–38.

Mayer, A.M.S. and M.T. Hamann. 2005. Marine pharmacology in 2001–2002: marine compounds with anthelmintic, antibacterial, anticoagulant, antidiabetic, antifungal, anti-inflammatory, antimalarial, antiplatelet, antiprotozoal, antituberculosis, and antiviral activities; affecting the cardiovascular, immune and nervous systems and other miscellaneous mechanisms of action. Comparative Biochemistry and Physiology, Part C 140: 265–286.

Mayer, A.M.S., A.D. Rodríguez, R.G.S. Berlinck and N. Fusetani. 2010. Marine pharmacology in 2007–8: Marine compounds with antibacterial, anticoagulant, antifungal, anti-inflammatory, antimalarial, antiprotozoal, antituberculosis, and antiviral activities; affecting the immune and nervous system, and other miscellaneous mechanisms of action. Comparative Biochemistry and Physiology Part C 153: 191–222.

Milledge, J.J. 2010. Commercial application of microalgae other than as biofuels: a brief review. Reviews in Environmental Sciences and Biotechnology 10(1): 31–41.

Mohamed, S., S.N. Hashim and H.A. Rahman. 2012. Seaweeds: a sustainable functional food for complementary and alternative therapy. Trends in Food Science & Technology 23(2): 83–96.

Morrondo, J.M.S. 2011. Cultivo de laminariales y acuicultura multitrófica. pp. 29–51. *In*: Ferreiro, U.V., M.I. Filgueira, R.F. Otero and J.M. Leal. (eds.). Macroalgas en la acuicultura multitrófica integrada peninsular. Valorización de su biomasa. Centro Tecnológico del Mar—Fundación CETMAR, Spain. ISBN: 978-84-615-4974-0.

Nakazawa, A., H. Matsuura, R. Kose, S. Kato, D. Honda, I. Inouye, K. Kaya and M.M. Watanabe. 2011. Optimization of culture conditions of the thraustochytrid *Aurantiochytrium* sp. strain 18W-13a for squalene production. Bioresource Technology 109: 287–291.

Naviner, M., J.P. Bergé, P. Durand and H. Le Bris. 1999. Antibacterial activity of the marine diatom *Skeletonema costatum* against aquacultural pathogens. Aquaculture 174(1): 15–24.

Osman, M.E.H., A.M. Abushady and M.E. Elshobary. 2010. *In vitro* screening of antimicrobial activity of extracts of some macroalgae collected from Abu-Qir bay Alexandria, Egypt. African Journal of Biotechnology 9(12): 7203–7208.

Pangestuti, R. and S.-K. Kim. 2011. Neuroprotective effects of marine algae. Marine Drugs 9(5): 803–818.

Patil, V., T. Källqvist, E. Olsen, G. Vogt and H.R. Gislerød. 2006. Fatty acid composition of 12 microalgae for possible use in aquaculture feed. Aquaculture International 15(1): 1–9.

Paz, B., A.H. Daranas, M. Norte, P. Riobó, J.M. Franco and J.J. Fernández. 2008. Yessotoxins, a Group of Marine Polyether Toxins: an Overview. Marine Drugs 6(2): 73–102.

Pereira, L. 2010. Littoral of Viana do Castelo—ALGAE. Uses in agriculture, gastronomy and food industry. Câmara Municipal de Viana do Castelo, Viana do Castelo, Portugal.

Pereira, L. 2011. A review of the nutrient composition of selected edible seaweeds. pp. 15–47. *In*: Pomin, V.H. (ed). Seaweed: Ecology, Nutrient Composition and Medicinal Uses. Nova Science Publishers Inc., Nova Iorque.

Pereira, L., A. Sousa, H. Coelho, A.M. Amado and P.J.A. Ribeiro-Claro. 2003. Use of FTIR, FT-Raman and C-NMR spectroscopy for identification of some seaweed phycocolloids. Biomolecular Engineering 20(4-6): 223–228.

Pereira, L., A.M. Amado, A.T. Critchley, F. van de Velde and P.J.A. Ribeiro-Claro. 2009a. Identification of selected seaweed polysaccharides (phycocolloids) by vibrational spectroscopy (FTIR-ATR and FT-Raman). Food Hydrocolloids 23(7): 1903–1909.

Pereira, L., A.T. Critchley, A.M. Amado and P.J.A. Ribeiro-Claro. 2009b. A comparative analysis of phycocolloids produced by underutilized versus industrially utilized carrageenophytes (Gigartinales, Rhodophyta). Journal of Applied Phycology 21(5): 599–605.

Pereira, L., S.F. Gheda and P.J.A. Ribeiro-Claro. 2013. Analysis by vibrational spectroscopy of seaweed polysaccharides with potential use in food, pharmaceutical, and cosmetic industries. Hindawi Publishing Corporation, International Journal of Carbohydrate Chemistry 537202: 1–7.

Pinto, I.S. and H. Abreu. 2011. Aquacultura multitrófica integrada—O que é? pp. 54–77. *In*: Ferreiro, U.V., M.I. Filgueira, R.F. Otero and J.M. Leal. Macroalgas en la acuicultura multitrófica integrada peninsular. Valorización de su biomasa. Centro Tecnológico del Mar—Fundación CETMAR, Spain. ISBN: 978-84-615-4974-0.

Poulet, S.A., A. Cueff, V.T. Wichard, J. Marchetti, C. Dancie and G. Pohnert. 2007. InXuence of diatoms on copepod reproduction. III. Consequences of abnormal oocyte maturation on reproductive factors in *Calanus helgolandicus*. Marine Biology 152(2): 415–428.

Prakash, S., S.L. Sasikala and V.H.J. Aldous. 2010. Isolation and identification of MDR-*Mycobacterium tuberculosis* and screening of partially characterized antimycobacterial compounds from chosen marine micro algae. Asian Pacific Journal of Tropical Medicine 3(8): 655–661.

Pratt, R., J.F. Oneto and J. Pratt. 1945. Studies on *Chlorella vulgaris*. X. Influence of the age of the culture on the accumulation of chlorellin. American Journal of Botany 32(7): 405–408.

Rawat, I., R.R. Kumar, T. Mutanda and F. Bux. 2010. Dual role of microalgae: phycoremediation of domestic wastewater and biomass production for sustainable biofuels production. Applied Energy 88(10): 3411–3424.

Reddy, L.H. and P. Couvreur. 2009. Squalene: a natural triterpene for use in disease management and therapy. Advanced Drug Delivery Reviews 61(15): 1412–1426.

Richards, R.G. and B.J. Mullins. 2012. Using microalgae for combined lipid production and heavy metal removal from leachate. Ecological Modelling 249: 59–67.

Romero-Lopez, J., V. Lopez-Rodas and E. Costas. 2012. Estimating the capability of microalgae to physiological acclimatization and genetic adaptation to petroleum and diesel oil contamination. Aquatic Toxicology 124–125: 227–237.

Rosaline, X.D., S. Sakthivelkumar, K. Rajendran and S. Janarthanan. 2012. Screening of selected marine algae from the coastal Tamil Nadu, South India for antibacterial activity. Asian Pacific Journal of Tropical Biomedicine 2(1): 140–146.

Scholz, B. and G. Liebezeit. 2012. Screening for biological activities and toxicological effects of 63 phytoplankton species isolated from freshwater, marine and brackish water habitats. Harmful Algae 20: 58–70.

Sousa, A.P.A., M.R. Torres, C. Pessoa, M.O. Moraes, F.D.R. Filho, A.P.N.N. Alves and L.V. Costa-Lotufo. 2006. *In vivo* growth-inhibition of sarcoma 180 tumor by alginates from brown seaweed *Sargassum vulgare*. Carbohydrate Polymers 69(1): 7–13.

Spolaore, P., C. Joannis-Cassan, E. Duran and A. Isambert. 2006. Commercial applications of microalgae. Journal of Bioscience and Bioengineering 101(2): 87–96.

Tubaro, A., P. Durando, G. Del Favero, F. Ansaldi, G. Icardi, J.R. Deeds and S. Sosa. 2011. Case definitions for human poisonings postulated to palytoxins exposure. Toxicon 57(3): 478–495.

Twiner, M., J.N. Rehmann, P. Hess and G.J. Doucette. 2008. Azaspiracid shellfish poisoning: A review on the chemistry, ecology, and toxicology with an emphasis on human health impacts. Marine Drugs 6(2): 39–72.

Vasconcelos, V., J. Azevedo, M. Silva and V. Ramos. 2010. Effects of marine toxins on the reproduction and early stages development of aquatic organisms. Marine Drugs 8(1): 59–79.

Vo, T.-S. and S.-K. Kim. 2010. Potential anti-HIV agents from marine resources: an overview. Marine Drugs 8(12): 2871–2892.

Wijesinghe, W.A.J.P. and Y.-J. Jeon. 2012. Biological activities and potential industrial applications of fucose rich sulfated polysaccharides and fucoidans isolated from brown seaweeds: A review. Carbohydrate Polymers 88(1): 13–20.

Xia, W., J. Li, M. Geng, X. Xin and J. Ding. 2005. Potentiation of T Cell function by a marine algae-derived sulfated polymannuroguluronate: *in vitro* analysis of novel mechanisms. Journal of Pharmacological Sciences 97(1): 107–115.

Yamamoto, T., I. Goto, O. Kawaguchi, K. Minagawa, E. Ariyoshi and O. Matsuda. 2008. Phytoremediation of shallow organically enriched marine sediments using benthic microalgae. Marine Pollution Bulletin 57(1-5): 108–115.

Yang, S., R.S.S. Wu and R.Y.C. Kong. 2002. Biodegradation and enzymatic responses in the marine diatom *Skeletonema costatum* upon exposure to 2,4-dichlorophenol. Aquatic Toxicology 59(3-4): 191–200.

Yue, C.-J. and Y. Jiang. 2009. Impact of methyl jasmonate on squalene biosynthesis in microalga *Schizochytrium mangrovei*. Process Biochemistry 44(8): 923–927.

Analysis by Vibrational Spectroscopy of Seaweed with Potential Use in Food, Pharmaceutical and Cosmetic Industries

Leonel Pereira[1],* and *Paulo J.A. Ribeiro-Claro*[2]

1 Introduction

1.1 Vibrational Spectroscopy

In order to determine the chemical nature of the compounds present in seaweed, vibrational spectroscopy arises as a useful tool, as it can reveal detailed information concerning the properties and structure of materials at a molecular level. Until now, this type of analysis required the extraction of polysaccharides and other compounds, through lengthy and complicated procedures. With the development of FTIR diffuse-reflectance spectroscopy (DRIFTS) it became possible to directly analyze ground, dried seaweed material (Chopin and Whalen 1993). On the other hand, the development of Raman Spectroscopy with affordable low energy lasers (FT-Raman) allowed the use of this complementary technique in the study of the same ground samples.

[1] IMAR-CMA (Institute of Marine Research) and MARE (Marine and Environmental Sciences Centre), Department of Life Sciences, FCTUC, University of Coimbra, 3001-455 Coimbra, Portugal.
Email: leonel.pereira@uc.pt
[2] Department of Chemistry—CICECO, University of Aveiro, 3810-193 Aveiro, Portugal.
* Corresponding author

Pereira and collaborators (2003, 2006, 2009, 2013) developed an analysis technique based on FTIR-ATR (attenuated total reflectance) and FT-Raman spectroscopy, which allowed for the accurate identification of diverse polysaccharides (namely the phycocolloids) and other natural compounds present in seaweeds.

1.1.1 Infrared spectroscopy

Infrared spectroscopy was until recently the most widely used vibrational technique for studying natural products. In the Fourier transform instruments, all frequencies are scanned simultaneously, making data collection extremely rapid, allowing to collect several scans in a short time, and resulting in an improved signal-to-noise ratio of final spectrum. In the spectrometer, an infrared source emits radiation with a range of frequencies that is then passed through the sample. Particular chemical bands in the sample absorb infrared radiation of specific frequencies from the beam (Fig. 1), and this is plotted as an absorbance spectrum against wave number.

This technique presents several advantages, the most important one being its sensitivity: the infrared absorption is efficient and a good infrared spectrum can be obtained from a small amount of sample.

Infrared absorption intensity depends on the change of dipole moment during the molecular vibration. In this way, the stronger IR signals are observed for vibrations involving polar functional groups (with permanent dipole moment), such as OH group (e.g., in water and carbohydrates)

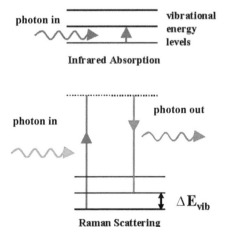

Figure 1. Diagram that illustrates infrared absorption and Raman scattering (Adapted from Pereira 2006).

Color image of this figure appears in the color plate section at the end of the book.

and C=O group (e.g., in aldehydes, ketones, and amides). This makes this technique quite suitable for studying these constituents and their intermolecular interactions (e.g., hydrogen-bonding) in organic and biological systems.

Usually, IR spectroscopy requires some sample preparation for transmission (e.g., dilution of sample in KBr pellets), but allows the use of pure samples, depending on the accessories used (ATR, IRRAS, DRIFTS).

1.1.2 Raman spectroscopy

In contrast with FTIR, the application of traditional Raman spectroscopy (Fig. 1) was limited until recently, due to the laser-induced fluorescence (strong background signal which is detected when some samples, such as biochemical compounds, are excited with visible lasers) and risk of sample destruction by the high energy of visible radiation. The use of Nd:YAG lasers operating at 1064 nm (infrared, lower energy radiation) has been generalized to decrease the fluorescence level and sample damage. Opto-electronic devices have progressed dramatically in the past decade as a consequence of major achievements in solid-state technology. As a result compact, efficient, and reliable diode lasers are now available from the visible to the infrared that have been demonstrated to work properly in Raman instruments in combination with suitable filter sets (Pereira 2006).

Raman spectroscopy comprises the family of spectral measurements made on molecular media based on inelastic scattering of monochromatic radiation. In a simplified description, part of the energy of the incident photon is left on the molecule, so that the scattered photon is of lower energy than the incident photon (Fig. 1). The energy difference between the incident and scattered photons is the exact measure of the energy difference between molecular vibrational levels—the same levels that are directly probed by infrared spectroscopy.

In Raman spectroscopy, the intensity of the radiation-matter interaction depends on the molecular polarizability change during the vibrational motion—which differs from the change of dipole moment that determines IR absorption. In this way, these two techniques provide complementary information: vibrational modes that are weak or not observed in one technique can usually be observed in the other.

1.2 Seaweed Polysaccharides

Many species of seaweed (marine macroalgae) are used as food and they have also found use in traditional medicine because of their perceived health benefits. Seaweeds are rich sources of sulphated polysaccharides, including some that have become valuable additives in the food industry because of their rheological properties as gelling and thickening agents (e.g., alginates, agar and carrageenan) (see Table 1). Sulphated polysaccharides are recognized to possess a number of biological activities

Table 1. Applications of Macroalgae Phycocolloids (adapted from van de Velde and de Ruiter 2002, Dhargalkar and Pereira 2005, Pereira 2004, 2008).

Use	Phycocolloid	Function
Food additives		
Baked food	Agar Kappa, Iota, Lambda	Improving quality and controlling moisture
Beer and wine	Alginate Kappa	Promotes flocculation and sedimentation of suspended solids
Canned and processed meat	Alginate Kappa	Hold the liquid inside the meat and texturing
Cheese	Kappa	Texturing
Chocolate milk	Kappa, lambda	Keep the cocoa in suspension
Cold preparation puddings	Kappa, Iota, Lambda	Thicken and gelling
Condensed milk	Iota, lambda	Emulsify
Dairy Creams	Kappa, iota	Stabilize the emulsion
Fillings for pies and cakes	Kappa	Give body and texture
Frozen fish	Alginate	Adhesion and moisture retention
Gelled water-based desserts	Kappa + Iota Kappa + Iota + CF	Gelling
Gums and sweets	Agar Iota	Gelling, texturing
Hot preparation flans	Kappa, Kappa + Iota	Gelling and improving taste
Jelly tarts	Kappa	Gelling
Juices	Agar Kappa, Lambda	Viscosity, emulsifier
Low calorie gelatins	Kappa + Iota	Gelling
Milk ice-cream	Kappa + GG, CF, X	Stabilize the emulsion and prevent ice crystals formation
Milkshakes	Lambda	Stabilize the emulsion
Salad dressings	Iota	Stabilize the suspension
Sauces and condiments	Agar Kappa	Thicken
Soymilk	Kappa + iota	Stabilize the emulsion and improve the taste
Cosmetics		
Shampoos	Alginate	Vitalization interface
Toothpaste	Carrageenan	Increase viscosity

Table 1. contd....

Table 1. contd.

Use	Phycocolloid	Function
Lotions	Alginate	Emulsification, elasticity and skin firmness
Lipstick	Alginate	Elasticity, viscosity
Medicinal and Pharmaceutical uses		
Dental mould	Alginate	Form retention
Laxatives	Alginate Carrageenan	Indigestibility and lubrication
Tablets	Alginate Carrageenan	Encapsulation
Metal poisoning	Carrageenan	Binds metal
HSV	Alginate	Inhibit virus
Industrial and Lab Uses		
Paints	Alginate	Viscosity and suspension, glazing
Textiles	Agar, Carrageenan	Sizing and glazing
Paper making	Alginate, Agar, Carrageenan	Viscosity and thickening
Analytical separation	Alginate, Carrageenan	Gelling
Bacteriological media	Agar	Gelling
Electrophoresis gel	Agar, Carrageenan	Gelling

Non-seaweed colloids: CF—Carob flour; GG—Guar gum; X—Xanthan

including anticoagulant, antiviral, and immune-inflammatory activities that might find relevance in nutraceutical/functional food, cosmetic and pharmaceutical applications (Pereira 2011, Table 2).

Some seaweeds produce hydrocolloids, associated with the cell wall and intercellular spaces. Members of the red algae (Rhodophyta) produce galactans (e.g., carrageenans and agars) and the brown algae (Heterokontophyta, Phaeophyceae) produce uronates (alginates) and other sulphated polysaccharides (e.g., fucoidan and laminaran) (Peat et al. 1958, Rinaudo 2002, Costa et al. 2010, Rioux et al. 2010, Ale et al. 2011, Jiao et al. 2011).

The different phycocolloids used in food industry as natural additives are (European codes of phycocolloids):

- Alginic acid–E400
- Sodium alginate–E401
- Potassium alginate–E402
- Ammonium alginate–E403
- Calcium alginate–E404
- Propylene glycol alginate–E405

Table 2. Summary of bioactive activity of some seaweed compounds (polysaccharides and phenolic compounds).

Category	Compounds	Seaweed source	Potential health benefit	Reference
Polyphenols	Flavonoids	*Palmaria palmata*	At high experimental concentrations that would not exist *in vivo*, the antioxidant abilities of flavonoids *in vitro* are stronger than those of vitamin C and E	Bagchi et al. 1999, Yuan et al. 2005
	Phlorotannins	Brown algae	Antioxidant activity of polyphenols extracted from brown and red seaweeds has already been demonstrated by *in vitro* assays; anti-inflammatory effect	Nakamura et al. 1996, Shin et al. 2006, Shibata et al. 2008, Wijesekara et al. 2010, Ngo et al. 2011
			Algicidal and bactericidal effect	Nagayama et al. 2002, 2003, Ngo et al. 2011
Polysaccharides and dietary fibers	Agars, carrageenans, ulvans and fucoidans	Red and brown algae	These polysaccharides are not digested by humans and therefore can be regarded as dietary fibers	Lahaye and Thibault 1990, Lahaye 1991, Costa et al. 2010
	Ulvan	*Ulva pertusa*	Antihyperlipidemic effects	Fujiwara-Arasaki et al. 1984, Pengzhan et al. 2003
	Carrageenan, fucoidan	Red algae (carrageenophytes), *Undaria pinnatifida*, brown algae	Antitumor and anti-viral	Haijin et al. 2003, Pengzhan et al. 2003, Choosawad et al. 2005, Nisizawa 2006, Hemmingson et al. 2006,Yuan and Song 2005, Yuan et al. 2010, Ngo et al. 2011
	Carrageenan (lambda, iota and nu variants)	Red algae (carrageenophytes)	Anti-viral, anti-HSV and anti-HIV	Spieler 2002, Smit 2004, Jiao et al. 2011

Table 2. contd....

Table 2. contd.

Category	Compounds	Seaweed source	Potential health benefit	Reference
	Fucoidan	Brown algae	Anticoagulant and antithrombotic activity	Smit 2004, Li et al. 2008, Ngo et al. 2011
			Antitumor and immunomodulatory activity	Choosawad et al. 2005, Li et al. 2008, Kim et al. 2010
			Antiviral and anti-HIV	Sugawara et al. 1989, Béress et al. 1993, Witvrouw and de Clercq 1997, Feldman et al. 1999, Smit 2004, Li et al. 2008, Cornish and Garbary 2010
	Fucoidan	*Fucus vesiculosus* *Saccharina japonica*	Hypolipidemic effect	Vázquez-Freire et al. 1996, Huang et al. 2010

Abbreviations: HIV—Human immunodeficiency virus; HSV—Herpes simplex virus

- Agar–E406
- Carrageenan–E407
- Semi-refined carrageenan or "processed *Eucheuma* seaweed"–E407A

1.2.1 Ulvan

Ulvan represents 8–29% of the algae dry weight and is produced by species belonging to the phylum Chlorophyta (green algae), mostly belonging to the class Ulvophyceae (Robic et al. 2009). It is mainly made up of disaccharide repeating sequences composed of sulfated rhamnose and glucuronic acid, iduronic acid, or xylose (Percival and McDowell 1967, Quemener et al. 1997). The two major repeating disaccharides are aldobiuronic acids designated as: type A, ulvanobiuronic acid 3-sulfate (A3s) and type B, ulvanobiuronic acid 3-sulfate (B3s) (Fig. 2). Partially sulfated xylose residues at O-2 can also occur in place of uronic acids (Fig. 2). Low proportions of galactose, glucose, mannose, and protein are also generally found in ulvan. Aditionally, minor repeat units have been reported that contain sulfated xylose replacing the iduronic acid or glucuronic acid (Lahaye and Robic 2007, Jiao et al. 2011).

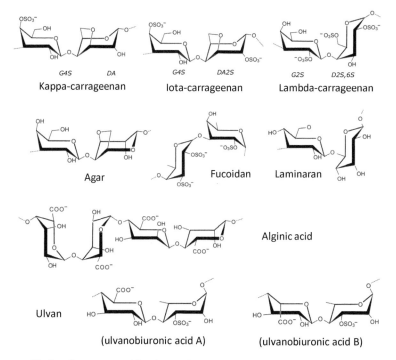

Figure 2. Idealized structures of the chemical units of kappa-, iota-, and lambda-carrageenan, agar, alginic acid, fucoidan, laminaran and ulvan.

1.2.2 Laminaran

Laminaran (Fig. 2) is a small glucan present in either soluble or insoluble form. The first form is characterized by complete solubility in cold water, while the other is only soluble in hot water (Kylin 1913, Chevolot et al. 2001). This polysaccharide is composed of D-glucose with β-(1,3) linkages, with β-(1,6) intra-chain branching (Barry 1939, Peat et al. 1958, Rioux et al. 2010).

1.2.3 Fucans

Fucans (Fig. 2) are sulphated polysaccharides that are composed of a fucose backbone. One of the best studied fucans from brown algae is fucoidan, which was first isolated by Kylin in 1913 (Kylin 1913). The fucoidan from *Fucus vesiculosus* has been available commercially for decades (Sigma-Aldrich Chemical Company, St. Louis). Early work on its structure showed that it contained primarily (1→2) linked 4-O-sulphated fucopyranose residues. However, 3-linked fucose with 4-sulphated groups was subsequently reported to be present on some of the fucose residues. Chevolot and colleagues reported that the fucoidan from *F. vesiculosus* and *Ascophyllum nodosum* contains a predominant disaccharide motif containing sulphate at the 2-position of the 3-linked fucose and sulphate groups on the 2- and 3-positions of the 4-linked fucose (Chevolot 2001).

1.2.4 Alginates

Alginic acid (Fig. 2) was discovered in 1883 by E.C.C. Stanford, a British pharmacist who called it algin. In seaweeds, algin is extracted as a mixed salt of sodium and/or potassium, calcium and magnesium. The exact composition varies with algal species. Since Stanford discovered algin, the name has been applied to a number of substances, e.g., alginic acid and all alginates, derived from alginic acid. The extraction process is based on the conversion of an insoluble mixture of alginic acid salts of the cell wall in a soluble salt (alginate) which is appropriate for water polysaccharide is derived from several genera of brown algae (e.g., mixed Fucales and Laminariales) that are utilized as raw materials by commercial alginate producers; these include *Macrocystis, Laminaria, Lessonia, Ascophyllum, Alaria, Ecklonia, Eisenia, Nereocystis, Sargassum, Cystoseira* and *Fucus*, with *Macrocystis pyrifera* and *Ascophyllum nodosum* being the principal sources of the world's alginate supply. The intercellular mucilage in these seaweeds has been regarded as the principal site of algin although it has also been found to occur in the cell walls. Alginic acid is a complex organic compound composed of D-mannuronic acid and L-guluronic acid monomers (Sartori et al. 1997, Rinaudo 2002, Pereira 2004).

1.2.5 Agar

Agar (Fig. 2) was the first colloid to be developed and has applications as a gelling agent for food and also as an inert support medium for microbial culture. This polysaccharide is the dried hydrophilic, colloidal substance extracted commercially from certain marine algae of the phylum Rhodophyta. The most important commercial agarophyte genera are *Gelidium*, *Pterocladiella*, *Gelidiella* and *Gracilaria*. Agar has also been found in species of *Ceramium*, *Phyllophora*, *Ahnfeltia*, *Campylaephora*, *Acanthopeltis* and *Gracilariopsis*. It is a polysaccharide, consisting primarily of D- and L-galactose units. About every tenth D-galactopyranose unit contains a sulphate ester group. Calcium, magnesium, potassium or sodium cations are also associated with the polysaccharide. Agar may be separated into two fractions. One is a neutral polymer, agarose, composed of repeating units, referred to as agarobiose, of alternating 1,3-linked β-D-galactopyranose and 1,4-linked 3,6-anhydro-α-L-galactopyranose. The second fraction has agaropectin, a more complicated structure. It contains residues of sulphuric, pyruvic, and uronic acids, in addition to D-galactose and 3,6-anhydro-L-galactose (Lahaye 2001b, Bixler and Porse 2011).

1.2.6 Carrageenan

Carrageenan and agar (Fig. 2) are the principal sulphated polysaccharides produced by red seaweeds (Rhodophyta); the main difference between the highly sulphated carrageenans from the less sulphated agars is the presence of D-galactose and anhydro-D-galactose in carrageenans and of D-galactose, L-galactose or anhydro-L-galactose in agars.

The structure of the various types of carrageenans is defined by the number and position of sulphate groups, the presence of 3,6-anhydro-D-galactose and conformation of the pyranosidic ring. There are about fifteen idealized carrageenan structures traditionally identified by Greek letters (Chopin et al. 1999).

The commercial carrageenans are normally divided into three main types: kappa, iota and lambda-carrageenan. Generally, seaweeds do not produce these idealized and pure carrageenans, but more likely a range of hybrid structures. The precursors (mu and nu), when exposed to alkali conditions, are modified into kappa and iota, respectively, through formation of the 3,6-anhydro-galactose bridge (Bixler and Porse 2011).

Different types of carrageenan are obtained from different species of the Gigartinales (Rhodophyta). Kappa-carrageenan is predominantly obtained by extraction from the cultivated tropical seaweed *Kappaphycus alvarezii* (know in the trade as "cottonii"). *Eucheuma denticulatum* (commonly referred to as "spinosum" in the trade) is the main species for the production of iota-carrageenan (Pereira 2013). Lambda-carrageenan is obtained from different species from the genera *Gigartina* and *Chondrus* (trade name "Irish moss") (van de Velde and de Ruiter 2002, Pereira 2013).

The rheological properties of the gelling carrageenans (e.g., kappa and iota) are quite distinct: the kappa-type forms gels that are hard, strong and brittle, whereas iota-carrageenan forms soft and weak gels. The common feature of these carrageenans is the anhydro-galactose bridge of the 4-linked galactose residue, respectively DA and DA2S, which adopts the $^1C_4$-chair conformation. This conformation is crucial for the formation of the helical structure and, thereby, for the ability to form a gel. Lambda-carrageenan and the precursors mu- and nu-carrageenan lack the 3,6-anhydro bridge and, therefore, the 4-linked residue adopts the $^4C_1$-chair conformation, which disturbs the helical conformation. Thus, lambda-carrageenan acts simply as a thickening agent (van de Velde and de Ruiter 2002).

1.3 Phenolic Compounds

Phenolic compounds are commonly found in plants, reportedly having several biological activities (see Table 2), including antioxidant properties. Earlier reports have revealed that seaweed extracts, especially polyphenols, have antioxidant activity (Lim et al. 2002, Kuda et al. 2005, Martins et al. 2013). Phenolic compounds are regarded for their important dietary roles as antioxidants and chemo preventive agents (Bravo 1998).

Antioxidants can be classified into two major groups; that is, enzymatic and nonenzymatic antioxidants. Some of these antioxidants are endogenously produced, including enzymes, low-molecular-weight molecules, and enzyme cofactors. Many nonenzymatic antioxidants are obtained from dietary sources. Dietary antioxidants can be classified into various classes, of which polyphenols is the largest one (Bunaciu et al. 2012).

2 Analysis of Seaweeds by Vibrational Spectroscopy

The combined use of FTIR-ATR (see Table 3 and 4) and FT-Raman (see Table 5) spectroscopy analysis can be used for the identification of main seaweed polysaccharides, namely, alginates, fucoidan, laminaran, agars, kappa-, iota-, and lambda-carrageenans (Pereira et al. 2009, 2013). Infrared spectroscopic techniques can also be used for the identification of phenolic content and total antioxidant capacity of some compounds from seaweeds (Bunaciu et al. 2012, Vijayabaskar and Shiyamala 2012, Rajauria et al. 2013).

Therefore, vibrational spectroscopy (FTIR-ATR and FT-Raman) is proposed as a useful tool for the cosmetic, pharmaceutical and food industry

Table 3. Peaks found in seaweed FTIR spectra with their attributed bonds (Adapted from Chopin et al. 1999, Pereira et al. 2009, Souza et al. 2012).

Wave numbers (cm⁻¹)	Bonds/Assignments
3500	O-H
2960	CH_2
2900–2920	C-H (good reference for total sugar content)
2845	$O-CH_3$ (shoulder on the band at 2920 in highly methylated agars)
1725	COOH
1690–1695	Amide I from proteins
1640–1650	H_2O and proteins CO-NH/amide II from proteins
1605	Carboxylate anion of pyruvate
1450	Ester-sulfate
1420	Amide III from proteins
1370–1320	Ester-sulfate
1210 <1240 <1260	S=O of ester-sulfate (good indicator for total sulfate content)
1180	P-O-C (alkyl substituent's suggesting organic phosphates)
1150	Ester-sulfate
1040–1080	Skeleton of galactans
1070	C-O of 3,6-anhydrogalactose (shoulder)
1065	Gelling type carrageenans
1040	C-O of ester-sulfate and hydroxyl
1037–1071	Symmetric C-O vibration associated with a $C-O-SO_3$ of heterofucans
1020	Non-gelling type carrageenans
1000–1200	Sulfates and floridean starch
970–975	Galactose: peak with alkali modified iota carrageenan, small peak with unmodified iota, and also present in agars
930–940	Vibrations of the C-O-C of 3,6-anhydrogalactose
905	$C-O-SO_4$ on C_2 of 3,6-anhydrogalactose (shoulder)
890–900	Unsulfated β-D-galactose (or with 6–0-methylgalactose or with pyruvate); agar specific band
867	$C-O-SO_4$ on C_6 of galactose (shoulder, indicates precursors)
845–850	$C-O-SO_4$ on C_4 of galactose/floridean starch
825–830	$C-O-SO_4$ on C_2 of galactose (narrow when xi-carrageenan present)
820	Galactose 6-sulfate
815–820	$C-O-SO_4$ on C_6 of galactose
805	$C-O-SO_4$ on C_2 of 3,6-anhydrogalactose
790	Characteristic of agar-type in second derivative spectra
730–750	C-S/C-O-C bending mode in glycosidic linkages of agars
717	Characteristic of agar-type in second derivative spectra/C-O-C bending mode in glycosidic linkages of agars
705	$C-O-SO_4$ on C_4 of galactose
580	S-O in sulfated galactans

Table 4. Identification of Carrageenan types by Infrared Spectroscopy (Adapted from Chopin et al. 1999, Pereira et al. 2009).

Wave numbers (cm^{-1})	Bond(s)/Group(s)	Letter Code	Type of carrageenan							
			Kappa (κ)	Mu (μ)	Iota (ι)	Nu (ν)	Beta (β)	Theta (θ)	Lambda (λ)	Xi (ξ)
1240–1260	S=O of sulphate esters		+	++	++	+++	-	++	+++	++
1070	C-O of 3,6-anhydrogalactose	DA	+	-	+	-	+	+	-	-
970–975	Galactose	G/D	+	s	+	s	+	+	-	-
930	C-O of 3,6-anhydrogalactose	DA	+	-	+	-	+	+	-	-
905	C-O-SO$_3$ on C$_2$ of 3,6-anhydrogalactose	DA2S	-	-	+	-	-	+	-	-
890–900	Unsulphated β-D-galactose	G/D	-	-	-	-	+	-	-	-
867	C-O-SO3 on C$_6$ of galactose	G/D6S	-	+	-	+	-	-	+	-
845	C-O-SO3 on C$_4$ of galactose	G4S	+	+	+	+	-	-	-	-
825–830	C-O-SO3 on C$_2$ of galactose	G/D2S	-	-	-	+	-	+	+	n
815–820	C-O-SO3 on C$_6$ of galactose	G/D6S	-	+	-	+	-	-	+	-
805	C-O-SO3 on C$_2$ of 3,6-anhydrogalactose	DA2S	-	-	+	-	-	+	-	-

-, absent; +, medium; ++, strong; +++, very strong; s, shoulder peak; n, narrow peak.

Table 5. Identification of Carrageenan Types by Raman Spectroscopy (adapted from Pereira 2006, Pereira et al. 2009).

Wave numbers (cm⁻¹)	Bond(s)/Group(s)	Letter code	Type of carrageenan							
			Kappa (κ)	Mu (μ)	Iota (ι)	Nu (ν)	Beta (β)	Theta (θ)	Lambda (λ)	Xi (ξ)
1240–1260	S=O of sulphate esters		++	++	++	+++	-	++	++	++
1075–1085	C-O of 3,6-anydrogalactose	DA	+++	-	+++	-	+	+	-	-
970–975	Galactose	G/D	+	+	s	s	+	+	-	-
925–935	C-O of 3,6-anydrogalactose	DA	+	-	+	-	+	+	-	-
905–907	C-O-SO$_4$ on C$_2$ of 3,6-anydrogalactose	DA2S	-	-	+	-	-	+	+	+
890–900	Un-sulphated β-D-galactose	G/D	-	-	-	-	+	-	-	-
867–871	C-O-SO$_4$ on C$_6$ of galactose	G/D6S	-	s	-	+	-	-	+	-
845–850	C-O-SO$_4$ on C$_4$ of galactose	G4S	++	+	++	+	-	+	+	+
825–830	C-O-SO$_4$ on C$_2$ of galactose	G/D2S	-	-	-	+	-	+	+	-
815–825	C-O-SO$_4$ on C$_6$ of galactose	G/D6S	-	s	-	s	-	-	+	+
804–808	C-O-SO$_4$ on C$_2$ of 3,6-anhydrogalactose	DA2S	-	-	++	-	-	+	-	-

–, absent; +, medium; ++, strong; +++, very strong; s, shoulder peak.

to check the phycocolloid quality of a raw seaweed material by a quick and non-destructive method (Pereira 2004, Pereira et al. 2009).

2.1 Polysaccharides from Brown Algae

The main polysaccharide found in studied brown seaweeds (Phaeophyceae) was alginate, a linear copolymer of mannuronic (M) and guluronic acid (G). Different types of alginic acid present different proportions and/or different alternating patterns of guluronic (G) and mannuronic (M) units. The presence of these acids can be identified from their characteristic bands in the vibrational spectra; in accordance with Mackie (1971) these phycocolloids show two characteristic bands in IR spectra: 808 cm^{-1}, assigned to M units, and 787 cm^{-1}, assigned to G units. However, Chandia and co-workers, in work with specimens of *Lessonia* genus, assign both bands to G units (Chandia et al. 2001, 2004). Filipov and Kohn (1974) propose that M/G ratios of the different samples can be estimated from the ratio of absorbance of the bands at 1320 and 1290 cm^{-1} in FTIR spectra. According Sakugawa and collaborators (2004) the M/G concentration ratio characterizing a certain alginate sample can be inferred from the relative intensity ratio of the two bands 1030/1080 cm^{-1}, in calcium alginate and 1019/1025 cm^{-1}, in manganese alginate. In accordance with the same authors, the absorbance at 1030 cm^{-1} directly reflects the change of mannurate concentration of calcium alginate and the 1025 cm^{-1} is attributed to the OH bending of guluronate (Sakugawa et al. 2004).

Alginate M/G ratio was tentatively estimated from the 1030/1080 cm^{-1} band ratio in infrared spectra, suggesting higher values of mannuronic than guluronic acid blocks (M/G> 1) in *Himanthalia elongata* (Pereira et al. 2013). However, the FTIR spectra of *Saccorhiza polyschides* show an intense broad band centered at 1025 cm^{-1}, indicating that the samples considered are particularly rich in guluronic acid. According to several works (Skriptsova et al. 2004, Torres et al. 2007, Sahayaraj et al. 2012, Pereira et al. 2013), the spectrum of *U. pinnatiffida* (old adult thallus) indicates that the relative amounts of both mannuronate and guluronate residues are similar.

The spectra presented by Pereira and co-workers (2013), suggesting higher values of guluronic than mannuronic acid blocks in *Padina pavonica* and similar amounts of both mannuronate and guluronate residues in *Sargassum vulgare*, are in accordance with other published works (Skriptsova et al. 2004, Torres et al. 2007, Sahayaraj et al. 2012).

Some brown algae, such *Saccorhiza polyschides* and *Undaria pinnatifida*, also exhibit a broad band around 1220–1260 cm^{-1}, assigned to the presence of sulphate ester groups (S=O) which is a characteristic component in fucoidan and sulphated polysaccharides other than alginate in brown seaweeds. *Padina pavonica* and *Sargassum vulgare* also exhibit a broad band in this region (around 1195–1237 cm^{-1} for *Padina* and 1210–1280 cm^{-1} for *Sargassum*) assigned to (S=O). However, *Sargassum vulgare* contains a larger amount of fucoidan than *Padina pavonica* (Pereira et al. 2013). According to Camara and co-workers (2011), characteristic sulfate absorptions were identified in the FTIR spectra of heterofucans: bands around 1239–1247 cm^{-1} for asymmetric S=O stretching vibration and bands around 1037–1071 cm^{-1} for symmetric C-O vibration associated with a C-O-SO$_3$ group. The peaks at 820–850 cm^{-1} were assigned to the bending vibration of C-O-S. However, *Sargassum vulgare* and *Padina pavonica* contain little amounts of laminaran (Peat et al. 1958, Nelson and Lewis 1974, Rioux et al. 2010, Jiao et al. 2011).

2.2 Identification of Ulvan

A typical infrared spectrum of ulvan shows strong absorbance at about 1650, 1250, and 1070 cm^{-1} and small ones at about 1400, 850, and 790 cm^{-1} (Ray and Lahaye 1995, Robic et al. 2009). Some of these bands are easily assigned to carboxylate groups and to sulfate esters. Carboxylate groups show two bands, an asymmetrical stretching band near 1650 cm^{-1} and a weaker symmetric stretching band near 1400 cm^{-1}, and sulfate esters show a major band at about 1250 cm^{-1}. Two other bands at 850 and 790 cm^{-1}, likely related to sugar cycles, are also observed. The 1200–1000 cm^{-1} region is dominated by sugar ring vibrations overlapping with stretching vibrations of (C–OH) side groups and the (C–O–C) glycosidic bonds vibration. All ulvan spectra presented a maximum absorption band at around 1055 cm^{-1} which are likely due to C–O stretching from the two main sugars, rhamnose and glucuronic acid: their individual IR spectra present a maximum at the same wavenumber, 1055 cm^{-1} (Robic et al. 2009).

2.3 Identification of Agar and Agarocolloids

Agars differ from carrageenans as they have the L-configuration for the 4-linked galactose residue; nevertheless, they have some structural similarities with carrageenans. The characteristic broad band of sulphate esters, between 1210 to 1260 cm^{-1} (Chopin et al. 1999), is much stronger in carrageenan than in agar. Especially in the anomeric region (700–950 cm^{-1}), agar and carrageenan show several similar bands (see Table 2). Thus, the strong IR band at 930 cm^{-1} assigned to the presence of 3,6-anhydro-galactose was common to agar and carrageenans; the band at 890 cm^{-1} corresponded to anomeric CH of β-galactopyranosyl residues and Raman bands at 770 and 740 cm^{-1} (strong in the FT-Raman spectra and weak in the FTIR-ATR) are assigned to the skeleton bending of pyranose ring (Matsuhiro 1996, Pereira et al. 2003), both in agar and carrageenans. Also, the bands at 1010–1030 cm^{-1}

may be assigned to C–O and C–C stretching vibrations of pyranose ring common to all polysaccharides. So, the main polysaccharide composition of *Gelidium corneum* and of *Pterocladiella capillacea* (Rhodophyta) is agar (Pereira et al. 2003, 2013).

A typical FT-Raman spectrum of agar shows a strong band centered at 837 cm^{-1}, which is absent in the FTIR spectra (Pereira et al. 2003, 2013). According to Matsuhiro (1996), this band is associated with the CH vibration coupled with C-OH related modes of α residues. Moreover, the spectral feature at 890 cm^{-1}, also particularly intense in the FT-Raman spectra, is mainly associated with vibrational modes of the β-galactose residues.

Laurencia obtusa (Rhodophyta) presents a complex agar-like sulfated galactan. These polysaccharides belong to the agar group, being agarose derivatives with a rather high content in sulfate groups and with a reduced amount of 3,6-anhydro-L-galactose residues (700–950 cm^{-1}) (Usov and Elashvili 1991, Pereira et al. 2013).

2.4 Identification of Carrageenan

In the works of Pereira and co-workers, the FTIR-ATR and FT-Raman spectra of *Kappaphycus alvarezii* were compared with those of commercial kappa-carrageenan (Pereira et al. 2003, 2013). The spectra of the ground seaweed show the same main features of commercial kappa-carrageenan: a strong Raman band at approximately 845 cm^{-1} (with moderate intensity in the IR spectrum), which is assigned to D-galactose-4-sulphate (G4S), and a relatively strong band at approximately 930 cm^{-1} in the FTIR-ATR spectra, weak in FT-Raman spectrum, indicating the presence of 3,6-anhydro-D-galactose (DA) (Pereira 2004, Pereira et al. 2009).

According to Pereira et al. (2003, 2013) the spectra of iota-carrageenan and of *Calliblepharis jubata* show bands at approximately 930 and 845 cm^{-1}, with the same intensity pattern as in kappa-carrageenan. However, an additional well-defined feature is visible around 805 cm^{-1} in both IR and Raman spectra. This band, indicating the presence of sulphate ester in the 2-position of the anhydro-D-galactose residues (DA2S), is a characteristic band of the iota-carrageenan (Pereira 2004, Pereira et al. 2009).

The vibrational spectra of lambda-carrageenan and ground *Chondrus crispus* tetrasporophytes indicate high sulphate content, evidenced from the presence of a broad band between 820 and 830 cm^{-1} in FTIR-ATR spectra (see Table 4 and 5). The *C. crispus* and lambda-carrageenan FT-Raman spectra show the two combined weak bands between 815 and 830 cm^{-1} (Pereira 2004, Pereira et al. 2009).

2.5 Identification of Phenolic Compounds

The great antioxidant capacity of the phenolic compounds, from certain seaweeds (e.g., *Turbinaria conoides, T. ornata, Himanthalia elongata*— Phaeophyceae; and *Gracilaria follifera*—Rhodophyta), when compared with the standard Gallic acid activity, was evidenced in some recent works (Meenakshi et al. 2009, Devi et al. 2011, Bunaciu et al. 2012, Vijayabaskar and Shiyamala 2012, Rajauria et al. 2013).

The FTIR spectrum of the standard Gallic acid contain ten major peaks, at approximately 3366, 3283, 3065, 2654, 1703, 1618, 1541, 1449, 1099 and 1026 cm^{-1}. The absorption peaks observed for hydroxyl groups (around 3300–3500 cm^{-1}) and aromatic ring (around 1450–1470 cm^{-1} and 2850–2960 cm^{-1}) in the spectra of *T. ornata* also suggest the presence of phenolic compounds (Vijayabaskar and Shiyamala 2012).

In FTIR analysis of seaweed (*T. conoides*) extracts, the characteristic absorption of sulphate and carboxyl groups were identified. The peaks at 3537–3396 and 3419 cm^{-1} and 3419 cm^{-1} were assigned to the stretching vibration of O-H of strong relative strength in axial position, the stretching vibration of C-H, and C=O stretching mode, respectively. Bands at 3536–3440 cm^{-1} correspond to stretching vibration of O-H (Devi et al. 2011).

3 Conclusion

In respect to the methodology used in phycocolloid analysis, the development of Fourier transform infrared spectroscopy (FTIR) and of Fourier transform laser Raman spectroscopy (FT-Raman) has produced great advances in structural study of polysaccharides (Chopin and Whalen 1993, Cáceres et al. 1996, Matsuhiro 1996, Pereira 2006, Pereira et al. 2009).

FT-Raman spectroscopy in the solid state gives well-defined characteristic spectra of phycocolloids (Matsuhiro 1996, Pereira et al. 2003, Pereira 2006, Pereira et al. 2009). The analysis of ground-dried seaweed by FT-Raman possesses all the great advantages of the analysis of ground algal material with FTIR-ATR, and is a rapid and simple technique; it requires only small amounts of non-manipulated samples, which allows the most accurate determination of the native composition of the algal polysaccharides and phenolic compounds (Pereira et al. 2009, Bunaciu et al. 2012, Pereira et al. 2013).

Acknowledgments

The authors acknowledge the financial support provided by the Portuguese Foundation for Science and Technology (FCT), the European Union, QREN,

FEDER through "Programa Operacional Factores de Competitividade (COMPETE)", IMAR-CMA, and Laboratório Associado Centro de Investigação em Materiais Cerâmicos e Compósitos, CICECO (FCOMP-01-0124-FEDER-037271—Ref. FCT PEst-C/CTM/LA0011/2013), for their general funding scheme.

References Cited

Ale, M.T., J.D. Mikkelsen and A.S. Meyer. 2011. Important determinants for fucoidan bioactivity: A critical review of structure-function relations and extraction methods for fucose-containing sulfated polysaccharides from brown seaweeds. Marine Drugs 9(10): 2106–2130.

Bagchi, M., M. Mark, W. Casey, B. Jaya, Y. Xumei, S. Sidney and D. Bagchi. 1999. Acute and chronic stress-induced oxidative gastrointestinal injury in rats, and the protective ability of a novel grape seed proanthocyanidin extract. Nutrition Research 19: 1189–1199.

Barry, V. 1939. Constitution of laminarin—isolation of 2, 4, 6-trimethylglucopyranose. Scientific Proceedings of Royal Dublin Society 22: 59–67.

Béress, A., O. Wassermann, T. Bruhn, L. Béress, E.N. Kraiselburd, L.V. Gonzalez, G.E. de Motta and P.I. Chavez. 1993. A new procedure for the isolation of anti-HIV compounds (polysaccharides and polyphenols) from the marine alga *Fucus vesiculosus*. Journal of Natural Products 56: 478–488.

Bixler, H.J. and H. Porse. 2011. A decade of change in the seaweed hydrocolloids industry. Journal of Applied Phycology 23: 321–335.

Bravo, L. 1998. Polyphenols: Chemistry, dietary sources, metabolism, and nutritional significance. Nutrition Reviews 56(11): 317–333.

Bunaciu, A.A., H.Y. Aboul-Enein and S. Fleschin. 2012. FTIR Spectrophotometric Methods used for antioxidant activity assay in medicinal plants. Applied Spectroscopy Reviews 47(4): 245–255.

Cáceres, P.J., C.A. Faundez, B. Matsuhiro and J.A. Vasquez. 1996. Carrageenophyte identification by second-derivative Fourier transform infrared spectroscopy. Journal of Applied Phycology 8: 523–527.

Camara, R.B.G., L.S. Costa, G.P. Fidelis, L.T.D.B. Nobre, N. Dantas-Santos, S.L. Cordeiro, M.S.S.P. Costa, L.G. Alves and H.A.O. Rocha. 2011. Heterofucans from the Brown Seaweed *Canistrocarpus cervicornis* with Anticoagulant and Antioxidant Activities. Marine Drugs 9: 124–138.

Chandia, N.P., B. Matsuhiro and A.E. Vasquez. 2001. Alginic acids in *Lessonia trabeculata*: characterization by formic acid hydrolysis and FT-IR spectroscopy. Carbohydrate Polymers 46: 81–87.

Chandia, N.P., B. Matsuhiro, E. Mejias and A. Moenne. 2004. Alginic acids in *Lessonia vadosa*: Partial hydrolysis and elicitor properties of the polymannuronic acid fraction. Journal of Applied Phycology 16: 127–133.

Chevolot, L., B. Mulloy, J. Ratiskol, A. Foucault and S. Colliec-Jouault. 2011. A disaccharide repeat unit is the major structure in fucoidans from two species of brown algae. Carbohydrate Research 330(4): 529–535.

Choosawad, D., U. Leggat, C. Dechsukhum, A. Phongdara and W. Chotigeat. 2005. Anti-tumour activities of fucoidan from the aquatic plant *Utricularia aurea* lour. Songklanakarin Journal of Science and Technology 27: 799–807.

Chopin, T. and E. Whalen. 1993. A new and rapid method for carrageenan identification by FT-IR diffuse-reflectance spectroscopy directly on dried, ground algal material. Carbohydrate Research 246: 51–59.

Chopin, T., B.F. Kerin and R. Mazerolle. 1999. Phycocolloid chemistry as taxonomic indicator of phylogeny in the Gigartinales, Rhodophyceae: A review and current developments using Fourier transform infrared diffuse reflectance spectroscopy. Phycological Research 47:167–188.

Cornish, M.L. and D.J. Garbary. 2010. Antioxidants from macroalgae: potential applications in human health and nutrition. Algae 25: 155–171.

Costa, L.S., G.P. Fidelis, S.L. Cordeiro, R.M. Oliveira, D.A. Sabry, R.G.B. Camara, L. Nobre, M. Costa, J. Almeida-Lima, E.H.C. Farias, E.L. Leite and H.A.O. 2010. Biological activities of sulfated polysaccharides from tropical seaweeds, Biomedicine & Pharmacotherapy 64(1): 21–28.

Devi, G.K., K. Manivannan, G. Thirumaran, F.A.A. Rajathi and P. Anantharaman. 2011. *In vitro* antioxidant activities of selected seaweeds from Southeast coast of India. Asian Pacific Journal of Tropical Medicine 4: 205–211.

Dhargalkar, V.K. and N. Pereira. 2005. Seaweed: promising plant of the millennium. Science and Culture 71: 60–66.

Feldman, S.C., S. Reynaldi, C.A. Stortz, A.S. Cerezo and E.B. Damont. 1999. Antiviral properties of fucoidan fractions from *Leathesia difformis*. Phytomedicine 6: 335–340.

Filippov, M.P. and R. Kohn. 1974. Determination of composition of alginates by infra-red spectroscopic methods. Chem. Zvesti 28: 817.

Fujiwara-Arasaki, T., N. Mino and M. Kuroda. 1984. The protein value in human nutrition of edible marine algae in Japan. Hydrobiologia 116/117: 513–516.

Haijin, M., J. Xiaolu and G. Huashi. 2003. A κ-carrageenan derived oligosaccharide prepared by enzymatic degradation containing anti-tumor activity. Journal of Applied Phycology 15: 297–303.

Hemmingson, J.A., R. Falshaw, R.H. Furneaux and K. Thompson. 2006. Structure and antiviral activity of the galactofucan sulfates extracted from *Undaria pinnatifida* (Phaeophyta). Journal of Applied Phycology 18: 185–193.

Huang, L., K. Wen, X. Gao and Y. Liu. 2010. Hypolipidemic effect of fucoidan from *Laminaria japonica* in hyperlipidemic rats. Pharmaceutical Biology 48: 422–426.

Jiao, G.L., G.L. Yu, J.Z. Zhang and H.S. Ewart. 2011. Chemical structures and bioactivities of sulfated polysaccharides from marine algae. Marine Drugs 9(2): 196–223.

Kim, K.J., O.H. Lee, H.H. Lee and B.Y. Lee. 2010. A 4-week repeated oral dose toxicity study of fucoidan from the sporophyll of *Undaria pinnatifida* in sprague-dawley rats. Toxicology 267: 154–158.

Kuda, T., M. Tsunekawa, H. Goto and Y. Araki. 2005. Antioxidant properties of four edible algae harvested in the Noto Peninsula, Japan. Journal of Food Composition and Analysis 18(7): 625–633.

Kylin, H. 1913. Biochemistry of sea algae. Zeitschrift für Physikalische Chemie 83: 171–197.

Lahaye, M. 1991. Marine algae as sources of fibres: determination of soluble and insoluble dietary fiber contents in some sea vegetables. Journal of the Science of Food Agriculture 54: 587–594.

Lahaye, M. 2001a. Chemistry and physico-chemistry of phycocolloids. Cahiers de Biologie Marine 42 (1-2): 137–157.

Lahaye, M. 2001b. Developments on gelling algal galactans, their structure and physico-chemistry. Journal of Applied Phycology 13: 173–184.

Lahaye, M. and J.F. Thibault. 1990. Chemical and physio-chemical properties of fibers from algal extraction by-products. pp. 68–72. *In*: Southgate, D.A.T., K. Waldron, I.T. Johnson and G.R. Fenwick (eds.). Dietary Fibre: Chemical and Biological Aspects. Royal Society of Chemistry, Cambridge.

Lahaye, M. and A. Robic. 2007. Structure and functional properties of Ulvan, a polysaccharide from green seaweeds. Biomacromolecules 8: 1765–1774.

Li, B., F. Lu, X. Wei and R. Zhao. 2008. Fucoidan: structure and bioactivity. Molecules 13: 1671–1695.

Lim, S.N., P.C.K. Cheung, V.E.C. Ooi and P.O. Ang. 2002. Evaluation of antioxidative activity of extracts from a brown seaweed, *Sargassum siliquastrum*. Journal of Agricultural and Food Chemistry 50(13): 3862–3866.

Lobban, C.S., D.J. Chapman and B.P. Kremer. 1988. Experimental Phycology: A Laboratory Manual Phycological Society of America. Cambridge University Press, Cambridge, 366 pp.

Mackie, W. 1971. Semi-quantitative estimation of composition of alginates by infra-red spectroscopy. Carbohydrate Research 20: 413–415.

Martins, C.D.L., F. Ramlov, N.P.N. Carneiro, L.M. Gestinari, B.F. dos Santos, L.M. Bento, C. Lhullier, L. Gouvea, E. Bastos, P.A. Horta and A.R. Soares. 2013. Antioxidant properties and total phenolic contents of some tropical seaweeds of the Brazilian coast. Journal of Applied Phycology 25(4): 1179–1187.

Matsuhiro, B. 1996. Vibrational spectroscopy of seaweed galactans. Hydrobiologia 327: 481–489.

Meenakshi, S., D. Manicka Gnanambigai, S. Tamil mozhi, M. Arumugam and T. Balasubramanian. 2009. Total flavanoid and *in vitro* antioxidant activity of two seaweeds from Rameshwaram Coast. Global Journal of Pharmacology 3: 59–62.

Nagayama, K., Y. Iwamura, T. Shibata, I. Hirayama and T. Nakamura. 2002. Bactericidal activity of phlorotannins from the brown alga *Ecklonia kurome*. Journal of Antimicrobial Chemotherapy 50: 889–893.

Nagayama, K., T. Shibata, K. Fujimoto, T. Honjo and T. Nakamura. 2003. Algicidal effect of phlorotannins from the brown alga *Ecklonia kurome* on red tide microalgae. Aquaculture 218: 601–611.

Nakamura, T., K. Nagayama, K. Uchida and R. Tanaka. 1996. Antioxidant activity of phlorotannins isolated from the brown alga *Eisenia bicyclis*. Fisheries Science 62: 923–926.

Nelson, T.E. and B.A. Lewis. 1974. Separation and characterization of the soluble and insoluble components of insoluble laminaran. Carbohydrate Research 33(1): 63–74.

Ngo, D.H., I. Wijesekara, T.S. Vo, Q.V. Ta and S.K. Kim. 2011. Marine food-derived functional ingredients as potential antioxidants in the food industry: An overview. Food Research International 44: 523–529.

Nisizawa, K. 2006. Seaweeds Kaiso—Bountiful harvest from the seas. *In*: Critchley, A., M. Ohno and D. Largo (eds.). World Seaweed Resources—An Authoritative Reference System: ETI Information Services Ltd. Hybrid Windows and Mac DVD-ROM. Amsterdam, The Netherlands. ISBN: 90-75000-80-4.

Peat, S., W.J. Whelan and H.G. Lawley. 1958. The structure of laminarin. Part I. The main polymeric linkage. Journal of the Chemical Society 724–728.

Pengzhan, Y., P.Z. Quanbin, L. Ning, X. Zuhong, W. Yanmei and L. Zhi'en. 2003. Polysaccharides from *Ulva pertusa* (Chlorophyta) and preliminary studies on their antihyperlipidemia activity. Journal of Applied Phycology 15: 21–27.

Percival, E.J.V. and R.H. McDowell. 1967. Chemistry and Enzymology of Marine Algal Polysaccharides. Academic Press, London, New York, 219 pp.

Pereira, L. 2004. Estudos em macroalgas carragenófitas (Gigartinales, Rhodophyceae) da costa portuguesa—aspectos ecológicos, bioquímicos e citológicos. PhD Thesis. Departamento de Botânica—FCTUC, Universidade de Coimbra, Coimbra, 293 pp.

Pereira, L. 2006. Identification of phycocolloids by vibrational spectroscopy. *In*: Critchley A.T., M. Ohno and D.B. Largo (eds.). World Seaweed Resources—An Authoritative Reference System. ETI Information Services Ltd, Amsterdam, The Netherlands, Hybrid Windows and Mac DVD-ROM. ISBN: 90-75000-80-4.

Pereira, L. 2008. As algas marinhas e respectivas utilidades, Monografias.com, 19 pp. http://br.monografias.com/trabalhos913/algas-marinhas-utilidades/algas-marinhas-utilidades.pdf.

Pereira, L. 2011. A Review of the nutrient composition of selected edible seaweeds. pp. 15–47. *In*: Pomin, V.H. (ed.). Seaweed: Ecology, Nutrient Composition and Medicinal Uses. Nova Science, New York, USA.

Pereira, L. 2013. Population Studies and Carrageenan Properties in Eight Gigartinales (Rhodophyta) from Western Coast of Portugal. The Scientific World Journal vol. 2013, Article ID 939830, 11 pp. doi:10.1155/2013/939830.

Pereira, L., A. Sousa, H. Coelho, A.M. Amado and P.J.A. Ribeiro-Claro. 2003. Use of FTIR, FT-Raman and ^{13}C-NMR spectroscopy for identification of some seaweed phycocolloids. Biomolecular Engineering 20: 223–228.

Pereira, L., A.M. Amado, A.T. Critchley, F. van de Velde and P.J.A. Ribeiro-Claro. 2009. Identification of selected seaweed polysaccharides (phycocolloids) by vibrational spectroscopy (FTIR-ATR and FT-Raman). Food Hydrocolloids 23(7): 1903–1909.

Pereira, L., F.S. Gheda and P.J.A. Ribeiro-Claro. 2013. Analysis by vibrational spectroscopy of seaweed polysaccharides with potential use in food, pharmaceutical and cosmetic industries. International Journal of Carbohydrate Chemistry vol. 2013, Article ID 537202, 7 pp. http://dx.doi.org/10.1155/2013/537202.

Quemener, B., M. Lahaye and C. Bobin-Dubigeon. 1997. Sugar determination in ulvans by a chemical-enzymatic method coupled to high performance anion exchange chromatography. Journal of Applied Phycology 9: 179–188.

Rajauria, G., A.K. Jaiswal, N. Abu-Gannam and S. Gupta. 2013. Antimicrobial, antioxidant and free radical-scavenging capacity of brown seaweed *Himanthalia elongata* from Western Coast of Ireland. Journal of Food Biochemistry 37(3): 322–335.

Ray, B. and M. Lahaye. 1995. Cell-wall polysaccharides from the marine green alga *Ulva* "rigida" (Ulvales, Chlorophyta). Extraction and chemical composition. Carbohydrate Research 274: 251–261.

Rinaudo, M. 2002. Alginates and carrageenans. Actualite Chimique (11-12): 35–38.

Rioux, L.-E., S.L. Turgeon and M. Beaulieu. 2010. Structural characterization of laminaran and galactofucan extracted from the brown seaweed *Saccharina longicruris*. Phytochemistry 71(13): 1586–1595.

Robic, A., D. Bertrand, J.F. Sassi, Y. Lerat and M. Lahaye. 2009. Determination of the chemical composition of ulvan, a cell wall polysaccharide from *Ulva* spp. (Ulvales, Chlorophyta) by FT-IR and chemometrics. Journal of Applied Phycology 21: 451–456.

Sahayaraj, K., S. Rajesh and J.M. Rathi. 2012. Silver nanoparticles biosynthesis using marine alga *Padina Pavonica* (Linn.) and its microbicidal activity. Digest Journal of Nanomaterials and Biostructures 7: 1557–1567.

Sakugawa, K., A. Ikeda, A. Takemura and H. Ono. 2004. Simplified method for estimation of composition of alginates by FTIR. Journal of Applied Polymer Science 93: 1372–1377.

Sartori, C., D.S. Finch, B. Ralph and K. Gilding. 1997. Determination of the cation content of alginate thin films by FTir spectroscopy. Polymer 38: 43–51.

Shibata, T., K. Ishimaru, S. Kawaguchi, H. Yoshikawa and Y. Hama. 2008. Antioxidant activities of phlorotannins isolated from Japanese Laminariaceae. Journal of Applied Phycology 20: 705–711.

Shin, H.C., H.J. Hwang, K.J. Kang and B.H. Lee. 2006. An antioxidative and anti-inflammatory agent for potential treatment of osteoarthritis from *Ecklonia cava*. Archives of Pharmacal Research 29: 165–171.

Skriptsova, A., V. Khomenko and I. Isakov. 2004. Seasonal changes in growth rate, morphology and alginate content in *Undaria pinnatifida* at the northern limit in the Sea of Japan (Russia). Journal of Applied Phycology 16(1): 17–21.

Smit, A.J. 2004. Medicinal and pharmaceutical uses of seaweed natural products: A review. Journal of Applied Phycology16: 245–262.

Souza, B.W.S., M.A. Cerqueira, A.I. Bourbon, A.C. Pinheiro, J.T. Martins, J.A. Teixeira, M.A. Coimbra and A.A. Vicente. 2012. Chemical characterization and antioxidant activity of sulfated polysaccharide from the red seaweed *Gracilaria birdiae*. Food Hydrocolloids 27: 287–292.

Spieler, R. 2002. Seaweed compound's anti-HIV efficacy will be tested in southern Africa. Lancet 359: 16–75.

Sugawara, I., W. Itoh, S. Kimura, S. Mori and K. Shimada. 1989. Further characterization of sulfated homopolysaccharides as anti-HIV agents. Cellular and Molecular Life Sciences 45: 996–998.

Torres, M.R., A.P.A. Sousa, E.A.T.S. Filho, F.M.B.B. Dirce, J.P.A. Feitosa, R.C.M. de Paula and M.G.S. Lima. 2007. Extraction and physicochemical characterization of *Sargassum vulgare* alginate from Brazil. Carbohydrate Research 342: 2067–2074.

Usov, A.I. and M.Y. Elashvili. 1991. Polysaccharides of Algae. 44. Investigation of sulfated galactan from *Laurencia nipponica* Yamada (Rhodophyta, Rhodomelaceae) using partial reductive hydrolysis. Botanica Marina 34: 553–560.

van de Velde, F. and G.A. de Ruiter. 2002. Carrageenan. pp. 245–274. *In*: Vandamme E.J., S.D. Baets and A. Steinbèuchel (eds.). Biopolymers, Vol. 6: Polysaccharides II, polysaccharides from eukaryotes. Wiley-VCH, Chichester, UK.

Vázquez-Freire, M.J., M. Lamela and J.M. Calleja. 1996. Hypolipidaemic activity of a polysaccharide extract from *Fucus vesiculosus* L. Phytotherapy Research 10: 647–650.

Vijayabaskar, P. and V. Shiyamala. 2012. Antioxidant properties of seaweed polyphenol from *Turbinaria ornata* (Turner) J. Agardh, 1848. Asian Pacific Journal of Tropical Biomedicine 2(1 Supplement): 90–98.

Wijesekara, I., N.Y. Yoon and S. Kim. 2010. Phlorotannins from *Ecklonia cava* (Phaeophyceae): Biological activities and potential health benefits. BioFactors 36: 408–414.

Williams, D.H. and I. Fleming. 1980. Spectroscopic Methods in Organic Chemistry, 3rd Edition. McGraw-Hill, London, 251 pp.

Witvrouw, M. and E. De Clercq. 1997. Sulfated polysaccharides extracted from sea algae as potential antiviral drugs. General Pharmacology: The Vascular System 29: 497–511.

Yuan, H. and J. Song. 2005. Preparation, structural characterization and *in vitro* antitumor activity of kappa-carrageenan oligosaccharide fraction from *Kappaphycus striatum*. Journal of Applied Phycology 17: 7–13.

Yuan, H., J. Song, X. Li, N. Li and S. Liu. 2010. Enhanced immunostimulatory and antitumor activity of different derivatives of kappa-carrageenan oligosaccharides from *Kappaphycus striatum*. Journal of Applied Phycology 23: 59–65.

Yuan, Y.V., D.E. Bone and M.F. Carrington. 2005. Antioxidant activity of Dulse (*Palmaria palmata*) extract evaluated *in vitro*. Food Chemistry 91: 485–494.

CHAPTER 8

Kappaphycus (Rhodophyta) Cultivation: Problems and the Impacts of Acadian Marine Plant Extract Powder

Anicia Q. Hurtado,[1,]* *Renata Perpetuo Reis,*[2,a]
Rafael R. Loureiro[2,b] *and Alan T. Critchley*[3]

1 Introduction

Kappaphycus is one of the most significant, economically valuable red seaweeds, cultivated in tropical and sub-tropical waters. This alga demands a relatively high market value globally, due to applications of the kappa carrageenan colloid that is industrially extracted from the biomass. Carrageenan is widely used in food, pharmaceuticals, nutraceuticals and for aquaculture applications.

The first successful commercial cultivation of *Kappaphycus* (previously called *Eucheuma*) was recorded from the southern Philippines. It took more than five years of field trials from 1967 to the early 70s in order to domesticate *Kappaphycus* for reliable commercial cultivation. The first commercial quantities of "cottonii" produced from extensive cultivation

[1] Integrated Services for the Development of Aquaculture and Fisheries, McArthur Highway, Tabuc Suba, Jaro 5000 Iloilo City, Philippines.
Email: anicia.hurtado@gmail.com
[2] Instituto de Pesquisas, Jardim Botânico do Rio de Janeiro, Rua Pacheco Leão, 915, Rio de Janeiro, RJ CEP 22460-030, Brazil.
[a] Email: rreis@jbrj.gov.br
[b] Email: rafael.rloureiro@gmail.com
[3] Acadian Seaplants Limited, 30 Brown Avenue, Dartmouth, NS B3B 1X8, Canada.
Email: Alan.Critchley@acadian.com
* Corresponding author

were obtained in 1974 with a total production of 8,000 t. Dramatic increases in production were achieved and the Philippines was the leading producer of *Kappaphycus* for 33 years until it was overtaken by Indonesia in 2008. It was in 1978, when *Kappaphycus* farming first saw successful adoption in Indonesia under the initiative of the Copenhagen Pectin Factory. Due to these successes, *Kappaphycus* farming has also grown commercially in East Africa, Fiji Is., India, Malaysia, Vietnam, Cambodia, Myanmar and southern China, although their volumes are minimal when compared to Indonesia and the Philippines. On the other hand, Latin American (Argentina, Colombia, Brazil, Equador, Venezuela, Mexico, and Peru) and the Caribbean countries (St. Lucia, Jamaica, and Panama) are even more recent entrants to *Kappaphycus* cultivation. Most information comes from published pilot-plot, demonstration farms and scientific studies, as such, only relatively small commercial quantities are produced presently.

The success of seaweed aquaculture in the Philippines and Indonesia has been dominated by various species and strains of two major genera: viz. *Kappaphycus alvarezii* (i.e., the '*cottonii*' type and source of kappa carrageenan) and *Eucheuma denticulatum* (the '*spinosum*' type and source of iota carrageenan). Once these species were domesticated, they comprised a relatively small genetic stock and their propagation has been mainly vegetative, i.e., repeated cutting of young branches from the same plant. The same method was practiced by the seaweed farmers from the start of farming and continues to present times. It is surprising that although there have been attempts to use cultivars developed from spores and tissue culture, the application of this relatively mainstream technology is still very much in its infancy, as applied to carrageenophyte cultivation. Furthermore, *in situ* proving and testing through field trials are still required to test the viability of the 'new plants' developed.

Not unlike the development of vegetatively developed terrestrial crops, the early years of commercial farming of *Kappaphycus* were beset with problems, of which 'ice-ice' malaise and macro-epiphytism were predominant. These problems still persist today. Epiphytic Filamentous Algae (EFA) infestations are a more recent problem which adversely affected the quality and quantity of the seaweed. One of the most notable negative effects of severe EFA infestation by *Neosiphonia* (a filamentous red alga), with a concomitant 'ice-ice' malaise incidence, is the scarcity, or the lack of availability of good quality propagules for re-planting purposes. Recent reports showed that dipping the cultivars in Acadian Marine Plant Extract Powder (AMPEP), a commercial extract of the brown seaweed *Ascophyllum nodosum* resulted in more vigorous 'seedlings' and increased resistance to 'ice-ice' malaise, epiphytes and particularly *Neosiphonia* infestation. Good aquaculture practices and improved and consistent management practices have, likewise, contributed to improved productivity and production.

The socio-economic impacts of seaweed farming on coastal communities have been overwhelmingly positive. Family-unit-type operations are far more advantageous than the initial corporate-owner-operator style farms, especially in remote areas where coastal communities are faced with a limited number of opportunities for alternative economic activities. There are few published articles, but several anecdotal reports state that seaweed farming has improved the economic status of coastal fisherfolk. However, the profitability of seaweed farming during 'ice-ice' outbreaks and *Neosiphonia* infestations, has been severely negatively impacted because the carrageenophyte crops have been damaged which in turn affected productivity and importantly the carrageenan quality and quantity, such that there have been significant downturns in availability of crops for commercial harvest.

A limited number of papers have shown the impact of 'ice-ice' outbreaks and *Neosiphonia* infestations on seaweed farming. Surface seawater temperature, salinity, water movement and light intensity, all acting in combination, play significant roles in *Kappaphycus* health and growth. It is not conclusively established if these changes are in fact related to global climate change in surface seawater temperature and/or other environmental and/or biological variables. The economic impact of the decline of *Kappaphycus* in some areas, especially the Philippines, requires more intensive and regional investigation to establish cause and effects. There is an important need to sustain carrageenophyte farming due to the many inter-linked economic and coastal development benefits which can be derived, if practiced and maintained in a sustainable manner. The key players along the value chain must work in harmony providing increased and sustained efforts to continue to provide the extraction industry with this high-value, natural colloid. If this is not the case, the hydrocolloid industry could switch to other sources of raw materials with similar or competing rheological properties.

The consumption and utilization of seaweed worldwide is associated with a myriad of products that generate near US$ 8 billion per year (FAO 2010). Almost 90 percent is food products for human consumption; the remainder is for the hydrocolloid industry focused on agar, carrageenan and alginates. Macroalgae for direct or indirect consumption (i.e., coloring, flavoring agents and biologically active compounds sold as dietary supplements) have been gaining market share mainly due to the shift of men and women of traditional uses of seaweed in their daily lives. A good example is the United States and certain countries of South America which in the last 15 years had an 83 percent increase in domestic consumption of products derived from algae, much of which has been provided by the popularization of Asian food (McHugh 2003, FAO 2010).

Seaweed cultivation has expanded over the years, fueled by the growing demand for macroalgal raw material by an industry that initially relied on the exploitation of natural beds (Pickering et al. 2007). This demand necessitates a search for new technologies to improve macroalgal production, seeking more effective ways to produce biomass and also more resistant individual cultivars (Zemke-White and Ohno 1999, Ask and Azanza 2002).

Records of the first "crops" of *K. alvarezii* date back to the 1960s in southern Mindanao, Philippines using seedlings derived from individuals collected from natural beds (Bindu and Levine 2011). After four decades, the cultivation of *K. alvarezii* and *Eucheuma denticulatum* (N.L. Burman, F.S. Collins and Hervey) is responsible for 88 percent of the world's production of kappa (*K. alvarezii*) and iota (*E. denticulatum*) carrageenan, producing 120,000 dry tons/year. The main producing countries are the Philippines, Indonesia and Tanzania (Ask 2001, Hayashi et al. 2010, Bindu and Levine 2011). However, *K. alvarezii* crops can be found in over 20 countries located at latitudes near 10° which allow ideal conditions for the cultivation of the species (Hayashi et al. 2010).

2 History of *Kappaphycus* Farming

In the mid 1960s, experimental farms for *Kappaphycus* cultivation were first established in southern Philippines by the Marine Colloids Philippines, University of Hawaii, Bureau of Fisheries and Aquatic Resources and the University of the Philippines. In 1972, commercial quantities of *Kappaphycus* (then called *Eucheuma* and "*cottonii*") were harvested from the cultivation areas. A continuous supply of dried '*cottonii*' has been exported (Doty and Alvarez 1981) from that time to the present day. Vegetative propagation has been the only widely used method employed for developing biomass for cultivation. Young, robust and healthy branches were cut from a selected 'mother plant' and then tied to soft plastic rope previously wound around a cultivation rope (i.e., monofilament rope, polyethylene rope, or flat binder). Five methods of culturing *Kappaphycus* were introduced in the Philippines in 1973 (Doty 1973, Ricohermoso and Deveau 1979): (1) off-bottom mono-line, presently called fixed-off bottom, (2) broadcast, (3) floating bamboo, (4) net system, and (5) the tubular net. However, the off-bottom monoline, single floating-raft and hanging long-lines (adapted in shallow waters) are the most popular methods adopted in most recent times, not only in the Philippines (Hurtado-Ponce et al. 1996, Hurtado and Agbayani 2000, Hurtado et al. 2008) but also in other countries such as Fiji (Luxton et al. 1987), Hainan, China (Wu et al. 1989), India (Mairh et al. 1995, Eswaran and Jha 2006, Kaliaperumal et al. 2008, Bindu and Levine 2011), Indonesia (Adnan and Porse 1987, Firdausy and Tisdell 1991, Luxton 1993), Kiribati

(Luxton and Luxton 1999), Madagascar (Mollion and Braud 1993), Tanzania (Lirasan and Twide 1993), and Vietnam (Ohno et al. 1996). Over time and with gains in experience, innovations were introduced, e.g., the free swing method (Fig. 1a; Hurtado et al. 2008), multiple, raft long-lines (Fig. 1b; Hurtado and Agbayani 2002, Hurtado 2007); and the 'spider web' (Fig. 1c; Hurtado 2007, Hurtado et al. 2008) which are adapted for practices in deeper waters. Table 1 summarizes the history of *Kappaphycus* farming and its introduction to other tropical and sub-tropical countries.

The Americas and the Caribbean countries are new entrants to *Kappaphycus* farming. Optimal conditions for farming can be found in these countries that could possibly provide a great potential, if correctly administered. However, the biggest concern of the introduction in these countries is the potential environmental risk of a non-native species. A series of bio-invasions have been attributed to the introduction of *K. alvarezii* for farming purposes in Hawaii (Rodgers and Cox 1999, Conklin and Smith 2005) with later reports in Venezuela (Barrios 2005) and India (Chandrasekaran et al. 2008). However, other researchers found that the 'invasiveness' of this species is restricted to sites that are already affected by anthropogenic environmental perturbations (Pickering et al. 2007). Moreover, the lack of culture and tradition of seaweed farming, and poor government incentives have unfortunately not led to a concerted effort to solve the problems. Thus academia has taken up the challenge of generating alternatives in biotechnology and simple reliable techniques to ensure that eucheumatoid cultivation in the Americas and Caribbean countries remains possible. Since these countries are new to *Kappaphycus* cultivation, farming practices of each country will be presented briefly.

2.1 Brazil

The exotic species *K. alvarezii* is the main macroalga that is commercially cultivated in Brazil (Pellizzari and Reis 2011, Góes and Reis 2011). Its success is explained by the relatively easy management techniques which are employed (i.e., fragmentation of the thallus) and the species' cultured resilience to local abiotic factors that resulted in high growth rates (Ask 1999, Ask and Azanza 2002).

The introduction of *K. alvarezii* into Brazil took place at Ubatuba Bay, São Paulo State by Dr. Édison José de Paula and his research group from University of São Paulo. It was responsibly introduced as a seedling clone from Japan which had originated in the Philippines. After a quarantine procedure of 10 months, a series of tests were conducted, not only with *K. alvarezii* (Paula et al. 2002), but also with *Kappaphycus striatum* (F. Schmitz) Doty ex P.C. Silva, which produced viable tetraspores and was removed from the sea in order to avoid environmental risks (Bulboa et al. 2007).

Figure 1. (a-c) Innovations in the cultivation of *Kappaphycus* in deeper waters (Photos courtesy of A.Q. Hurtado): (a) free swing; (b) multiple raft long line; (c) spider web.

Table 1. History of *Kappaphycus* farming and its introduction to other tropical and sub-tropical countries.

Year of introduction	Year of commercialization	Country	Culture technique	Reported by
1967	1971	Philippines (Sulu archipelago)	off-bottom monoline, broadcast, floating bamboo, net and tubular	Doty 1973, Doty and Alvarez 1981, Parker 1974
			off-bottom monoline, hanging long-line	Neish and Barraca 1978, Ricohermoso and Deveau 1979
1969	1978	Indonesia	floating rafts	Soerjodinoto 1969, Soegiarto 1979, Adnan and Porse 1987
1976	1982	Fiji Islands	off-bottom monoline	Booth et al. 1983, Luxton et al. 1987, Prakash 1990
1977	1986	Kiribati	off-bottom monoline, long line	Why 1985, Uan 1990
1978	1978	Sabah, Malaysia	off-bottom monoline	Doty 1980
1984	1989	Solomon Islands	off-bottom monoline	Smith 1990, Kronen et al. 2013 (in press)
1985		Micronesica		Doty 1985
1985	1989	Zanzibar & Pemba Is, Tanzania, East Africa	off-bottom monoline	Mshigeni 1985
				Eklund and Patterson 1992, Lirasan and Twide 1993
				Msuya et al. 2007, Shechambo et al. 1996
1985	1986	Hainan, China	hanging long line	Wu et al. 1989
1989/1995	1998	Tamil Nadu, India	bag and net methods	Mairth et al. 1995, Krishnan and Narayana 2013 (in press)
			raft method	
1990	1991	Madagascar	off-bottom monoline	Mollion and Braud 1993

Table 1. contd....

Table 1. contd.

Year of introduction	Year of commercialization	Country	Culture technique	Reported by
1991	Data not available	Colombia	floating rafts	Rincones 2006
1993	1998	Vietnam	off-bottom monoline (lagoon)	Ohno et al. 1995, 1996
			pond cultivation	
			raft cultivation (off-shore)	
			fixed-off bottom	Pham 2009, Le 2011
			floating rafts	
1995	1998	Brazil	floating rafts	Paula et al. 2002, Castelar et al. 2009
1996	Data not available	Kenya	netting bags	
1996	Data not available	Venezuela	floating rafts	Rincones and Rubio 1999, Smith and Rincones 2006
1999	2000	Cambodia	hanging long line	Cambodia Ministry of Fisheries
1999	2002	Gulf of Mexico	fixed off-bottom, floating system	Munoz et al. 2004
2002	2008	Panama (Colon)	floating rafts	Batista 2006, 2009
2007	2008	Myanmar	hanging long line	The Myanmar Times, July 23–29, 2008
2010	Data not available	Equador	off-bottom monoline (enclosed tanks)	Equator Ministry of Fisheries and Aquiculture
2011	2011	Saint Lucia, Saint Martin & Grenadines	floating rafts; off-bottom	Rincones and Sepulveda, pers. communication

The first commercial farming in Brazil was located at Ilha Grande Bay, Rio de Janeiro State by Miguel Sepúlveda, in 1998, after a quarantine period in tanks at the University of Santa Catarina University, using seedlings which were brought from Venezuela (Castelar et al. 2009). In 2004, another farm was installed in Sepetiba Bay, Rio de Janeiro State, by Sete Ondas Biomar Cultivo de Algas Marinhas Ltda., in an enterprise that occupied over 20 ha; which became the biggest farming site for *Kappaphycus* in Latin America. However, this enterprise was not long-lived and shut down operations in 2008 due to high operational costs. After that, small farms were transferred to the southern coast of Sepetiba Bay (Góes and Reis 2011). Presently, most of the commercial cultivations are at Ilha Grande Bay, to the south of Rio de Janeiro State, and belong to private companies with no attached government initiatives.

In Brazil, the cultivation techniques used for *K. alvarezii* are comprised of floating rafts made with PVC tubes (Figs. 2a-e), differing from the most widely used techniques in cultivation of this species in other countries, i.e., the off-bottom systems (Hayashi et al. 2010). The adaption of these Brazilian techniques is mainly due to higher water turbidity and muddy bottom substrata at the cultivation sites, making the off-bottom techniques impossible.

Góes and Reis (2011) discussed the use of a novel culture technique, i.e., the tubular net (Fig. 2b). The floating rafts of PVC are retained; however tubular networks stuffed with *K. alvarezii* seedlings replaced the traditional 'tie-tie'. This technique showed no difference in daily growth rates of seedlings or carrageenan yield and quality when compared to the 'tie-tie', but saved time and had lower costs and provided ease of management.

2.2 Colombia

Kappaphycus alvarezii was introduced in Colombia in 1991, through an FAO funded program supervised by Prof. Germnán Bula Meyer. The Cuban seedlings, which originated from Venezuela, were quarantined and acclimatized in tanks before being used for cultivation (R.E. Rincones, personal communication).

Presently, branches of the original FAO program, now funded by the non-governmental organization Fundación Terrazul at Cabo de La Vela in the La Guarija Peninsula, works with Native American Wayuu communities on experimental farms using floating rafts and tubular nets in unused shrimp breeding tanks. However, these attempts to create a semi-closed cultivation system, to lower the risks of bio-invasions, seemed to be ineffective since *K. alvarezii* was reported to occur along the La Guarija coastline since 2004 (Rincones 2006).

Figure 2. (a-e) Cultivation methods in Brazil, South America (Photos courtesy of RP Reis and R.R. Loureiro): (a) PVC raft; (b) tubular net; (c-d) stuffing of seedling into the tubular net; (e) tubular net with *Kappaphycus* 'seedlings'.

Color image of this figure appears in the color plate section at the end of the book.

2.3 Equador

Kappaphycus seedlings were introduced by a Brazilian seaweed consultant into Ecuador to be part of an enterprise called Ecuaalgas S.A. which was supported by the local government through the Ministry of Fisheries and Aquiculture (MAGAP). The cultivation is still restricted to the enclosed disused shrimp breeding tanks where their acclimation and quarantine process took place, being set in the off-bottom method.

In these tanks, *K. alvarezii* is afflicted by 'ice-ice' and consequent epiphyte infestation creating major seedling losses with no apparent solution (Sepúlveda, personal communication).

2.4 Mexico

Muñoz et al. (2004) proposed the cultivation of *K. alvarezii* at the Yucatan Peninsula obtaining very promising initial results. This activity did not receive government support, even though the project was aimed towards a sustainable cultivation program conducted by fishermen and coastal communities through social cooperatives.

The seedlings used were originally from Brazil and were quarantined following the rules of the Official Mexican Norm (NOM). The farming practice was mostly the off-bottom cultivation technique; there were no reports of problems with either herbivores or epiphyte infestations. The main cause of crop losses in this region seemed to be due to 'ice-ice', which was most frequently observed when the seawater temperatures exceeded 30°C. This is possibly a problem that could be solved by making use of floating rafts (Robledo, personal communication).

2.5 Panama

Alongside Brazil and the Caribbean countries of Saint Lucia, Saint Martin and Grenadines, Panama is one of the few that has commercial farms of *K. alvarezii* in the Americas having exported its production to Europe and Asia (Vega and Rincones, personal communication).

K. alvarezii seedlings, from Venezuela, were originally introduced by the Smithsonian Tropical Research Institute group with no previous quarantine procedure. The material was for experimental purposes in the Colón Province, the location of the only commercial site in Panama; activities were supported by private companies, such as Panamá Seafarms and Bansistemas S.A., overseen by the local government through the Panama Authority of Aquatic Resources (ARAP). Moreover, in the Bocas del Toro Province, experimental studies were being conducted using the same floating rafts and 'tie-tie' system used in the Colón Peninsula farms (Batista 2006, 2009).

When the seedlings were originally brought to the Colón Peninsula, infestations of epiphytes (i.e., red, green and brown macroalgae) seemed to be a major concern negatively influencing *K. alvarezii* daily growth rates and causing crop losses (Batista 2009). Today, there are no reports of epiphytes, however, the occurrence of 'ice-ice' and losses due to herbivores seem to be of great concern to the local farmers (R.E. Rincones, personal communication).

2.6 Saint Lucia, Saint Martin and Grenadines

With no apparent government support, and relying on private companies and non-governmental organizations, such as Ashton Multipurpose Cooperative AMCO and the Sustainable Grenadines Project, these Caribbean countries were one of the few to develop fully functional *K. alvarezii* farms, with the original stock coming from Venezuela. These seedlings were introduced in 2011, with no apparent quarantine procedure (R.E. Rincones and Sepúlveda, personal communication).

A myriad of available cultivation techniques were employed and varied according to the farming site, and ranging from floating rafts to off-bottom long-lines (both of which utilized 'tie-tie'). There are no reports of epiphyte infestations. 'Ice-ice' and herbivores are issues in these countries, as in most of the Americas, and continue to be the biggest impediment to productivity (R.E. Rincones, personal communication).

2.7 Venezuela

K. alvarezii was legally introduced to Venezuela; the seedlings originated from the Philippines in 1996. A six week acclimation and quarantine process was applied. The project had full government support, through the National Institute of Agropecuary Investigations (INIA) and the Rural Development, Innovation and Capacitation Foundation, Venezuela conducted experimental studies with *K. alvarezii* in the Araya Peninsula and Cubagua Isle (Rincones and Rubio 1999, Smith and Rincones 2006).

Those studies provided the information necessary for the government to permit the farming of an exotic species along the Araya Peninsula (Sucre State), using floating rafts with tubular nets and 'tie-tie'; techniques varied depending on the cultivation site. Tubular nets were more common in places with the greatest water motion so as to avoid crop losses. So far no reports of complications due to epiphyte infestations have been were reported from the region, while a small degree of 'ice-ice' and herbivore grazing were the only problems reported (R.E. Rincones, personal communication).

3 Global Production of '*Cottonii*' and '*Spinosum*'

The first successful commercial cultivation of *Kappaphycus* (previously called *Eucheuma*) was recorded from the southern Philippines. It took more than five years of field trials from 1967 to the early 70s in order to domesticate *Kappaphycus* for reliable commercial cultivation. The first commercial quantities of '*cottonii*' produced from extensive cultivation, were obtained in 1974 with a total production of 8,000 t. Thereafter, the Philippines continued to cultivate and export dried *Kappaphycus* to the present. Figure 3 shows the production of '*cottonii*' (*Kappaphycus*) and '*spinosum*' (*Eucheuma*) from 2002–2009. The '*cottonii*' production of the Philippines first showed signs of decline in 2003. Production was regained the following year, however, it steadily decreased thereafter. From the start of commercial seaweed farming, the Philippines retained the status of number one producer in the world for '*cottonii*'—*Kappaphycus*. However, at the same time, Indonesia's proximity to the Philippines made for easy and rapid transfer of seaweed farming technology and seedlings. Through the initiative of the Copenhagen Pectin Factory in 1978, the introduction and successful adoption of *Kappaphycus* farming in Indonesia was made possible. The geographic location of Indonesia, occupying about 65 percent of the total shoreline within the 10° latitude of the Coral Triangle creates an ideal environment for the adoption of seaweed farming in coastal waters. The Philippines has

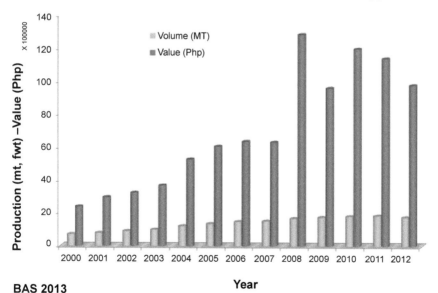

BAS 2013

Figure 3. Production (dwt, MT) of 'cottonii' in the Philippines from 2000–2012.

only 15 percent of the total area of the Coral Triangle and the remaining 20 percent is divided among Malaysia, Papua New Guinea, Timor Leste and the Solomon Islands (I.C. Neish, personal communication). The ideal geographic location of Indonesia, coupled by a receptive, iterative approach to seaweed farm development, accompanied by decentralized policies of the Indonesian government, in concert with traditional 'adat' forms of village government, paved the way for the rapid increase in Indonesian production of cultivated seaweeds (I.C. Neish, personal communication). In addition, strong market linkages and the all important assistance by the Business Development Services (BDS), played significant roles in the rapid and extensive development of seaweed farms. The combined effects were such that by 2008, Indonesia became the world's number one producer of *Kappaphycus*.

Since *Kappaphycus* had also been transplanted to a number of other tropical and sub-tropical countries from as early as the mid'80s, commercial farming has also become a successful enterprise, after several years of research and trial farming in East Africa, Fiji Is., India, Malaysia, Vietnam, Cambodia, Myanmar and southern China. However, their volumes, even today, are minimal when compared to Indonesia and the Philippines (Fig. 4). On the other hand, some Latin American countries (e.g., Colombia, Brazil, Equador, Venezuela, Mexico and Peru) and the Caribbean countries

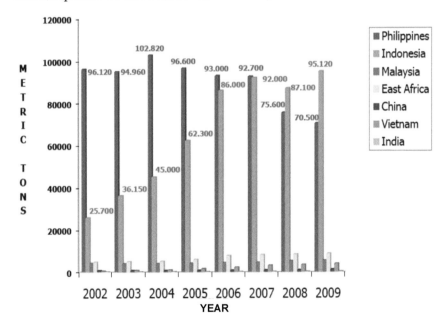

Figure 4. Global production of 'cottonii' (dwt, MT) from the major producing countries. *Color image of this figure appears in the color plate section at the end of the book.*

(e.g., St. Lucia, Jamaica and Panama) are even more recent entrants to those countries practicing *Kappaphycus* cultivation. Most information on these activities comes from published pilot-plot, demonstration farms and scientific studies; as such, only small commercial quantities are produced presently, but the scale could increase if there is sufficient demand for the supply of raw materials.

4 Two Major Problems of *Kappaphycus* Cultivation

4.1 Lack of Availability of Good Quality Propagules (=Seedlings)

Successful seaweed aquaculture in the Philippines and Indonesia has been dominated by a relatively small number of species and, in particular, strains of just two major genera: viz. *Kappaphycus alvarezii* and *K. striatum* (Figs. 5a-b; i.e., the *'cottonii'* type and source of kappa carrageenan) and *Eucheuma denticulatum* (Fig. 6; the *'spinosum'* type and source of iota carrageenan). After these species were domesticated, they formed a relatively limited genetic stock and their subsequent multiplication has been exclusively through vegetative propagation, i.e., relatively simple, repeated cutting of young branches from the same plant. The same method of cropping and ensuring the availability of new material for the next harvest has been practiced and perpetuated by the seaweed farmers to the present time. It is somewhat surprising that although there have been attempts to use

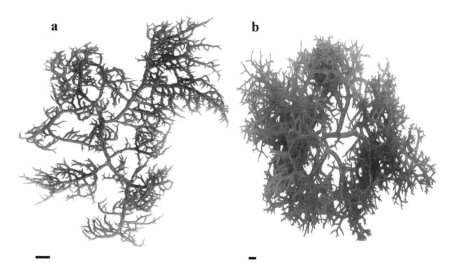

Figure 5. Two major species of *Kappaphycus*, 'cottonii' type (a) *alvarezii* and (b) *striatum*, commonly used in cultivation (bar = 1 cm, Photos courtesy of A.Q. Hurtado).

Color image of this figure appears in the color plate section at the end of the book.

Figure 6. 'Spinosum' type *Eucheuma denticulatum* (bar = 1 cm, Photo courtesy of A.Q. Hurtado).

Color image of this figure appears in the color plate section at the end of the book.

cultivars, as developed from spore production (see: Azanza-Corrales et al. 1996, Azanza and Aliaza 1999, Bulboa et al. 2007, 2008, Luhan and Sollesta 2010) and also tissue culture (see: Dawes and Koch 1991, Dawes et al. 1993, 1994, Hurtado and Cheney 2003, Muñoz et al. 2006, Hayashi et al. 2007, Hurtado and Biter 2008, Hurtado et al. 2009, Yunque et al. 2011, Yong et al. 2011), the application of these relatively mainstream technologies are still very much in their infancy, as applied to carrageenophyte cultivation. Furthermore, *in situ* proving and testing through field trials are still required to test the viability of the 'new plants' developed. As a consequence, the simple method of vegetative propagation prevails; unfortunately this technique has only perpetuated the limited genetic stock of the same limited number of strains or cultivars.

4.2 Occurrence of Disease and Epiphytes

Not unlike the development of vegetatively developed terrestrial crops, the early years of commercial farming of *Kappaphycus* were beset with problems, of which 'ice-ice' malaise and macro-epiphytism were predominant. These problems still persist today. Epiphytic filamentous algae (EFA) infestation was first described by Ask and Azanza (2002) as numerous species of filamentous algae that attach to the cortical layer of the host thalli. Recent EFA infestation is dominated by the filamentous red alga *Neosiphonia* previously called *Polysiphonia*, which adversely affected the quality and quantity of the seaweed. The most notable negative effects of severe EFA infestation by *Neosiphonia* with a concomitant 'ice-ice' malaise incidence is the resultant scarcity of good quality propagules for re-planting purposes.

'Ice-ice' Malaise

The early years of commercial farming of *Kappaphycus* in the Philippines and elsewhere were beset with problems, of which 'ice-ice' (Fig. 7) and macro-epiphytism (Fig. 8) were predominant. The outbreaks of 'ice-ice' malaise were most pronounced in areas with lower water quality, slow water exchange (almost stagnant), low salinity and extreme (low or high) temperatures (Doty 1987, Largo et al. 1995b). The 'ice-ice' phenomenon appeared with a greening of the normally pink/red tissue followed by whitening, like 'ice' of necrotic tissue as decay of the surface tissue lead to softening of entire or partial branches, which finally resulted in fragmentation of the thallus and considerable loss of biomass. Bacteria (e.g., *Pseudomonas*, *Cytophaga-Flavobacterium* complex, *Vibrio*, *Xanthomonas* and *Achromobacter*) have been isolated from the affected areas of tissue (Uyengco 1977, Uyengco et al. 1981, Largo et al. 1995a, Butardo et al. 2003), but no specific causative bacterial species have been linked to the 'ice-ice' malaise. Hence, it is interpreted as an indication of "stress" rather than a pathogen-induced disease (i.e., the bacteria are secondary rather than primary, causative instigators of the disease). It is likely that the bacteria

Figure 7. A branch of *Kappaphycus* with an 'ice ice' (bar = 1 cm, Photo courtesy of A.Q. Hurtado).
Color image of this figure appears in the color plate section at the end of the book.

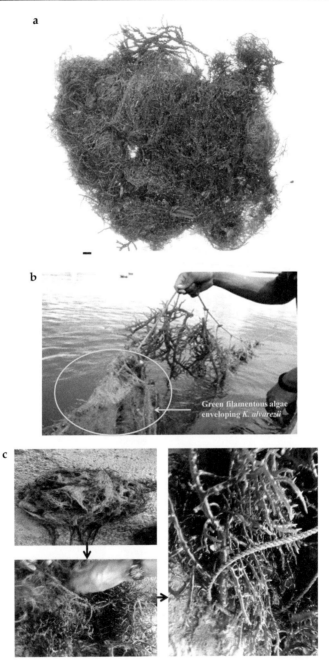

Figure 8. (a-c) Examples of macro-epiphytes (Photos courtesy of A.Q. Hurtado).

Color image of this figure appears in the color plate section at the end of the book.

are able to effectively invade the alga only at periods when the (stressed) seaweed is less able to produce and release extracellular products which may be inhibitory to the biofilm, fouling microorganisms (Uyengco et al. 1981).

The incidence and degree of infection of 'ice-ice' was observed to vary significantly between color morphotypes of *K. alvarezii* and *E. denticulatum*, time (season) and geographical location (Tisera and Naguit 2009). *Eucheuma denticulatum* was found to be more resistant to 'ice-ice' than *K. alvarezii* in Negros Oriental and Zamboanga del Norte, Philippines. High water temperature and low salinity during April and October gave rise to the higher incidence of 'ice-ice', as claimed by the authors.

Most reports on the incidence and degree of infection of 'ice-ice' by *K. alvarezii* and *E. denticulatum* were mainly due to the interplay of ecological factors and secondary infections by pathogenic marine bacteria (Uyengco 1977, Uyengco et al. 1981, Doty 1987, Largo et al. 1995a,b) and most recently by marine fungi (Table 2; Solis et al. 2010). The ability of marine fungi (e.g., *Aspergillus ochraceus*, *A. terreus* and *Phoma* sp.) to produce seaweed-degrading enzymes, i.e., carrageenases and cellulase, and the preference of these fungi for carrageenan as a carbon source, supports their potential as causative agents of 'ice-ice' malaise (Table 2; Solis et al. 2010).

An understanding of the significance of the 'ice-ice' problem should be viewed not only from the interplay of the ecological or environmental conditions of where the seaweed is grown, but more importantly, on the underlying cell physiology and cell wall responses to the surrounding environment. The early works of Collen et al. (1995), Mtolera et al. (1995, 1996) and Pedersen et al. (1996) on *E. denticulatum*, *E. platycladum* and other marine algae are the only known reports which demonstrated the role of water quality in the production of stress-induced hydrogen peroxide (H_2O_2) and volatile halogenated compounds by seaweeds susceptible to 'ice-ice'. The production of H_2O_2, as an oxidative burst, is probably a part of the chemical defense mechanisms of *E. denticulatum* (Collen et al. 1995). A burst of H_2O_2 might mimic the algal responses to the grazing activity of fish or invertebrates. Mtolera et al. (1995) showed that continuous production of H_2O_2 in *Eucheuma* increased with increasing irradiance and pH. In the field, *Eucheuma* spp. are rarely found subjected to epiphytes and pathogens (Mtolera et al. 1996), which is perhaps due to the production of H_2O_2; this may explain how the surface of the alga is maintained clean of epiphytes and pathogens. H_2O_2 produced in seaweeds induces other reactions inside the cell and cell walls, and the formation of other strong oxidants such as hypochlorite, hypobromite and mono-chloroamine; these reactions can take place with the help of halogenated peroxides that are present in the algae (Collen et al. 1995). Seawater contains halide ions, in high concentrations and these may serve as substrates in halogenating reactions. A burst of hydrogen peroxide from *E. denticulatum* may create a release of HOCl, HOBr and

Table 2. List of marine-derived fungus isolated from healthy and infected *K. alvarezii* and *K. striatum* farmed in Calatagan, Batangas, Philippines (orange and green varieties) (Solis et al. 2010).

Taxon	No. of isolated MDF strains						Total no. of strains
	Kappaphycus alvarezii		*Kappaphycus striatum* (orange variety)		*Kappaphycus striatum* (green variety)		
	Healthy host	Infected host	Healthy host	Infected host	Healthy host	Infected host	
Scopulariopsis brumptii Salv.-Duval	0	0	1	0	0	0	1
Cladosporium sp. 1	0	0	1	0	0	0	1
Phoma nebulosa (Pers.) Mont.	0	0	1	0	0	0	1
Cladosporium sp. 2	1	0	0	4	5	0	10
Phoma lingam (Tode) Desm.	0	3	0	3	3	0	9
Aspergillus terreus Thom	0	1	0	3	2	0	6
Eurotium sp.	0	1	0	8	0	0	9
Phoma sp.	0	6	0	8	0	0	14
Aspergillus sydowii (Bainier et sartory) Thom et Church	4	4	0	2	0	0	10
Curvularia intermedia Boedijn	0	9	0	9	0	0	18
Cladosporium sp. 3	0	12	1	2	0	0	15
Fusarium sp.	0	7	0	7	5	0	19
Fusarium solani (Mart.) Sacc.	1	7	0	0	0	1	9
Aspergillus ochraceus G. Wilh.	0	0	0	6	0	0	6
Aspergillus flavus Link	0	0	5	1	0	1	7
Penicillium sp.	2	0	0	0	0	0	2
Penicillium purpurogenum Stoll	0	0	0	1	0	0	1
Engyodontium album (Limber) de Hoog	1	1	4	0	0	0	6
Total	9	51	13	54	15	2	144

chloramines in response to certain pathogenic microorganisms or physical damage by grazers and, as such, could be a very efficient chemical defense (Pedersen et al. 1996). However, the defense mechanism of H_2O_2 production can, in itself, be damaging to the alga particularly if the algal density is high (compared to the water volume into which the strong oxidants diffuse), as might be found in a high density seaweed farm. Field observations indicate that *Kappaphycus alvarezii* is more easily subjected to 'ice-ice' and epiphyte infestation as compared to *E. denticulatum* (Hurtado, personal observation). This was noted especially in high density cultivation areas, particularly where support structures such as bamboo rafts and spider webs were in close proximity to one another. Such intensive arrangements may impede water movement, and act as substrata for the development of epiphytic communities. Further research into these preliminary observations will be required to develop management action plans for healthier cultivation practices. Healthy *Kappaphycus* or *Eucheuma* thalli/seedlings have a greater chance of maintaining efficient chemical defense mechanisms when exposed to lower water quality, with slow exchange, low salinity and low or high temperatures (Uyengco et al. 1981, Largo et a. 1995a,b, Tisera and Naguit 2009). It is ironic that the intensively cultivated stressed seaweeds may become victim to their own defense mechanisms, such that they may eventually perish.

The earliest practice of cultivating *Kappaphycus* in shallow areas, especially in seagrass beds (Doty and Alvarez 1981), proved to be detrimental, both to the seagrass bottom and the cultured seaweed. The interference of seagrass with the uptake of CO_2 by *Eucheuma* resulted in poor growth (Collen et al. 1995). In addition, sometimes, the sea grasses were cut during construction of the seaweed farm to accommodate the stakes and cultivation lines; this in turn had negative impacts on the growth and population diversity of the associated benthic ecosystem for the support of fish nurseries, invertebrates and zooplankton.

Epiphytes

The characteristics of epiphytes and the extent of their damage at the host/ epiphyte interface was classified into five groups (B. Kloareg, personal communication, Leonardi et al. 2006):

1. **Type I**: Epiphytes weakly attached to the surface of the host and not associated with any host tissue damage. The contact between the host and the epiphyte was so close that the interface was indistinct
2. **Type II**: Epiphytes strongly attached to the surface of the host but not associated with any host tissue damage

3. **Type III**: Epiphytes penetrate the outer layer of host cell wall without damaging the cortical cells
4. **Type IV**: Epiphytes penetrate the outer layer of the host cell wall, and are associated with the host's cortical disorganization
5. **Type V**: Epiphytes invade the tissue of the host, growing intercellularly, and are associated with destruction of cortical and (in some cases) medullary cells

The question is posed as to, at which point does an epiphyte, which invades the tissues of the host, become a parasite?

Epiphytes and grazers create large problems for seaweed farmers in general. Hard substrata are often limiting in the sea, and seaweed thalli are attractive surfaces for the settlement of epiphytic algae and animals. Epiphytism may be problematic since it can deprive the cultured seaweed of sufficient irradiance required for photosynthesis, as well as decrease the amount of available nutrients and carbon dioxide, or dissipation of oxygen. Taken together, these factors may reduce growth rate and quality of the cultivated material. *In vitro*, Loureiro et al. (2010) observed that the use of AMPEP was efficient in improving the daily growth rate of *K. alvarezii* and eliminated some epiphytes like *Ulva* and *Cladophora*. In Brazilian commercial cultivations, the same response was observed using AMPEP to improve growth and carrageenan yield (Reis, personal field and laboratory observations). The most common macro-epiphytes of *Kappaphycus* are the green (e.g., *Ulva*) and red (e.g., *Acanthophora, Hypnea, Hydrocanthus, Chondrophycus*) algae which can entirely envelope the surface of the thalli (see Figs. 8a-c). The negative effects of macro-epiphytism bring more working hours to the farmers, requiring the removal of the attached seaweeds. The poor practice of some farmers to just throw away the removed epiphytised plants within the existing farms, brings more problems to the entire cultivation area. The discarded covered crop can be carried by the current from one area to another only to act as a vector for the dispersal of the epiphytes; the problem becomes a vicious circle. To combat this problem, the pruned seaweed segments must be brought on land, sundried and perhaps even used as soil conditioners for agricultural or horticulture plants.

Endophytic Filamentous Algae (EFA) (Fig. 9) infestation is a more recent problem that adversely affected the quality and quantity of *K. alvarezii* commercial production. In Brazilian cultivation the main problem is with the calcareous organisms that can sink the rafts and consequently increase costs to clean the rafts (Marroig and Reis 2011). The most notable adverse effect of severe EFA infestation was by *Neosiphonia*, with a concomitant 'ice-ice' malaise incidence, which resulted in the scarcity or even the lack of available good quality propagules for re-planting purposes. Furthermore, there were signs of slowing growth rates which decreased the capacity for

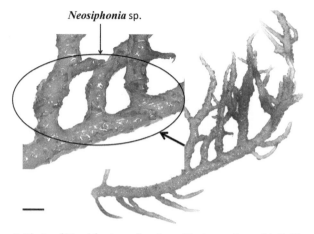

Figure 9. Photo of *Neosiphonia* sp. (bar 1 cm, Photo courtesy of A.Q. Hurtado).

regeneration, leading to a significant decrease in both carrageenan yield and also molecular weight of the carrageenan extracted (Mendoza et al. 2002). A similar situation was reported in *Chondrus crispus*, a carrageenophyte of temperate waters when infected with the endophytic green alga (*Acrochaete heteroclada* and *A. operculata*). Sporophytes of *Chondrus crispus* became completely endophytised, resulting in severe damage of the host tissue, with secondary bacterial infections, and eventual disintegration and death of the thallus (Correa and McLachlan 1991, 1992, 1994, Bouarab et al. 2001, Potin et al. 2002, Weinberger et al. 2005).

The survey work of Hurtado (2005a) revealed that the earliest report of the red algal epiphyte *Neosiphonia* (first recorded as *Polysiphonia*), infestation in the Philippines was in 1994 in Sitangkai, Tawi-Tawi, Philippines. The author reported that *Neosiphonia* infestations (Table 3a) also occurred in Iloilo and Antique (1997), Zamboanga City (1998), Calaguas Is., Camarines Norte (2000), Zamboanga del Norte, Palawan and Quezon (2002), and Bohol (2003). These early incidences of *Neosiphonia* infestation did not concern the Philippine seaweed industry until a severe infestation was experienced in Calaguas Is., Camarines Norte in 2000 (Critchley et al. 2004). As a result, a high percentage of *Neosiphonia* infestation in combination with 'ice-ice' occurred due to overcrowding in the cultivation areas and localized slow water movement (Hurtado et al. 2006, Hurtado and Critchley 2006). The problem of *Neosiphonia* infestation became so severe in some areas as to limit cultivation activities. The Sulu Archipelago, a region of large areas of ideal sites for *Kappaphycus* farming, was severely affected by this outbreak. This resulted in cultivation activities relocating to areas which remained unaffected, especially in the Visayas and Luzon. Seaweed farmers in Sitangkai allotted an exclusion area in order to promote the recovery

Table 3a. Seasonality of good growth of *K. alvarezii*, 'ice-ice' and *Neosiphonia* incidence in some major producing areas in the Philippines (Critchley et al. 2004, Hurtado 2005, Hurtado and Critchley 2006, Hurtado et al. 2006, Hurtado 2008).

	J	F	M	A	M	J	J	A	S	O	N	D
Bohol												
good growth	■	■	■						■	■	■	■
'ice-ice'		▦	▦									
Neosiphonia sp.			░	░								
Camarines Norte												
good growth	■	■	■						■	■	■	■
'ice-ice'		▦	▦	▦	▦						▦	
Neosiphonia sp.			░	░								
Palawan												
good growth	■	■	■	■					■	■	■	■
'ice-ice'			▦	▦								
Neosiphonia sp.			░	░								
Panay												
good growth	■	■	■								■	■
'ice-ice'			▦	▦	▦							
Neosiphonia sp.			░	░								
Zamboanga del Norte												
good growth	■	■	■	■					■	■	■	■
'ice-ice'				▦	▦							
Neosiphonia sp.				░	░							
Sibutu, Tawi-Tawi												
good growth	■	■	■	■			■	■	■	■	■	■
'ice-ice'				▦	▦							
Neosiphonia sp.			░	░	░							
Sorsogon												
good growth	■	■	■	■							■	■
'ice-ice'			▦	▦								
Neosiphonia sp.								░	░			

Table 3b. Seasonality of best growth of *K. alvarezii*, 'ice-ice' and *weeds* (epiphytes) incidence in some major producing areas in Indonesia (I.C. Neish, Personal communication).

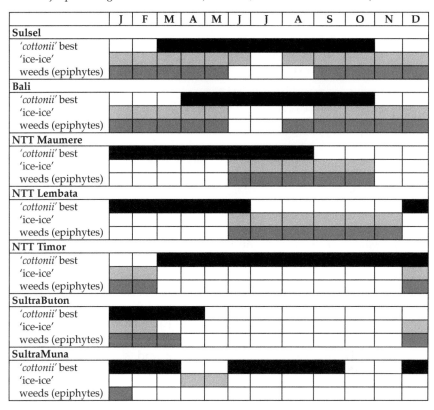

Table 3c1. Farmers' perception of seasonal problems at Muugoni, Unguja, Zanzibar. A black box denotes the time farmers recognize the problem: (1) katika/ngeupe = 'ice-ice', (2) nyuzi/chafu = EFA and nuisance algae, (3) wimbi = waves. The Table also shows the monsoon, precipitation and temperature readings in the area (Davis 2011).

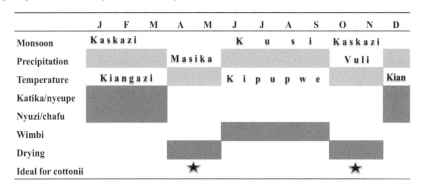

Table 3c2. Farmers' perceptions of seasonal problems in Jambiani, Unguja, Zanzibar. A black box denotes the time farmers recognize the problem. (1) katika/nyuepe = 'ice-ice', (2) mashava = EFA and nuisance algae, (3) wimbi = waves, (4) unumba = sea urchins. The Table also shows the monsoon, precipitation and temperature readings in the area (Davis 2011).

	J	F	M	A	M	J	J	A	S	O	N	D
Monsoon	Kaskazi					K	u	s	i	Kaskazi		
Precipitation				Masika						Vuli		
Temperature	Kiangazi					K	i	p	u	p	w e	Kian
Katika/nyeupe												
Nyuzi/chafu												
Wimbi												
Drying												
Ideal for cottonii				★						★		

of *Kappaphycus* from 'ice-ice' and *Neosiphonia* (A.Q. Hurtado, personal observation). Indonesia had a lot more areas suitable for *Kappaphycus* available than the Philippines; hence, the farmers transferred their farms to places which were still free from *Neosiphonia* infestation. In areas which have a history of a number of years of cultivation, 'ice-ice' and *Neosiphonia* (Table 3b; I.C. Neish, personal communication) infestation problems have also occurred; one such example is Tanzania (Tables 3c1-2; Davis 2011).

The cultivation of *Kappaphycus alvarezii* and *K. striatum* has been extensive in the Southeast Asian region, notably the Philippines, Indonesia and Malaysia. It is in these regions where *Neosiphonia* infestation and 'ice-ice' have been most severely experienced. *K. alvarezii* cultivation in Sabah, Malaysia was reported to be susceptible to *Neosiphonia* infestation during the dry months (March–June and September–November, Vairappan 2006). Here, the infective organism was dominated by *Neosiphonia savatieri* (Hariot) M.S. Kim *et* I.K. Lee, with 80–85 percent cover of *Kappaphycus* during the peak season. The author further claimed, that the emergence of epiphytes in late February/March coincided with large increases in salinity (from 28–34 ppt) and temperature (from 27–31°C). However, the opposite was observed in the second occurrence of epiphytes and 'ice-ice' in September–November when both salinity and water temperature decreased from 29–27 ppt and 30–25°C. Five notable epiphytes (Fig. 10) were recorded in Malaysia in the following order of abundance: *Neosiphonia savatieri* > *N. apiculata* > *Ceramium* sp. > *Acanthophora* sp. > *Centroceras* sp. (Vairappan 2006). *Neosiphonia apiculata* earlier identified as *Polysiphonia* sp. in the Philippines (Hurtado et al. 2006, Hurtado and Critchley 2006) was also recorded in Indonesia, Philippines and Tanzania (Vairappan et al. 2008).

The growth and development of *Neosiphonia savatieri* (Figs. 11A-D), as described by Vairappan (2006) started as black spots in the surface cuticle on *K. alvarezii*. Tissue cross-sections of the black spots revealed the presence of epiphyte tetraspores embedded between the outer cortical cells. These black spots increased in size, disturbing the host tissue and the vegetative epiphyte emerged after 3–4 weeks and grew to become reproductively mature. The

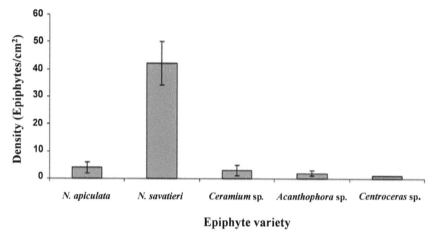

Epiphyte variety

Figure 10. Five notable epiphytes recorded in Malaysia (Vairappan 2006).

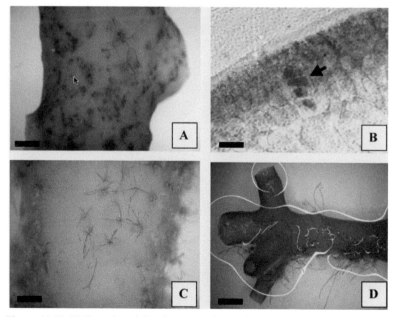

Figure 11. (A-D) Growth and development of *Neosiphonia savatieri* (Vairappan 2006).

presence of a secondary bacterial infection was reported from Scanning Electron Microscopy (Figs. 12a-f), and was observed to further contribute to the disintegration of the infected tissue, leading to fragmentation of the thallus and loss of material from the cultivation lines (Vairappan et al. 2008).

Unfortunately, there is paucity of information on how the endophytes first attach to and subsequently penetrate the host's tissues. Epiphytic filaments may enter the host's internal tissues through wounds (Apt 1988), an observation which was not seen in *Undaria pinnatifida* (Gauna et al. 2009) and *K. alvarezii* (Vairappan 2006, Hurtado and Critchley 2006). Spores of *Neosiphonia* may behave like the spores of *Laminarionema elsbetiae* (Peters

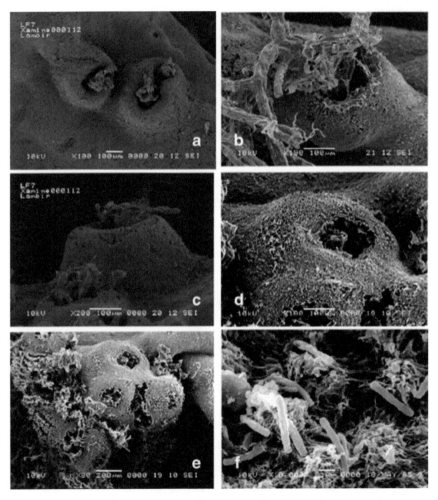

Figure 12. (a-f) Secondary bacterial infection in *Kappaphycus* through an SEM photography (Vairappan et al. 2008).

and Ellerstsdottir 1996) as specialized infective agents that attach to, and penetrate, the host surface and require neither specific wounds nor an opening for a successful invasion of the host.

The reduction of biomass in cultivated *K. alvarezii* populations could be possibly explained by the infection of *N. savatieri* if there was also damage to the oxygen evolving-complex (OECs), which could thereby decrease active reaction centers (RCs) and the plastoquinone (PQ) pool resulting on a significant reduction in the Performance Index (PI) of Phytosystem II (PSII) further leading to reduced photosynthetic activity of *K. alvarezii* (Pang et al. 2011). Such a phenomenon occurred in southern China when the *K. alvarezii* farms were infested with *N. savatieri*. Excessive growth of *N. savatieri* on the surface of the *K. alvarezii* reduced water motion nearby, as well as the gas and nutrient exchange between the *K. alvarezii* and the seawater. Thus, *N. savatieri* limited N, P, CO_2, O_2 and exchange of mineral elements required for growth. At times of heavy infestation, the host *K. alvarezii* becomes nutrient-deficient, resulting in stunted growth while *N. savatieri* has sufficient nutrient and gaseous exchange and proliferates rapidly. Likewise, light becomes limiting to *K. alvarezii* due to the endophytic *Neosiphonia*. Though both seaweeds produce oxygen through photosynthesis during daytime, oxygen consumption through respiration during night-time becomes insufficient especially for the *K. alvarezii*.

5 The Impact of AMPEP on the Growth and Vigor of *K. alvarezii*

Many efforts have been made to prevent 'ice-ice' and solve the problems of declining cultivated biomass. One such effort was the development of cultivation methods using the fluctuation system, instead of the off-bottom method, traditionally used in Asian countries. Whereas the off-bottom method can be accessed only during low tides, the technique of deep water, floating lines, can be accessed consistently, and is not dependent on tides. However, the material cost of operating in deep water is more expensive and the use of boats for crop management is an additional requirement. These costs can be a challenge for poor coastal communities, which constitute the majority of *K. alvarezii* producers (Msuya and Salum 2007).

Seaweed and seaweed extract have been used extensively in agricultural crops such as lettuce (Crouch et al. 1990), tomatoes (Crouch and van Staden 1992), potatoes (Kowalski et al. 1999), barley (Rayorath et al. 2008), canola (Ferreira 2002) bean, wheat, and maize (Blunden et al. 1997), spinach (Fan et al. 2011), carrot (Jayaraj et al. 2008), cucumber (Jayaraj et al. 2011) and horticulture as a source of nutrients, bioactive compounds and biostimulants. The report of Craigie (2011) on seaweed extract stimuli in plant science and agriculture presented a comprehensive account of the many potential applications of seaweed extract. *Ascophyllum nodosum*, a brown alga abundant in the temperate waters of the North Atlantic (Ugarte et al. 2006, Ugarte and Sharp 2001) is one of the economically important seaweeds which are used in the production of liquid or powder extracts (Craigie 2011).

There are only few reports on the use of seaweed extract, as applied to aquatic plants or seaweeds. Kelpak and Acadian Marine Explant Powder (AMPEP) extracted from *Ecklonia maxima* and *Ascophyllum nodosum*, respectively, are the only known seaweed extracts used in the cultivation of another seaweed. Kelpak has been used effectively for the tips of *Gracilaria gracilis* when grown under laboratory conditions, and for *Ulva* grown on a pilot-scale, as a natural feed to abalone, using a 1:2500 concentration in combination with the effluents of turbot fish, demonstrated the highest growth, suggesting that Kelpak could be used in commercial seaweed mariculture (Roberstson-Anderson et al. 2006).

5.1 Development of Kappaphycus Varieties Plantlets Through Tissue Culture Techniques

Acadian Marine Plant Extract Powder (AMPEP) extracted from *A. nodosum* has the following properties (Table 4). Extract from *A. nodosum* either in powder or liquid form has been used extensively for the benefit of land plants. However, recently, AMPEP has been found to work efficiently and economically for the benefit of seaweeds (Hurtado et al. 2009, Loureiro et al. 2010, Borlongan et al. 2011, Yunque et al. 2011). The use of AMPEP in the tissue culture of three different varieties of *K. alvarezii* and one variety of *K. striatum* proved to be an efficient medium for the regeneration of young plants, for the production of seed stock for nursery and out-planting purposes (Hurtado et al. 2009). The authors further claimed that the addition of Plant Growth Regulators (i.e., Phenyl Acetic Acid (PAA) + zeatin) to AMPEP hastened shoot formation of the explants, as compared to AMPEP when used singly at lower concentrations (0.001–0.10 mg/L). To prove further that AMPEP was an efficient and economical major source of culture media in tissue culture techniques, optimal media concentrations with, or without PGRs, pH-temperature combinations, and explant density: media volume combinations were determined to improve the production of *Kappaphycus* plantlets. *Kappaphycus alvarezii* var. tambalang purple (PUR), Kapilaran brown (KAP), Vanguard brown (VAN), Adik-Adik (AA), Tungawan green (TGR), and *K. striatum* var. Sacol green (GS) were used as explants. Based on the shortest period for shoot emergence and the economical use of AMPEP, the optimum enriched media were 3.0 mg/L AMPEP and 0.1 mg/L AMPEP+PGR 1 mg/L each PAA and zeatin for PUR;

Table 4. Physical and chemical properties of Acadian Marine Plant Extract Powder (AMPEP) extracted from the brown alga *Ascophyllum nodosum.*

Physical data	
Appearance	Brownish-black crystals
Odor	Marine odor
Solubility in water	100%
pH	10.0–10.5
Typical analysis	
Maximum moisture	6.50%
Organic matter	45–55%
Ash (Minerals)	45–55%
Total nitrogen (N)	0.8–1.5%
Available phosphoric acid (P2O5)	1–2%
Soluble potash (K2O)	17–22%
Sulfur (S)	1–2%
Magnesium (Mg)	0.2–0.5%
Calcium (Ca)	0.3–0.6%
Sodium (Na)	3–5%
Boron (B)	75–100 ppm
Iron (Fe)	75–250 ppm
Manganese (Mn)	5–20 ppm
Copper (Cu)	1–5 ppm
Zinc (Zn)	25–50 ppm
Carbohydrates	Alginic acid, mannitol, laminarin

1.0 mg/L AMPEP+PGR for KAP and GS, 0.1 mg/L AMPEP+PGR for VAN; and 3.0 mg/L AMPEP and 0.001 mg/L AMPEP+PGR for AA and TGR. Results showed that the addition of the PGR to low concentrations of AMPEP hastened shoot formation. pH–temperature combinations for the most rapid shoot formation were determined for the brown (KAP) and purple (PUR) color morphotypes of *K. alvarezii* var. tambalang and the green morphotype of *K. striatum* var. sacol (GS) cultured in 1.0 mg/L AMPEP+PGR. The brown morphotype produced the most number of shoots at pH 7.7 at 20°C after as short as 20 days. Purple *K. alvarezii* showed an increased shoot formation at pH 6.7 at 25°C and the green *K. striatum* morphotype at pH 8.7 at 25°C. These results simply indicated that each variety responded differently to the pH-temperature combinations tested. The optimal number of explants added to the culture media was also determined for Tungawan green (TGR), Brown (KAP), and Tambalang purple (PUR) varieties of *K. alvarezii* in 1.0 mg/L AMPEP+PGR. The number of explants and the volume of the culture media combination were also tested. The highest average number of shoots formed occurred in two explants: 1 mL culture media (2:1) for KAP and PUR (35 percent and 17 percent, respectively) and 1 explant: 2 mL culture media for the TGR (100 percent) with a range of 0.5–3.0 mm shoot length after 40 days in culture. The earliest shoot formation was observed after 21 days for the brown and 9 days for both the green and purple color morphotypes of *Kappaphycus*, in all densities investigated. This indicated that within the range tested, the density of explants did not have a significant effect on the rate of shoot formation but did influence the average number generated from the culture. The rate of production of new and improved *Kappaphycus* explants for a commercial nursery stock was improved through the use of AMPEP with optimized culture media pH, temperature, and density conditions (Yunque et al. 2011). These findings parallel those of Rayorath et al. (2008) on the effect of *A. nodosum* extract to induce amylase activity, independent of gibberellic acid (GA_3) which may act in concert with GA-dependent amylase production leading to enhanced germination and seedling vigor in barley. Similar effects of *A. nodosum* extract have been reported for the model plant *Arabidopsis thaliana* during early root and shoot formation (Rayorath et al. 2008).

The viability of laboratory generated, and hatchery reared, plantlets of different varieties of *Kappaphycus* developed from tissue culture techniques, using AMPEP for out-planting purposes is encouraging as demonstrated from the successful initial works at Southeast Asian Fisheries Development Center, Aquaculture Department, Tigbauan, Iloilo, Philippines (Figs. 13a-d) and the National Seaweed Technology Development Center, Cabid-

Figure 13. (a-d) Plantlet regenerants using tissue culture techniques at SEAFDEC-AQD, Tigbauan, Iloilo, Philippines (bar = 1 cm, Photos courtesy of A.Q. Hurtado).

an, Sorsogon, Philippines (Figs. 14a-d). Mass production of plantlets of *Kappaphycus* varieties using AMPEP is thus encouraged to develop good quality propagules for commercial cultivation; this should be scaled-up to warrant pilot-testing and demonstration farming of the 'new and improved' *Kappaphycus* plantlets.

Figure 14. (a-d) Plantlet regenerants using tissue culture at NSTDC Cabid-an Sorsogon, Philippines (bar = 1 cm, Photos courtesy of I.T. Capacio).

Color image of this figure appears in the color plate section at the end of the book.

5.2 Growth of Kappaphycus *in Commercial Nurseries and Mariculture*

The use of AMPEP in commercial nurseries in Mindanao and Palawan, Philippines proved to accelerate the growth of *K. alvarezii* var. tambalang and *K. striatum* var. duyan-duyan. The study of Hurtado et al. (2012) showed that the average monthly growth rate of *K. alvarezii* (yellowish brown) ranged from 0.8 to 6.7 percent/day between July and January, respectively (Fig. 15). Significant differences (P<0.05) were observed between the various months of culture. Declining growth rates were observed from February to July while increasing growth rates were observed from August to October, with a slight decrease in November (but reaching its highest growth rate in January). The authors further reported that dipping the three color morphotypes of *K. alvarezii* for 30 min at the lower AMPEP concentrations of 0.01–0.1 g/L resulted in higher growth rates compared to the controls. The active compounds within the seaweed extract must be applied as small doses to be effective, as also claimed by Robertson-Anderson et al. (2006). The proliferation of young multiple shoots (Fig. 16) of the three color morphotypes of *K. alvarezii* and the yellowish brown

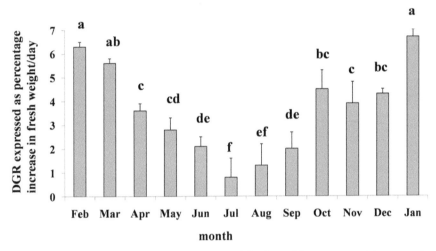

Figure 15. Average (± SD) daily growth grate of *K. alvarezii* in a commercial nursery in Zamboanga (Hurtado et al. 2012).

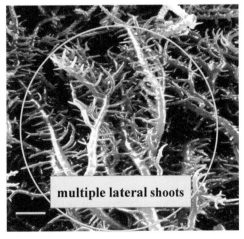

Figure 16. Multiple shoots in *K. alvarezii* treated with AMPEP (Photo courtesy of A.Q. Hurtado).

Color image of this figure appears in the color plate section at the end of the book.

K. alvarezii, grown in a commercial nursery, after 10–14 days of field growth could be attributed to the presence of growth elicitors which stimulated auxin-like activity in the seaweed. This further promoted shoot growth and paralleled the results of the study of Rayorath et al. (2008b), wherein the components of the commercial *A. nodosum* extract modulated the concentration and localization of auxins, which could account, at least in part, for the enhanced plant growth of the model plant *Arabidopsis thaliana.* Hurtado et al. (2012) further stated that though it was beyond the scope of the their study to speculate on the active compounds or mode of action of AMPEP, the enhanced plant growth effects of the extract treatment on the color morphotypes of *K. alvarezii* could be correlated with 'auxin-like', 'gibberellin-like' elicitors and primers (Rayorath et al. 2008), precursors of ethylene and betaine (Mckinnon et al. 2010) and cytokinin-like effects which are potentially involved in enhancing plant growth responses (Crouch and van Staden 1992, Wally et al. 2013).

The use of AMPEP in commercial nurseries prevented the worst of the deleterious effects of macro-epiphytism. Though *K. alvarezii* was enveloped with the deleterious *Ulva,* in the trials, no traces of pits or cavities were observed on the surface of the cultured seaweed (Hurtado et al. 2012), in fact the surface of the seaweed remained smooth and clear and the whole thallus appeared to be brittle. Severe macro-epiphytism, especially during periods of intermittent rains and sunny days, is a serious problem for the seaweed farmers. Sometimes when macro-epiphytes are dense, they are not readily removed, and therefore can remain attached to *K. alvarezii* for several days; fragmentation of the cultured seaweed was seen to be the ultimate result. Such phenomena were observed with *K. alvarezii* seedlings which had not been dipped in AMPEP. At this time, only few seaweed farmers use AMPEP in seaweed farming, but the number is expected to grow. In experiments using AMPEP in Brazilian commercial cultivation, it was observed that the rafts near the samples treated with AMPEP had less biofouling (R. Marroig, personal communication).

In the commercial mariculture of *K. striatum* (barako) in Panabulon, Guimaras Is., Philippines, the use of 1 mg/L AMPEP was sufficient to increase the monthly biomass by 2.5–3.5x the original weight (Data not shown). Further, consistent monthly dipping of the seaweed into a solution of 1 mg/L AMPEP increased the resistance to 'ice-ice' and *Neosiphonia* infestation for almost 18 months, without any change in the source of propagules (E. Ferrer, personal communication). The usual practice of seaweed farmers in the Philippines is to use only the material from the same propagules over 2–3 cycles of production. After such time the plant was deemed to be no longer viable due to aging, 'ice-ice' and/or *Neosiphonia* infestations and new sources of seedlings had be obtained. However, the erratic weather conditions experienced in 2011 in the Philippines, in the western Visayas in particular and the continuous vegetative cutting of the same seaweed stock perhaps contributed to weakening and loss of vigor of the commercial seaweed, which after 18 months became susceptible to *Neosiphonia* infestation brought on by environmental stresses.

The report of Borolongan et al. (2011) showed that dipping the cultivars of *Kappaphycus* in AMPEP induced more vigor and increased resistance to 'ice-ice' malaise and *Neosiphonia* infestation when the seedlings were experimentally cultivated at four different depths. The authors claimed that the use of AMPEP significantly increased ($P<0.05$) the growth rate of TUNG (Tungawan) and GTAM (Giant Tambalang) *Kappaphycus* varieties tested, but also decreased the percent occurrence of *Neosiphonia* sp. The percent occurrence of *Neosiphonia* sp. infection (6–50 percent at all depths) of both the *Kappaphycus* varieties tested (Figs. 17a-b) with AMPEP treatment, was significantly lower than the controls (i.e., 10–75 percent at all depths). Both the growth rate of the cultivated seaweed and the percent occurrence of the epiphytes decreased as the cultivation depth increased. Plants dipped in AMPEP, and suspended at the surface, had the highest growth rates (i.e., 4.1 percent, TUNG; 3.1 percent, GTAM) after 45 days. Those which were not dipped in AMPEP had the highest percentage occurrence of *Neosiphonia* infection (viz. 70–75 percent). The occurrence of the *Neosiphonia* infestation was found to correlate with changes in irradiance and salinity at the depths observed (Table 5). These results suggested that both varieties of *K. alvarezii*, used in this study, had the fastest growth rate when grown immediately at the surface. However, in order to minimize damage caused by the occurrence of epiphytic *Neosiphonia*, it was recommended that *K. alvarezii* should be grown within a depth range of 50–100 cm. These observations are important for the improved management of *Kappaphycus* for commercial farming. Furthermore, the use of AMPEP treatments for enhanced growth and reduction of *Neosiphonia* sp. infections parallels the results obtained in agricultural crops such as spinach (Fan et al. 2011), carrot (Jayaraj et al. 2008) and greenhouse cucumber (Jayaraj et al. 2011).

The administration of AMPEP extract to *K. alvarezii* under laboratory conditions reduced the effects of the oxidative burst (production of hydrogen peroxide) which may be extremely aggressive for an individual and its epiphytes (Loureiro et al. 2012). The bleaching of the non-corticated portions of *Polysiphonia subtilissima* thalli, which were co-cultivated as simulated epiphytes with AMPEP treatment, confirmed that the reaction was from hydrogen peroxide effects. The use of AMPEP acted as "seaweed vaccine", eliciting activation of the natural defenses of *K. alvarezii* against pathogens and thereby ameliorated the negative effects of long term exposure to oxidative bursts. The effectiveness of the commercial *Ascophyllum* extract was also confirmed on *K. alvarezii* thalli that were protected from bleaching, as compared with their control. Also *in situ*, when *K. alvarezii* was submitted

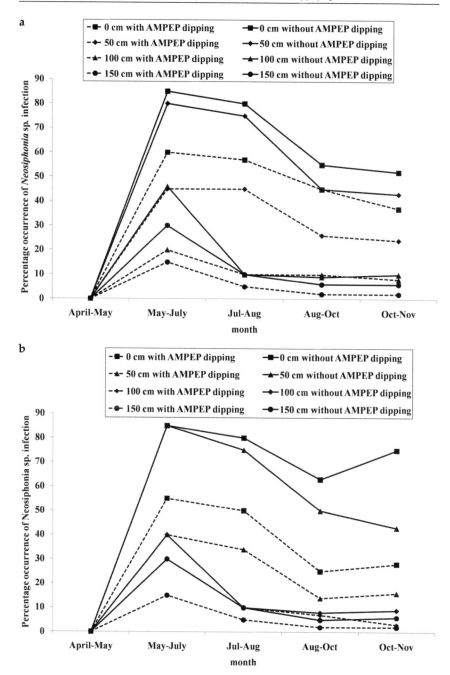

Figure 17. (a-b) % occurrence of *Neosiphonia* in *K. alvarezii* (giant tambalang) (Borlongan et al. 2011).

Table 5. Correlation coefficient between percent occurrence of *Neosiphonia* sp. and some environmental parameters in two varieties of *K. alvarezii* dipped and not-dipped in AMPEP (Borlongan et al. 2011).

Varieties of *K. alvarezii*	Irradiance μmol photons m^{-2}s^{-1}	Salinity ppt	Temperature °C	Turbidity NTU
Tungawan with AMPEP dipping	1.00	–0.99	0.21	–0.03
Tungawan without AMPEP dipping (control)	0.94	–0.95	0.46	–0.21
Giant tambalang with AMPEP dipping	0.97	–0.98	0.36	–0.14
Giant tambalang without AMPEP dipping (control)	0.94	–0.96	0.47	–0.25

to lethal temperatures that occur in some countries (16±1°C) where this species was introduced, the seedlings treated with AMPEP can be used as a preventive action for the cultivation of the seedlings in tanks and in the sea in periods of low temperatures since higher growth rates and carrageenan yield were obtained (Loureiro et al. 2013). They suggested that the better growth was due to the presence of betaines (that promote protection of the disruption of electron transport by the photosystem II and degradation of pigments) and cytokinin (an elicitor activity in the AMPEP extract and its antioxidant properties). Probably, the proline acted as mediator of osmotic adjustments, stabilizer of proteins and membranes, inducer of osmotic stress related genes, and source of reduction equivalents during stress recovery. Moreover, the elevated presence of LPCs significantly increased the concentration of soluble sugars in the cytosol in response to freezing stress, showing the mode of action of LPCs in protecting the tested samples at low temperatures and their effects in addition to priming cold response genes. Also the exudation of H_2O_2 (preventing effective action of catalase—CAT and the induction of ascorbate peroxidase gene—APX) may be responsible for the delay in responses appropriate to CAT and APX which guarantees an acclimatization in the long term, permitting *K. alvarezii* (Barros et al. 2006) to increase its growth rate under adverse environmental conditions combined with favorable compounds present in AMPEP and its antioxidant action.

In relation to the increased carrageenan yield on AMPEP treated samples, it is known that AMPEP act as defense-compound of macroalgae in the elicitation and priming in response to abiotic stresses. Since their daily growth rate was unaffected, it seems that the chlorophyll content was less affected. Nair et al. (2012) attest that AMPEP has a positive effect on reducing the expression of chlorophyllase under chilling stress.

6 Socio-economic Impact of Seaweed Diseases in *Kappaphycus* Cultivation

The socio-economic impacts of seaweed farming on coastal communities have been overwhelmingly positive (Alih 1990, Firdausy and Tisdell 1991, Samonte et al. 1993, Hurtado-Ponce et al. 1996, Crawford 1999, Hurtado et al. 2001, Hurtado and Agbayani 2002, Sievanen et al. 2005, Msuya et al. 2007, Hurtado et al. 2008, Hurtado 2013). Family-type operations are far more advantageous than the older style, initially corporate-owner-operator style farms, especially in remote areas where coastal communities are faced with a limited number of economic alternatives. The seaweed industry in Zamboanga Peninsula and elsewhere served as an alternative activity to farmers in their quest for survival, security and development (Jain 2006). Whether the industry can ultimately liberate the farmers from the trap of poverty depends largely on the support and attention that can be provided to the poor farmers by the government, local and international non-government organizations, big export traders, and also the local and multi-national corporations engaged in carrageenan processing. Seaweed farming provides an income that can help alleviate the hardship faced by this vulnerable group. The income from seaweed farming also reduces the impact of lack of food and malnutrition. The concept that the seaweed industry is the farmer's alternative to poverty holds true, depending on the adequacy of income achieved to allow the farmers access to the basic and prime necessities for survival security and empowerment (Jain 2006).

Prior to engaging in seaweed farming, many of the community or island people were marginal fishermen living below the poverty level. With sheer diligence and hard work, the life of the seaweed farmers became more meaningful. Incomes derived from the sale of seaweeds, allowed many farmers to experience substantial improvements in their standard of living, such as education for their children to college level, introduced improvements to their dwellings (from light materials to concrete houses), enhanced diets, increased purchasing power of material goods (e.g., appliances, vehicles, communications devices (mobile)) phones, and also access to political leadership in the community, improved banking credibility and social acceptance in the community (Cooke 2004, Hurtado 2005a). Women in seaweed farming are doubly respected in their communities, being engaged in an income-earning activity and at the same time efficient housewives (Aming 2004, Hurtado 2005a,b).

The most recent report on the socio-economic dimensions of seaweed aquaculture (Hishamunda and Valderrama 2013) is a collection of six contributions from six countries: Indonesia (Neish 2013), the Philippines (Hurtado 2013), Tanzania (Msuya et al. 2013), India (Krishnan et al. 2013), the Solomon Islands (Kronen et al. 2013) and Mexico (Robledo et al. 2013).

Each contributor emphasized that there is durability even under internal-chain stress; stability in the face of internal value-chain shocks, robustness under external value-chain stresses, resilience in the face of external value-chain shocks, and sustainability of value-chain functions over time and generations of farmers. Though the profitability of seaweed farming during 'ice-ice' outbreaks and *Neosiphonia* infestations seemed to reduce significantly since the carrageenophyte crops are damaged, amidst these circumstances, the seaweed farmers are resilient enough to bounce back. The seaweed farmers are still in the business after more than 40 years of providing the raw material for the processing of carrageenan. They were able to cope with, and recovered from stresses and shocks, maintained and enhanced the capabilities and assets, while not undermining the natural resources as reflected from their many years of seaweed farming and improved family life, both socially and economically. Until the present day, seaweed aquaculture is a sustainable activity (Hurtado 2013).

7 Global Future of Seaweed Farming in Relation to Diseases Brought by Climate Change

A limited number of papers have shown the impact of 'ice-ice' outbreaks and *Neosiphonia* infestations on seaweed farming. Surface seawater temperature, salinity, pH, water movement and light intensity all act in combination to play significant roles in *Kappaphycus* health and growth. It is not conclusively established if these changes are in fact related to global climate change in surface seawater temperature and/or other environmental/biological variables. There is an imperative need to pursue further research studies on the physiological and ecological processes involved between *K. alvarezii* and its environment, especially in relation to its productivity. The initial move to use the commercial extract from *A. nodosum* as biostimulant for growth, health and vigor is very encouraging. However, a thorough investigation to elucidate the mechanism(s) by which AMPEP exerts effects on enhancing disease resistance through the induction of defense genes or enzymes in *K. alvarezii* needs to be undertaken. When this aspect is fully understood, then possibly, the production of *K. alvarezii* will increase. On the other hand, the quest for faster growing and disease resistant strain(s), with ever high yielding carrageenan *K. alvarezii* must continue through natural selection, not through genetic modification.

The economic impact of the decline of *Kappaphycus* cultivation in some areas, especially the Philippines, requires more intensive and regional investigation due to its economic importance. There is an important need to sustain carrageenophyte farming due to the innumerable economic and coastal development benefits which one derives, if practiced and maintained in a sustainable manner. The key players along the value chain must work in harmony, providing increased and sustained effort to continue to provide the industry with this high-value natural colloid. If this is not the case, industry could switch to other sources of hydrocolloid with similar or competing rheological properties.

References Cited

Adnan, H. and H. Porse. 1987. Culture of *Eucheuma cottonii* and *Eucheuma spinosum* in Indonesia. Hydrobiologia 151/152: 355–358.

Alih, E.M. 1990. Economics of seaweed (*Eucheuma*) farming in Tawi-Tawi Islands, Philippines. *In*: Hirano, R. and I. Hanyu (eds.). Asian Fisheries Forum 2: 249–252.

Aming, N.A. 2004. Participation of Filipino Muslim women in seaweed farming in Sitangkai, Tawi-Tawi, Philippines. PhD dissertation, University of the Philippines in Los Baños, Laguna, Philippines.

Apt, K.E. 1988. Etiology and development of hyperplasia induced by *Streblomena* sp. (Phaeophyta) on members of the Laminariales (Phaeophyta). Phycologia 24: 28–34.

Ask, E.I. 1999. *Cottonii* and *Spinosum* Cultivation Handbook. FMC Food Ingredients Division, Philadelphia, 52 pp.

Ask, E.I. 2001. Creating sustainable commercial *Eucheuma cultivation* industry: the importance and necessity of human factor. *In*: Chapman A.R.O., R.J. Anderson, V.J. Vreeland and I.R. Davison (eds.). Proceedings 17th International Seaweed Symposium, Cape Town South Africa. 13–18.

Ask, E.I. and R.V. Azanza. 2002. Advances in cultivation technology of commercial eucheumatoid species: a review with suggestions for future research. Aquaculture 206: 257–277.

Azanza, R.V. and T.T. Aliaza. 1999. *In vitro* carpospore release and germination in *Kappaphycus* (Doty) Doty from Tawi-Tawi, Philippines. Botanica Marina 42: 281–284.

Azanza-Corrales, R., T.T. Aliaza and M.N.E. Montaño. 1996. Recruitment of *Eucheuma* and *Kappaphycus* on a farm in Tawi-Tawi, Philippines. Hydrobiologia 326/327: 235–244.

Barrios, J.E. 2005. Spread of exotic algae *Kappaphycus alvarezii* (Gigartinales: Rhodophyta) in the northeast region of Venezuela. Boletin del Instituto Oceanografico de Venezuela 44(1): 29–34.

Barros, M.P., O. Necchi Jr., P. Colepicole and M. Pedersen. 2006. Kinetic study of the plastoquinone pool availability correlated with H_2O_2 release in seawater and antioxidant responses in the red alga *Kappaphycus alvarezii* exposed to single or combined high light, chilling and chemical stresses. Biochimica et Biophysica Acta 1757: 1520–1528.

Batista, G. 2006. Seaweed resources of Panama. *In*: Critchley, A.T., M. Ohno and D.B. Largo (eds.). World Seaweed Resources: An Authoritative Reference System. DVD University of the Netherlands and UNESCO.

Batista, G. 2009. Cultivo eco-sostenible de *Kappaphycus alvarezii* en Panamá. Ph.D. Thesis, Universidad Las Palmas de Gran Canaria, España.

Bindu, M.S. and I.A. Levine. 2011. The commercial red seaweed *Kappaphycus alvarezii*—an overview on farming and environment. Journal of Applied Phycology 23: 789–796.

Blunden, G.T., T. Jenkins and Y.W. Liu. 1997. Enhanced chlorophyll levels in plants treated with seaweed extract. Journal of Applied Phycology 8: 535–543.

Booth, W.E., R.K. Solly and C.P. Wright. 1983. Preliminary attempts to culture *Eucheuma striatum* Schmitz (Rhodophyta) in Fiji waters. Fiji Agricultural Journal 45: 55–63.

Borlongan, I.A.G., K.R. Tibubos, D.A.T. Yunque, A.Q. Hurtado and A.T. Critchley. 2011. Impact of AMPEP on the growth and occurrence of epiphytic *Neosiphonia* infestation on two varieties of commercially cultivated *Kappaphycus alvarezii* grown at different depths in the Philippines. Journal of Applied Phycology 23: 615–621.

Bouarab, K., P. Potin, F. Weinberger, J.A. Correa and B. Kloareg. 2001. The *Chondrus crispus* *Acrochaete operculata* host-pathogen association, a novel model in glycobiology and applied phycopathology. Journal of Applied Phycology 13: 185–193.

Bulboa, C., E.J. de Paula and F. Chow. 2007. Laboratory germination and sea out-planting of tetrraspore progeny from *Kappaphycus striatum* (Rhodophyta) in subtropical waters of Brazil. Journal of Applied Phycology 19: 357–363.

Bulboa, C., E.J. de Paula and F. Chow. 2008. Germination and survival of tetraspores of *Kappaphycus alvarezii* var. *alvarezii* (Solieriaceae, Rhodophyta) introduced in sub-tropical waters of Brazil. Phycological Research 56: 39–45.

Bulboa, C.R. and E.J. de Paula. 2007. Introduction of non-native species of *Kappaphycus* (Rhodophyta) in sub-tropical waters: comparative analysis on growth rates of *Kappaphycus alvarezii* and *Kappaphycus striatum in vitro* and in the seas in south-eastern Brazil. Phycological Research 53: 183–188.

Butardo, V.M., E.T. Ganzon-Fortes, V. Silvestre, M.B.B. Bacano-Maningas, M.N.E. Montaño and A.O. Lluisma. 2003. Isolation and classification of culturable bacteria associated with ice-ice disease in *Kappaphycus* and *Eucheuma* (Rhodophyta, Solieriaceae). Philippine Scientist 40: 223–237.

Castelar, B., R.P. Reis, A.L. Moura and R. Kirk. 2009. Invasive potential of *Kappaphycus alvarezii* off the south coast of Rio de Janeiro State, Brazil: a contribution to environmentally secure cultivation in the tropics. Botanica Marina 52: 283–289.

Chandrasekaran, S., N.A. Nagendran, D. Pandiaraja, N. Krishnankutty and B. Kamalakannan. 2008. Bioinvasion of *Kappaphycus alvarezii* on corals in the Gulf of Mannar, India. Current Science 94: 1167–1172.

Collen, J., M. Mtolera, K. Abrahamsson, A. Semesi and M. Pedersen. 1995. Farming and physiology of the red algae *Eucheuma*: growing commercial importance in East Africa. Ambio 24: 497–501.

Conklin, E.J. and J.E. Smith. 2005. Abundance and spread of the invasive red algae, *Kappaphycus* spp., in Kane'ohe Bay, Hawai'i and an experimental assessment of management options. Biological Invasions 7: 1029–1039.

Cooke, F.M. 2004. Symbolic and social dimensions in the economic production of seaweed. Asia Pacific Viewpoint 45: 387–400.

Correa, J.A. and J.L. McLachlan. 1991. Endophytic algae of *Chondrus crispus* (Rhodophyta) 3. Host specificity. Journal of Applied Phycology 27: 448–459.

Correa, J.A. and J.L. McLachlan. 1992. Endophytic algae of *Chondrus crispus* (Rhodophyta). IV. Effects on the host following infections by *Acrochaete operculata* and *A. heteroclada* *(Chlorophyta)*. Marine Ecology Progress Series 81: 73–87.

Correa, J.A. and J.L. McLachlan. 1994. Endophytic algae of *Chondrus crispus* (Rhodophyta). V. Fine structure of the infection by *Acrochaete operculata* (Chlorophyta). European Journal of Phycology 29: 33–47.

Craigie, J.S. 2011. Seaweed extract stimuli in plant science and agriculture. Journal of Applied Phycology 23: 371–393.

Crawford, B. 1999. Seaweed farming: an alternative livelihood for small-scale fishers? Working Paper. Coastal Resources Center, University of Rhode Island, 10 pp.

Critchley, A.T., D.B. Largo, W. Wee, G. Bleicher-L'Honneur, A. Hurtado and J. Shubert. 2004. A preliminary summary of *Kappaphycus* farming and the impacts of epiphytes. Japan Journal of Phycology 52: 231–232.

Crouch, I.J. and J. van Staden. 1992. Effect of seaweed concentrate on the establishment and yield of greenhouse tomato plants. Journal of Applied Phycology 4: 291–296.

Crouch, I.J., R.P. Beckett and J. van Staden. 1990. Effect of seaweed concentrate on the growth and mineral nutrition of nutrient stressed lettuce. Journal of Applied Phycology 2: 269–272.

Davis, E. 2011. Seaweed farmer education: Is it enough to sustain the industry? Analyzing the status of stakeholder investment in Muungoni and Jambiani, Unguja. ISP Collection. Paper 1195. 41 pp.

Dawes, C.J. and E.W. Koch. 1991. Branch micropropagule and tissue culture of the red algae *Eucheuma denticulatum* and *Kappaphycus alvarezii* farmed in the Philippines. Journal of Applied Phycology 3: 247–257.

Dawes, C.J., G.C. Trono Jr. and A.O. Lluisma. 1993. Clonal propagation of *Eucheuma denticulatum* and *Kappaphycus alvarezii* for Philippine seaweed farms. Hydrobiologia 260-261: 379–383.

Dawes, C.J., A.O. Lluisma and G.C. Trono Jr. 1994. Laboratory and field growth studies of commercial strains of *Eucheuma denticulatum* and *Kappaphycus alvarezii* in the Philippines. Journal of Applied Phycology 6: 21–24.

Doty, M.S. 1973. Farming the red seaweed, *Eucheuma*, for carrageenans. Micronesica 9: 59–73.

Doty, M.S. 1987. The production and use of *Eucheuma*. *In*: Doty, M.S., J.F. Caddy and B. Santelices (eds.). Case Study of Seven Commercial Seaweed Resources. FAO Fish. Tech. Paper 281: 123–161.

Doty, M.S. and V.B. Alvarez. 1981. *Eucheuma* farm productivity. International Seaweed Symposium 8: 688–691.

Eklund, S. and P. Pettersson. 1992. Mwani is money: the development of seaweed farming and its socio-economic effects in the village of Paje. Department of Social Anthropology, Stockholm University. (Development Studies Unit Working Paper No. 24.

Eswaran, K. and B. Jha. 2006. Commercial cultivation of *Kappaphycus alvarezii* in India. Training manual, National Training Workshop on Seaweed Farming and Processing for Food, The Thassim Beevi Abdul Kader College for Women, Kilakarai, India, 40–46.

Fan, D., D.M. Hodges, J.Z. Zhang, C.W. Kirby, X.H. Ji, S.J. Locke, A.T. Critchley and B. Prithiviraj. 2011. Commercial extract of the brown seaweed *Ascophyllum nodosum* enhances phenolic antioxidant content of spinach (*Spinacia aloracea* L.) which protects *Caenorhabditis elegans* against oxidative and thermal stress. Food Chemistry 124: 195–202.

FAO. 2010. The state of world's fisheries and aquaculture. Technical Paper. Food and Agriculture Organization of the United Nations, Rome, 218 pp.

Ferreira, M.I. 2002. The efficacy of liquid extract on the yield of canola plants. South African Journal of Plant Soil 19: 159–161.

Firdausy, C. and C. Tisdell. 1991. Economic returns from seaweed (*Eucheuma cottonii*) farming in Indonesia. Asian Fisheries Science 4: 61–73.

Gauna, M.C., E.R. Parodi and E.J. Caceres. 2009. Epi-endophytic symbiosis between *Laminariocolax aecidioides* (Ectocarpales, Phaeophyceae) and *Undaria pinnatifida* (Laminariales, Phaeophyceae) growing on Argentinian coasts. Journal of Applied Phycology 21: 11–18.

Góes, H.G. and R.P. Reis. 2011. An initial comparison of tubular netting versus tie–tie methods of cultivation for *Kappaphycus alvarezii* (Rhodophyta, Solieriaceae) on the south coast of Rio de Janeiro State, Brazil. Journal of Applied Phycology 23: 607–613.

Hayashi, L., N.S. Yokoya, D.M. Kikuchi and E.C. Oliveira. 2007. Callus induction and micropropagation improved by colchicine and phytoregulators in *Kappaphycus alvarezii* (Rhodophyta, Solieriaceae). Journal of Applied Phycology 20: 653–659.

Hayashi, L., A.Q. Hurtado, F.E. Msuya, G. Bleicher-Lhonneur and A.T. Critchley. 2010. A review of *Kappaphycus* farming: prospects and constraints. pp. 255–279. *In*: Israel, A., R. Einav and J. Seckback (eds.). Seaweeds and their Role in Changing Global Environments. Springer, New York.

Hishamunda, N. and D. Valderrama. 2013. Social and Economic Dimensions of Seaweed Cultivation: Assessment and Case Studies. FAO Fisheries and Aquaculture Technical Paper 580. Rome, FAO. 204 pp.

Hurtado, A.Q. 2005a. Effect of epiphytes on the productivity of *Kappaphycus*: impact of diseases on the economics of producing and processing *Kappaphycus*. Terminal Report submitted to Degussa Texturant System (now Cargill Texturizing Solutions), France.

Hurtado, A.Q. 2005b. The role of women and children in seaweed farming activities in the Philippines. Report submitted to International Finance Corporation (IFC), USA.

Hurtado, A.Q. 2007. Establishment of Seaweed Nurseries in Zamboanga City, Philippines. Terminal Report submitted to International Finance Corporation (IFC, USA) and Asian Development Bank (ADB, Phil.).

Hurtado, A.Q. 2013. Socio-economic dimensions of seaweed farming in the Philippines. *In*: Hishamunda, N. and D. Valderrama (eds.). Social and Economic Dimensions of Seaweed Cultivation: Assessment and Case Studies. FAO Fisheries and Aquaculture Technical Paper 580. Rome, FAO. pp. 91–113

Hurtado, A.Q. and R.F. Agbayani. 2000. The farming of the seaweed *Kappaphycus*. SEAFDEC Aquaculture Department Extension Manual No. 32.

Hurtado, A.Q. and R.F. Agbayani. 2002. Deep-sea farming of *Kappaphycus* using multiple raft long-line method. Botanica Marina 45: 438–444.

Hurtado, A.Q. and D.P. Cheney. 2003. Propagule production of *Eucheuma denticulatum* (Burman) Collins et Harvey by tissue culture. Botanica Marina 46: 338–341.

Hurtado, A.Q. and A.T. Critchley. 2006. Seaweed industry of the Philippines and the problem of epiphytism in *Kappaphycus* farming. pp. 21–38. *In*: Phang, S.M., A.T. Critchley and P. Ang Jr. (eds.). Advances in Seaweed Cultivation and Utilization in Asia. University of Malaya Maritime Research Centre, Kuala Lumpur, Malaysia.

Hurtado, A.Q. and A.B. Biter. 2008. Plantlet regeneration of *Kappaphycus alvarezii* var. adik-adik by tissue culture. Journal of Applied Phycology 19: 783–786.

Hurtado, A.Q., R.F. Agbayani, R. Sanares and M.T.R. de Castro-Mallare. 2001. The seasonality and economic feasibility of cultivating *Kappaphycus alvarezii* in Panagatan Cays, Caluya, Antique, Philippines. Aquaculture 199: 295–310.

Hurtado, A.Q., A.T. Critchley, A. Trespoey and G. Bleicher-L'Honneur. 2006. Occurrence of *Polysiphonia* epiphytes in *Kappaphycus* farms at Calaguas Is., Camarines Norte, Philippines. Journal of Applied Phycology 18: 301–306.

Hurtado, A.Q., A.T. Critchley and G. Bleicher-L'Honneur. 2008. *Kappaphycus 'cottonii'* farming. (Revised ed.) Baupte, France.

Hurtado, A.Q., D.A. Yunque, K. Tibubos and A.T. Critchley. 2009. Use of Acadian marine plant extract powder from *Ascophyllum nodosum* in tissue culture of *Kappaphycus* varieties. Journal of Applied Phycology 21: 633–639.

Hurtado, A.Q., M. Joe, R.C. Sanares, D. Fan, B. Prithiviraj and A.T. Critchley. 2012. Investigation of the application of Acadian Marine Plant Extract Powder (AMPEP) to enhance the growth, phenolic content, free radical scavenging, and iron chelating activities of *Kappaphycus* Doty (Solieriaceae, Gigartinales, Rhodophyta). Journal of Applied Phycology 24: 601–611.

Hurtado-Ponce, A.Q., R.F. Agbayani and E.A.J. Chavoso.1996. Economics of cultivating *Kappaphycus alvarezii* using fixed-off bottom line and hanging-long line methods in Panagatan Cays, Caluya, Antique Philippines. Journal of Applied Phycology 105: 105–109.

Jain, S.A.M. 2006. The Seaweeds Industry, Exploring an Alternative to Poverty. Department of Trade and Industry, Philippines.

Jayaraj, J., A. Wan, M. Rahman and Z.K. Punja. 2008. Seaweed extract reduces foliar fungal disease on carrot. Crop Protection 10: 1360–1366.

Jayaraj, J., J. Norrie and Z.K. Punja. 2011. Commercial extract from brown seaweed *Ascophyllum nodosum* reduces fungal disease in greenhouse cucumber. Journal of Applied Phycology 23: 353–361.

Kaliaperumal, N., J.R. Ramalingan and B. Sulochanan. 2008. Growth and production of carrageenophyte *Kappaphycus alvarezii* in pilot scale cultivation in the Gulf of Mannar and Palk Bay. Seaweed Research Utilization 30: 57–66.

Kowalski, B., A.K. Jager and J. van Staden. 1999. The effect of seaweed concentrate on the *in vitro* growth and acclimatization of potato plantlets. Potato Research 42: 131–139.

Krishnan, M. and K. Narayana. 2013. Socio-economic dimensions of seaweed farming in India. *In*: Hishamunda, N. and D. Valderrama (eds.). Social and Economic Dimensions of Seaweed Cultivation: Assessment and Case Studies. FAO Fisheries and Aquaculture Technical Paper 580. Rome, FAO. 163–185.

Kronen, M., B. Ponia, T. Pickering, A. Teitelbaum, A. Meloti, J. Kama, P. Kenilolerie, S. Diake and J. Ngwaerobo. 2013. Socio-economic dimensions of seaweed farming in the Solomon Islands. *In*: Hishamunda, N. and D. Valderrama (eds.). Social and Economic Dimensions of Seaweed Cultivation: Assessment and Case Studies. FAO Fisheries and Aquaculture Technical Paper 580. Rome, FAO. 147–161.

Largo, D.B., K. Fukami and T. Nishijima. 1995a. Occasional pathogenic bacteria promoting ice-ice disease in the carrageenan-producing red algae *Kappaphycus alvarezii* and *Eucheuma denticulatum* (Solieriaceae, Gigartinales, Rhodophyta). Journal of Applied Phycology 7: 545–554.

Largo, D.B., K. Fukami, T. Nishijima and M. Ohno. 1995b. Laboratory-induced development of the ice-ice disease of the farmed red algae *Kappaphycus alvarezii* and *Eucheuma denticulatum* (Solieriaceae, Gigartinales, Rhodophyta). Journal of Applied Phycology 7: 539–543.

Le, Q. 2011. Poverty reduction from seaweed Kappaphycus alvarezii cultivation. Vietfish International. 8(3) May-Jun issue.

Leonardi, P.I., A.B. Miravalles, S. Faugeron, V. Flores, J. Beltran and J. Correa. 2006. Diversity, phenomenology and epidemiology of epiphytism in farmed *Gracilaria chilensis* (Rhodophyta) in northern Chile. European Journal of Phycology 41: 247–257.

Lirasan, T. and P. Twide. 1993. Farming *Eucheuma* in Zanzibar, Tanzania. Hydrobiologia 260/261: 353–355.

Loureiro, R.R., R.P. Reis and A.T. Critchley. 2010. *In vitro* cultivation of three *Kappaphycus alvarezii* (Rhodophyta, Areschougiaceae) variants (green, red and brown) exposed to a commercial extract of the brown alga *Ascophyllum nodosum* (Fucaceae, Ochrophyta). Journal of Applied Phycology 22: 101–104.

Loureiro, R.R., R.P. Reis, F.D. Berrogain and A.T. Critchley. 2012. Extract powder from the brown alga *Ascophyllum nodosum* (Linnaeus) Le Jolis (AMPEP): a 'vaccine-like; effect on *Kappaphycus alvarezii* (Doty) Doty ex P.C. Silva. Journal of Applied Phycology 24: 427–432.

Loureiro, R.R., R.P. Reis and R.G. Marroig. 2013. Effect of the commercial extract of the brown alga *Ascophyllum nodosum* Mont. on *Kappaphycus alvarezii* (Doty) Doty ex P.C. Silva *in situ* submitted to lethal temperatures. Journal of Applied Phycology.

Luhan, M.R.J. and H. Sollesta. 2010. Growing the reproductive cells (carpospores) of the seaweed, *Kappaphycus striatum*, in the laboratory until outplanting in the field and maturation to tetrasporophyte. Journal of Applied Phycology 22: 579–585.

Luxton, D.M. 1993. Aspects of the farming and processing of *Kappaphycus* and *Eucheuma* in Indonesia. Hydrobiologia 260/261: 365–371.

Luxton, D.M. and P.M. Luxton. 1999. Development of commercial *Kappaphycus* production in the Line Islands, Central Pacific. Hydrobiologia 398/399: 477–486.

Luxton, D.M., M. Robertson and M.J. Kindley. 1987. Farming of *Eucheuma* in the south Pacific islands of Fiji. Hydrobiologia 151/152: 359–362.

Mairh, O.P., S.T. Zodape, A. Tewari and M.R. Rajyaguru. 1995. Culture of marine red alga *Kappaphycus striatum* (Schmitz) Doty on the Saurashtra region, west coast of India. Indian Journal of Marine Science 24: 24–31.

Marroig, R. and R.P. Reis. 2011. Does biofouling influence *Kappaphycus alvarezii* (Doty) Doty ex Silva farming production in Brazil? Journal of Applied Phycology 23: 925–931.

McHugh, D.J. 2003. A guide to the seaweed industry. FAO Fisheries Technical Paper. Food and Agriculture Organization of the United Nations, Rome.

Mckinnon, S.L., D. Hiltz, R. Ugarte and C.A. Craft. 2010. Improved methods of analysis for betaines in *Ascophyllum nodosum* and its commercial seaweed extracts. Journal of Applied Phycology 22: 489–494.

Mendoza, W.G., M.N.E. Montano, E.T. Ganzon-Fortes and R.D. Villanueva. 2002. Chemical and gelling profile of *ice-ice* infected carrageenan from *Kappaphycus striatum* (Schmitz) Doty "sacol" strain (Solieriaceae, Gigartinales, Rhodophyta). Journal of Applied Phycology 14: 409–418.

Mollion, J. and J.P. Braud. 1993. A *Eucheuma* (Solieriaceae, Rhodophyta) cultivation test on the south-west coast of Madagascar. Hydrobiologia 260/261: 373–378.

Mshigeni, K.E. 1985. Pilot seaweed farming in Tanzania: progress report and future trends. Dar es Salaam, University of Dar es Salaam.

Msuya, F. 2013. Socio-economic dimensions of seaweed farming in Tanzania. *In*: Hishamunda, N. and D. Valderrama (eds.). Social and Economic Dimensions of Seaweed Cultivation: Assessment and Case Studies. FAO Fisheries and Aquaculture Technical Paper 580. Rome, FAO. 115–146.

Msuya, F.E. and D. Salum. 2007. Effect of cultivation duration, seasonality, nutrients, air temperature and rainfall on carrageenan properties and substrata studies of the seaweeds *Kappaphycus alvarezii* and *Eucheuma denticulatum* in Zanzibar, Tanzania. WIOMSA/MARG I n° 2007-06. 36.

Msuya, F.E., M.S. Shalli, K. Sullivan, B. Crawford, J. Tobey and A.J. Mmochi. 2007. A Comparative Economic Analysis of Two Seaweed Farming Methods in Tanzania. The Sustainable Coastal Communities and Ecosystems Program. Coastal Resources Center, University of Rhode Island and the Western Indian Ocean Marine Science Association.

Mtolera, M.S.P., J. Collen, M. Pedersen and A.K. Semesi. 1995. Destructive hydrogen peroxide production in *Eucheuma denticulatum* (Rhodophyta) during stress caused by elevated pH, high light intensities and competition with other species. European Journal of Phycology 30: 289–297.

Mtolera, M.S.P., J. Collen, M. Pedersen, A. Ekdahl, K. Abrahamsson and A.K. Semesi. 1996. Stress-induced production of volatile halogenated organic compounds in *Eucheuma denticulatum* (Rhodophyta) caused by elevated pH and high light intensities. European Journal of Phycology 31: 89–95.

Muñoz, J. and D. Sahoo. 2007. Impact of large scale *Kappaphycus alvarezii* cultivation in coastal water of India. Proceedings XIXth International Seaweed Symposium, Kobe, Japan. (Abstract only).

Muñoz, J., Y. Freile-Pelegrin and D. Robledo. 2004. Mariculture of *Kappaphycus alvarezii* (Rhodophyta, Solieriaceae) colour strains in tropical waters of Yucatan, Mexico. Aquaculture 239: 161–177.

Muñoz, J., A.C. Cahue-López, R. Patiño and D. Robledo. 2006. Use of plant growth regulators in micropropagation of *Kappaphycus alvarezii* (Doty) in airlift bioreactors. Journal of Applied Phycology 18: 209–218.

Nair, P., S. Kandasamy, J. Zhang, X. Ji, C. Kirby, B. Benkel, M.D. Hodges A.T. Critchley, D. Hiltz and B. Prithiviraj. 2012. Transcriptional and metabolic analysis of *Ascophyllum nodosum* mediates freezing tolerance in *Arabidopsis thaliana*. BMC Genomics 13: 643.

Neish, I.C. 2013. Socio-economic dimensions of seaweed farming in Indonesia. pp. 1–81. *In*: Hishamunda, N. and D. Valderrama (eds.). Social and Economic Dimensions of Seaweed Cultivation: Assessment and Case Studies. FAO Fisheries and Aquaculture Technical Paper 580. Rome, FAO. 61–89.

Neish, I.C. and R.T. Barraca. 1978. A survey of *Eucheuma* farming practices in Tawi-Tawi (The Sitangkai, Sibutu, and Tumindao Region). Marine Colloids (Philippines).

Neish, I.C., A.Q. Hurtado, B. Julianto and D. Saragih. 2009. Good aquaculture practices for *Kappaphycus* and *Eucheuma*. A compilation of nine training modules for seaweed farmers. SEAPlant.net Monograph no. HB2F 0909 V4 GAP, Sulawesi, Selatan, Indonesia.

Ohno, M., H.Q. Nang, N.H. Dinh and V.D. Triet. 1995. On the growth of the cultivated *Kappaphycus alvarezii* in Vietnam. Japan Journal of Phycology (Sorui) 43: 19–22.

Ohno, M., H.Q. Nang and S. Hirase. 1996. Cultivation and carrageenan yield and quality of *Kappaphycus alvarezii* in the waters of Vietnam. Journal of Applied Phycology 8: 431–437.

Pang, T., J. Liu, Q. Liu and W. Lin. 2011. Changes of photosynthetic behaviors in *Kappaphycus alvarezii* infected by epiphyte. eCAM. DOI:10.1155/2011/658906.

Pang, T., J. Liu, Q. Lin and L. Zhang. 2012. Impacts of glyphosate on photosynthetic behaviors in *Kappaphycus alvarezii* and *Neosiphonia savatieri* detected by JIP-text. Journal of Applied Phycology 24: 467–473.

Parker, H.S. 1974. The culture of the red algal genus *Eucheuma* in the Philippines. Aquaculture 3: 431–437.

Paula, E.J., R.T.L. Pereira and M. Ohno. 2002. Growth rate of the carrageenophyte *Kappaphycus alvarezii* (Rhodophyta, Gigartinales) introduced in subtropical waters of Sao Paolo State, Brazil. Phycological Research 50: 1–9.

Pedersen, M., J. Collen, K. Abrahamsson, M. Mtolera, A. Semesi and G. Garcia Reina. 1996. The ice-ice disease and oxidative stress of marine algae. Current Trends in Marine Botanical Research in the East African Region 11–24.

Pellizzari, F. and R.P. Reis. 2011. Seaweed cultivation on the Southern and Southeastern Brazilian Coast. Revista Brasileira de Farmacognosia (Brazilian Journal of Pharmacognosy) 21(2): 305–312.

Peters, A.F. and E. Ellerstdottir. 1996. New record of the kelp endophyte *Laminarionema elsbetiae* (Phaeophyceae, Ectocarpales) at Helgoland and its life history in culture. Nova Hedwigia 62: 341–349.

Pickering, T.D., P. Skelton and J.R. Sulu. 2007. International introduction of commercially harvested alien seaweeds. Botanica Marina 50: 338–350.

Potin, P., K. Bouarab, J.P. Salaun, G. Pohnert and B. Kloareg. 2002. Biotic interactions of marine algae. Current Opinion in Plant Biology 5: 308–317.

Prakash, J. 1990. Fiji. pp. 1–9. *In*: Adams, T. and R. Foscarini (eds.). Proceedings of the Regional Workshop on Seaweed Culture and Marketing. Suva, Fiji, November 1989, South Pacific Aquaculture Development Project, Food and Agriculture Organization of the United Nations. GCP/RAS/116/JPN.

Rayorath, P.M., W. Khan, R. Palanisamy, S.L. MacKinnon, R. Stefanova, S.D. Hankins, A.T. Critchley and B. Prithiviraj. 2008. Extracts of the brown seaweed *Ascophyllum nodosum* induce gibberellic acid (GA3)-independent amylase activity in barley. Journal of Plant Growth Regulation 27: 370–379.

Ricohermoso, M.A. and L.E. Deveau. 1979. Review of commercial propagation of *Eucheuma* (Florideophyceae) clones in the Philippines. *In*: Jensen, A. and J.R. Stein (eds.). International Seaweed Symposium 9: 525–531.

Rincones, R.E. 2006. The Jimoula initiative. *In*: Critchley, A.T., M. Ohno and D.B. Largo (eds.). World Seaweed Resources: An Authoritative Reference System. formato DVD University of the Netherlands and UNESCO.

Rincones, R.E. and J.N. Rubio. 1999. Introduction and commercial cultivation of the red alga *Eucheuma* in Venezuela for the production of phycocolloids. World Aquaculture Magazine 30: 57–61.

Robertson-Anderson, D.V., D. Leitao, J.J. Bolton, R.J. Robertson, A. Njobeni and K. Ruck. 2006. Can kelp extract (KELPAK) be useful in seaweed mariculture? Journal of Applied Phycology 18: 315–329.

Robledo, D., E. Gasca-Leyva and J. Fraga. 2013. Socio-economic dimensions of seaweed farming in Mexico. *In*: Hishamunda, N. and D. Valderrama (eds.). Social and Economic Dimensions of Seaweed Cultivation: Assessment and Case Studies. FAO Fisheries and Aquaculture Technical Paper 580. Rome, FAO. 185–204.

Rodgers, S.K. and E.F. Cox. 1999. Rate of spread of introduced rhodophytes *Kappaphycus alvarezii, Kappaphycus striatum,* and *Gracilaria salicornia* and their current distributions in Kane'ohe Bay, O'ahu, Hawai'i. Pacific Science 53: 232–241.

Rusell, D.J. 1982. Introduction of *Eucheuma* to Fanning Atoll, Kiribati, for the purpose of mariculture. Micronesica 18: 35–44.

Samonte, G.P.B., A.Q. Hurtado-Ponce and R.D. Caturao. 1993. Economic analysis of bottom line and raft monoline culture of *Kappaphycus alvarezii* var. *tambalang* in western Visayas, Philippines. Aquaculture 110: 1–11.

Shechambo, F., Z. Ngazy and F. Msuya. 1996. Socio-economic impacts of seaweed farming in the east coast of Zanzibar, Tanzania. Report submitted to the Canadian International Development Agency (CIDA), Institute of Marine Science, University of Dar es Salaam, Zanzibar Tanzania, IMS 1997/06.

Sievanen, L., B. Crawford, R. Pollnac and C. Lowe. 2005. Weeding through assumptions of livelihood approaches in ICM: Seaweed farming in the Philippines and Indonesia. Ocean and Coastal Management 48: 297–313.

Smith, A. and R.E. Rincones. 2006. The seaweed resources of the Caribbean. In: Critchley, A.T., M. Ohno and D.B. Largo (eds.). World Seaweeds Resources, DVDROM.

Smith, M.T. 1990. Solomon Islands. pp. 21–24. In: Adams, T. and R. Foscarini (eds.). Proceedings of the Regional Workshop on Seaweed Culture and Marketing. South Pacific Aquaculture Development Project, Food and Agriculture Organization of the United Nations. GCP/RAS/116/JPN.

Soegiarto, A. 1979. Indonesian seaweed resources: their utilization and management. Proceedings of International Seaweed Symposium 9: 463–469.

Soetjodinoto. 1969. Is the cultivation of seaweed (*Eucheuma spinosum* and *Eucheuma edule*) in Indonesia technically possible and economically justified? IPFC/C68/TECH 21 at 13th Session, IPFC, Brisbane, Australia, October 1968.

Solis, M.J.L., S. Draeger and T.E.E. dela Cruz. 2010. Marine-derived fungi from *Kappaphycus alvarezii* and *K. striatum* as potential causative agents of ice-ice disease in farmed seaweeds. Botanica Marina 53: 587–594.

The Myanmar Times. 2008. July 23–29.

Tisera, W. and M.R.A. Naguit. 2009. Ice-ice disease occurrence in seaweed farms in Bais Bay, Negros Oriental and Zamboanga del Norte. The Threshold 4: 1–16.

Trono, G.C. Jr. 1974. *Eucheuma* farming in the Philippines. U.P. Natural Science Research Center, Quezon City, Philippines.

Uan, J. 1990. Kiribati. pp. 10–15. In: Adams, T. and R. Foscarini (eds.). Proceedings of the Regional Workshop on Seaweed Culture and Marketing. South Pacific Aquaculture Development Project/Food and Agriculture Organization of the United Nations, 14–17 November 1989, Suva, Fiji.

Ugarte, R.A. and G. Sharp. 2001. A new approach to seaweed management in Eastern Canada: The case of *Ascophyllum nodosum*. Canadian Biology Marine 42: 63–70.

Ugarte, R.A., G. Sharp and B. Moore. 2006. Changes in the brown seaweed *Ascophyllum nodosum* (L. Le Jol.) Plant morphology and biomass produced by cutter rakes harvests in southern New Brunswick, Canada. Journal of Applied Phycology 18: 351–359.

Uyengco, F.R. 1977. Microbiological studies of diseased *Eucheuma* sp. and other seaweeds. National Seaweeds Symposium. Metro Manila, Philippines.

Uyengco, F.R., L.S. Saniel and G.S. Jacinto.1981. The "ice-ice" problem in seaweed farming. Proceedings International Seaweed Symposium 10: 625–630.

Vairappan, C.S. 2006. Seasonal occurrences of epiphytic algae on the commercially cultivated red alga *Kappaphycus alvarezii* (Solieriaceae, Gigartinales, Rhodophyta). Journal of Applied Phycology 18: 611–617.

Vairappan, C.S., C.S. Chung, A.Q. Hurtado, F. Msuya, G. Bleicher-L'Honneur and A.T. Critchley. 2008. Distribution and malaise of epiphyte infection in major carrageenophyte-producing farms. Journal of Applied Phycology 20: 477–483.

Wally, O.S.D., A.T. Critchley, D. Hiltz, J.S. Craigie, X. Han, L.I. Zaharia, S.R. Abrams and B. Prithiviraj. 2013. Regulation of phytohormone biosynthesis and accumulation in *Arabidopsis* following treatment with commercial extract from the marine macroalga *Ascophyllum nodosum*. Journal of Plant Growth Regulation 32: 324–339.

Weinberger, F., G. Pohmert, M.L. Berndt, K. Bouarab, B. Kloareg and P. Potin. 2005. Apoplastic oxidation of L-asparagine is involved in the control of the green algal endophyte *Acrochaete operculata* Correa and Nielsen by the red seaweed *Chondrus crispus* Stackhouse. Journal of Experimental Botany 56: 1317–1326.

Why, S. 1985. *Eucheuma* farming in Kiribati. South Pacific Commission, Noumea, Fisheries 17/Wp 19.

Wu, C., J. Li, E. Xia, Z. Peng, S. Ta, J. Li, Z. Wen, X. Huang, Z. Cai and G. Chen. 1989. On the transplantation and cultivation of *Kappaphycus alvarezii* in China. China Journal of Oceanology and Limnology 7: 327–334.

Yong, W.T.L., S.H. Ting, W.L. Chin, K.F. Rodrigues and A. Anton. 2011. *In vitro* micropropagation of *Euhceuma* seaweeds. IPCBEE 7: 58–60.

Yunque, D.A.T., K.R. Tibubos, A.Q. Hurtado and A.T. Critchley. 2011. Optimization of culture conditions for tissue culture production of young plantlets of carrageenophyte *Kappaphycus*. Journal of Applied Phycology 23: 433–438.

Zemke-White, W.L. and M. Ohno. 1999. World seaweed utilization: An end-of-century summary. Journal of Applied Phycology 11: 369–376.

CHAPTER 9

Marine Algae and the Global Food Industry

Maria Helena Abreu,[1,]* *Rui Pereira*[1] *and Jean-François Sassi*[2,3]

1 Introduction

According to the 2012 World Hunger map produced by the Food and Agricultural Organization (FAO) http://www.fao.org/fileadmin/templates/hunger_portal/docs/poster_web_001_WFS.pdf), 868 million people (12.5% of the world's population) are undernourished, especially in the regions of Africa and West Asia. At the same time, non-balanced diets and overeating trigger important obesity issues in several other countries (FAO 2013a). Adding to this, the world population continues to increase and additional food sources are required.

The basis for a balanced human body and mind development is good nutrition. According to Burlingame and Dernini (2010), sustainable diets are "...diets with low environmental impacts which contribute to food and nutrition security and to healthy life for present and future generations. Sustainable diets are protective and respectful of biodiversity and ecosystems, culturally acceptable, accessible, economically fair and affordable; nutritionally adequate, safe and healthy, while optimizing natural and human resources". Marine algae can be part of those diets, as will be shown in this review.

[1] Algaplus-Produção e comercialização de algas e seus derivados, Lda., Travessa Alexandre da Conceição, S/N, 3830-196 Ílhavo, Portugal.
[2] CEVA—Centre d'Etude et de Valorisation des Algues, Presqu'île de Pen Lan 22610 Pleubian, France.
[3] Present address: CEA—Centre de Cadarache 13108 Saint Paul Lez Durance, France.
* Corresponding author: htabreu@algaplus.pt

This chapter will focus on marine macroalgae (seaweeds), as microalgae used for human food are mostly produced in freshwater systems (e.g., *Spirulina* Turbin (ex Gomont) or *Chlorella* Beyerink). Many seaweed species have been consumed for centuries in Japan, China and Korea, partly for their nutritional and healing properties and, as it was recently found, it may all have started in Chile (Dillehay et al. 2008). The current search for a healthier and more environmental friendly lifestyle has led to an increasing interest of including seaweed in western diets (Hotchkiss and Trius 2007, Mithrill et al. 2013). Seaweeds health benefits go well beyond balanced nutrition, including for instance, lowering diabetes risk (Nwosu et al. 2011), preventing cardiovascular diseases (D'Orazio et al. 2012) or even lowering breast cancer occurrence (Teas et al. 2013), which puts them in the "superfoods list" (Hotchkiss and Trius 2007, Jaspers and Holmer 2013).

According to the latest FAO statistics, over 20 million tons of seaweeds were produced globally in 2011 from aquaculture (96% of total production), with ca. 75% of that biomass going to the food market. Currently, nearly all the seaweed biomass found on the international market (99.6%) originates from 8 countries: China, Indonesia, the Philippines, the Republic of Korea, Democratic People's Republic of Korea, Japan, Malaysia and the United Republic of Tanzania (FAO 2013b). Although more than 10,000 species of macroalgae are currently reported to exist (Guiry and Guiry 2013), only five to six seaweed genera make nearly 100% of the total biomass produced (Table 1). Worldwide it is estimated that nearly 200 species are consumed, however European regulation only considers 22 edible seaweed species (Table 2). That list may be extended if any of the EU countries is able to demonstrate the consumption of a species before 1997 (EC 258/97) or if a "novel-food" solicitation procedure is carried out.

Seaweed farming can be considered as a key environmental service for coastal areas since seaweeds uptake inorganic nutrients from the water where they grow, a natural process known as bioremediation, which allows lowering the risk of coastal eutrophication (Chopin et al. 2008). This has been done in a natural and unintentional way for many years in China and Japan with the massive cultivation of *Saccharina japonica* (Areschoug) C.E. Lane, C. Mayes, Druehl & G.W. Saunders and *Pyropia* J. Agardh (formerly known as *Porphyra*) species (He et al. 2008) in areas where heavy breeding of bivalves coexists. In recent years, planned Integrated Multi-Trophic Aquaculture (IMTA) systems have been essayed to promote higher seaweed yields, among other environmental and economical benefits. *Saccharina latissima* (Linnaeus) C.E. Lane, C. Mayes, Druehl & G.W. Saunders (Chopin 2012), *Palmaria palmata* (Linnaeus) Weber & Mohr and *Chondrus crispus* Stackhouse (Matos et al. 2006, Corey et al. 2013), as well as several *Gracilaria* Greville (Buschmann et al. 2008, Abreu et al. 2011), *Porphyra* C. Agardh (Chopin et al. 1999) and *Ulva* Linnaeus (Msuya and Neori 2008)

Table 1. Main seaweed genera produced (aquaculture) and their associated value in 2011 (source: FAO 2013b).

Scientific name (common name)	Production (metric tons)	Value ($USD in millions)	Main Producers
Eucheuma J. Agardh/ *Kappaphycus* Doty (eucheuma)	6,989,862	1,425.4	Indonesia/Philippines
Saccharina/Laminaria (kombu)	5,257,201	263.3	China
Gracilaria (ogonori)	2,215,695	625.2	China/Indonesia
Pyropia/Porphyra (nori)	1,645,341	963.2	China/Japan/Korea
Undaria Suringar (wakame)	1,754,504	729.0	Korea/China
Other species*	3,048,513	1,272.6	China
TOTALS	20,911,116	5,278.7	

*Refers to "aquatic plant nei" in FAO statistics; it is believed that most of this production also corresponds to *Gracilaria* species in China.

species are some of the edible seaweeds already tested in these systems. Seaweed exploitation can thus be considered as a sustainable activity with low impact on global marine biodiversity, fulfilling the requirements laid down by FAO (2013a) for a sustainable diet. Nonetheless, in some areas (e.g., Europe) seaweed production is still dependent on harvesting from wild resources and regulations are often non-existent or inadequate (see reports in www.netalgae.eu). With increasing demand for this natural resource in the western world, sustainable production systems with high quality control patterns will have to be assured. Land-based production of seaweed species destined for this market can be one of the solutions (Pereira et al. 2012a).

After production (wild harvest or aquaculture), the majority of seaweeds destined for the food market (not including polysaccharide extraction activities) are washed to remove debris and epiphytes, dried/ dehydrated and packaged in whole, flakes or powder format. This process is widely used because according to producers, it retains the organoleptic features of the product and reduces shipping costs.

2 Nutritional Value

A healthy diet contains a balanced and adequate combination of macronutrients (carbohydrates, fats and protein) and essential micronutrients (vitamins and minerals). According to FAO's latest report (2013a) on the status of the world's food and agriculture sectors, three of the most common

Table 2. Edible seaweed species[1] according to EU regulation and their nutritional composition[2] per a 6 g portion relative to the Guideline Daily Amount (GDA, FoodDrinkEurope) and Recommended Daily Intake (RDI, EC100/2008).

Seaweed species / Trade name	Energetic value kcal (%)	Protein g (%)	Carbohydrates g (%)	Lipids g (%)	Fibre g (%)	Salt g (%)
Ascophyllum nodosum / Rockweed, egg wrack	7.6 (0.4)	0.4 (0.9)	3.6 (1.3)	0.2 (0.2)	2.5 (10.1)	0.4 (6.5)
Fucus spp. / Bladder wrack	6.1 (0.3)	0.4 (0.8)	3.7 (1.4)	0.1 (0.1)	2.8 (11.2)	0.6 (9.8)
Himanthalia elongata (Linnaeus) S.F. Gray / Sea spaghetti	10.7 (0.5)	0.6 (1.2)	3.5 (1.3)	0.2 (0.2)	1.8 (7.3)	0.2 (3.6)
Undaria pinnatifida / Wakame	6.6 (0.3)	0.9 (1.7)	2.7 (1.0)	0.2 (0.2)	2.3 (9.1)	0.3 (5.5)
Laminaria digitata (Hudson) J.V. Lamouroux / Kombu	8.4 (0.4)	0.5 (1.0)	3.4 (1.3)	0.1 (0.1)	2.0 (8.0)	0.3 (5.8)
Saccharina japonica / Kombu, Japanese kelp	9.3 (0.5)	0.4 (0.9)	3.6 (1.3)	0.2 (0.2)	2.1 (8.3)	0.4 (6.4)
Saccharina latissima / Sugar kelp, Royal kombu	9.6 (0.5)	0.7 (1.4)	3.3 (1.2)	0.1 (0.1)	1.7 (6.9)	0.5 (9.0)
Alaria esculenta / Atlantic wakame, Wing kelp	6.9 (0.3)	0.7 (1.5)	3.8 (1.4)	0.1 (0.1)	3.0 (12.0)	0.3 (4.8)
Palmaria palmata / Dulse, Dilisk	10.1 (0.5)	1.0 (2.1)	3.0 (1.1)	0.1 (0.1)	1.7 (6.9)	0.2 (4.0)
Porphyra/Pyropia spp. / Nori, Laver	10.6 (0.5)	1.7 (3.3)	2.8 (1.0)	0.1 (0.1)	2.1 (8.2)	0.3 (4.9)
Chondrus crispus / Lichen, Irish moss	11.1 (0.6)	0.9 (1.9)	3.3 (1.2)	0.2 (0.2)	1.8 (7.3)	0.6 (9.5)
Gracilaria spp. / Ogonori, Wart weed	9.9 (0.5)	0.8 (1.7)	3.5 (1.3)	0.1 (0.1)	2.1 (8.2)	0.6 (10.1)
Enteromorpha (now *Ulva*) / Gut laver; Aonori	8.7 (0.4)	0.5 (1.0)	3.0 (1.1)	0.2 (0.3)	2.0 (8.0)	0.8 (12.9)
Ulva spp. (Sea lettuce)	7.9 (0.4)	1.0 (2.0)	2.9 (1.1)	0.1 (0.1)	2.1 (8.4)	0.3 (5.0)

[1]Not listed here but also accepted as edible: *Mastocarpus stellatus* (Stackhouse) Guiry and *Phymatolithon calcareum* (Pallas) W.H.Adey % D.L. McKibbin[2] Values from CEVA (www.ceva.fr).

micronutrient deficiencies refer to vitamin A, iron (anaemia) and iodine. Zinc, selenium and vitamin B_{12} are also recognized as important for human health.

Seaweeds are already considered a key element in healthy diets (Baik et al. 2013, Mithrill et al. 2013). They are characterized by high contents of fibres, minerals and essential vitamins coupled with low fat content, low salt and rich protein fraction (Tables 2 and 3). Several studies have looked

Table 3. Distinctive trace elements (minerals and vitamins) and amino acids for selected edible seaweed species.

Seaweed species	Trace Elements	Amino acids
Ochrophyta		
Fucus spp. and *A. nodosum*	Iodine, Calcium, Potassium; Vitamin C and E	Arginine, aspartic acid, glutamic acid
H. elongata	Iron, Phosphorus, Magnesium, Potassium, Calcium; Vitamins A, C	Aspartic acid, glutamic acid, lysine
U. pinnatifida	Iodine, Calcium, Potassium, Magnesium; Vitamins A, B9	Arginine, Aspartic acid, Glutamic acid, Leucine
Alaria, Laminaria and *Saccharina* species	Iodine, calcium, magnesium, iron; Vitamins A, B, C, E	Alanin, Aspartic acid, Glutamic acid, Leucine, Taurine
Rhodophyta		
P. palmata	Calcium, Potassium, magnesium, Iron; Vitamin C	Alanin, Arginine, Aspartic acid, Glutamic acid, Glycine, Leucine, Methionine, Phenylalanine, Valine
Porphyra/Pyropia spp.	Iron, Manganesium, Phosphorous, Vitamins C, B-complex, A, E	Alanine, Aspartic acid, Arginine, Glycine, Isoleucine, Leucine, Lysine, Proline, Valine, Taurine
C. crispus	Zinc, Vitamin C	Alanin, Arginine, Aspartic acid, Glutamic acid, amide,
Gracilaria spp.	Potassium, Iodine, Vitamins C	Glycine, Arginine, Alanine, Glutamic acid
Chlorophyta		
Ulva spp.	Magnesium, Calcium, Iron; Vitamins C, E, B12, D	Alanine, Aspartic acid, Glutamic acid, Glycine, Leucine, Lysine, Phenylalanine, Valine, Tyrosine

*Data from www.ceva.fr, Rouxel and Crouan 1995, Fleurence 1999, Niwa et al. 2003, Bocanegra et al. 2009, Cofrades et al. 2010, Larrea-Marin et al. 2010, Holdt and Kraan 2011, Pereira 2011, Tabarsa et al. 2012, Jaspars and Holmer 2013.

at the nutritional composition of edible seaweeds, many considering the same species but all relevant since seaweed composition changes with species, habitat and environmental conditions during growth, harvest time and processing methods (Bocanegra et al. 2009, Larrea-Marin et al. 2010, Patarra et al. 2010, Stengel et al. 2011, Madden et al. 2012 and many others). Major reviews on the nutritional composition of seaweeds have recently been done (MacArtain 2007, Holdt and Kraan 2011, Pereira 2011), providing a very detailed characterization on the nutritional composition of the most consumed seaweeds worldwide. The nutritional information in this chapter is mostly based on those works and on the data analysis available online at CEVA (Centre d'Étude et de Valorisation des Algues; http://www.ceva.fr), from which we have extracted the data regarding the species currently accepted for human consumption in Europe (Table 2). On average, Japanese people eat around 4–6 g of seaweed (dry matter equivalent) per day (MacArtain 2007, Chandini et al. 2008). This quantity is comparable to lettuce consumption in the French average diet. Putting this into perspective, we can probably expect that a committed western seaweed consumer would eat maybe 1 to 2 grams of seaweed (dry matter equivalent) a day. This kind of information is hard to provide accurately, but it may be important for companies to advise consumers on the individual portion size, thus preventing misinterpretations regarding nutrient daily intakes.

2.1 Macronutrients—Fibre, Proteins, Lipids

Seaweeds have high water content, with values ranging from around 60% up to 94% of the total fresh biomass. However, compared to fresh vegetables, the water content of seaweeds is either comparable or lower. The major and structural components of seaweed tissue are the polysaccharides that can make up to 76% of seaweeds dry weight (Holdt and Kraan 2011). Many polysaccharides are species-specific and essentially consist in cell wall polysaccharides (celluloses and hemicelluloses), storage polysaccharides (starch in green algae, laminaran in brown algae and floridean starch in red algae) and also intercellular fractions (Bocanegra et al. 2009). These last polysaccharides consist of alginates and fucoidan for Ochrophyta, carrageenan, agar and porphyran for Rhodophyta and ulvan for Chlorophyta species. Some of these compounds drive the major industrial activity related with seaweed: phycocolloid extraction (see relevant section below). They also function as soluble dietary fibres and promote high satiety levels (thus used for dietetic purposes) with positive impacts in gut health (Jaspers and Holmers 2013). A large share of the population in the western world does not meet the recommended daily dietary fibre intake of 24 g. In general, the average total fibre content in seaweed (up to 63%, according to Holdt and Kraan 2011) surpasses that of terrestrial foodstuffs, sometimes doubling the amount found in fruits, vegetables and cereals promoted today for their fibre content (MacArtain 2007). Considering Table 2, *Ascophyllum nodosum* (Linnaeus) Le Jolis, *Fucus Linnaeus* spp. and *Alaria esculenta* (Linnaeus) Greville can provide over 10% of the Guideline Daily Amount (GDA), determined by FoodDrinkEurope (former Confederation of Food and Drink Industries of the EEC, CIAA).

Algal protein content is generally higher in red and green algae (10–47% dw) when compared with brown algae (3–16% dw) and it is much greater than that found in high-protein leguminous seeds such as soybean (Fleurence 1999). *Porphyra* and *Palmaria* species are within the top choice for seaweed-based protein sourcing due to their high protein content (47% and 30%, respectively). Protein levels are highly affected by the nutrient availability in the water, mainly nitrogen and thus can be very different according to the seaweeds production conditions (Galland-Irmouli et al. 1999). In *Gracilaria*, for instance, protein content normally ranges from 5% to 23% (Holdt and Kraan 2011); however when cultured in nitrogen-enriched waters, like for instance within IMTA systems, protein values may rise up to average values of 38% and even to maximum records of 50% (Abreu et al. 2011). Specific proteins like phycoerythrin (part of the phycobilliproteins) and mycosporine-like aminoacids have high antioxidant and anti-UV properties and are used by the seaweeds to harvest light or protect themselves from high irradiances (Figueroa et al. 2003, Pereira 2011).

The nutritional value of proteins referred to as "amino acid score" is evaluated based on the composition of essential amino acids. Most seaweed score above 60, higher than in cereals and vegetables; some may even go close to 100 (nori and wakame), matching the scores of animal foods (Holdt and Kraan 2011). Aspartic and glutamic acids together generally compose a large part of the amino acid fraction in seaweed (brown > green > red) with specific properties for flavour; glutamic acid is a main trigger in the "umami" flavours, now famous around the world (MacArtain 2007). The free amino acids (FAAs) fraction, composed by alanine and taurine, among others, also give seaweeds their flavours, specifically the "nori taste". The most relevant amino acids for each species are listed in Table 3. *In vitro* protein digestibility tests in red and green seaweed showed values between 80% and 95% (Fleurence 1999). Proteins from brown seaweeds have the worst digestibility score, probably due to the antagonistic presence of polyphenols (Holdt and Kraan 2011).

Lipids are a concentrated source of energy and are essential to human diet. Most of the people get these compounds from vegetable oils (70%), followed by animal fats and a minimum from marine oils (mostly fish oils) (Holdt and Kraan 2011). Overall, seaweeds have very low lipid contents but still higher than the lipid levels found in several common terrestrial vegetables. Normal values can go up to 4.5% (dry weight) (Dawczynski et al. 2007, Bocanegra et al. 2009, Holdt and Kraan 2011). The lipid fraction of algae consists primarily of essential polyunsaturated fatty acids (PUFA), and is generally expected to be higher in colder water species or during cold months (Holdt and Kraan 2011). Seaweeds are starting to be considered as a potential source of PUFA, an alternative to fish and fish oils (Kumari et al. 2010, Pereira et al. 2012b).

Algae are particularly rich in n-3 PUFA that can account for 50% of the total fatty acid content (Dawczynski et al. 2007, Bocanegra et al. 2009, Pereira et al. 2012b). Red algae are richer in eicosapentaenoic acid (EPA) and arachidonic acid (Chen and Chou 2002). Green seaweeds have hexadecatetraenoic acid as well as oleic and palmitic acids and brown species have relatively high levels of palmitic and oleic acids. The advised n-6/n-3 ratio for a healthy diet is 5:1 but most western food products have a 15–17:1 ratio, suggesting a deficiency in n-3. As such, algae consumption has immense potential as a way to balance the n-6/n-3 ratio in diets (Bocanegra et al. 2009, Pereira et al. 2012b).

Besides fatty acids, the unsaponifiable fraction of marine algae contains sterols, terpenoids, tocopherols as well as carotenoids such as β-carotene (pro-vitamin A activity), lutein and violaxanthin in red and green seaweeds and fucoxanthin in brown seaweeds (Holdt and Kraan 2011). This lipid fraction of seaweed exhibits strong bioactivity potentials that are discussed in other chapters of this book. Just as an example, Lopes et al. (2011) analyzed the sterols present in several seaweed species, suggesting that the consumption of green and brown seaweed would increase phytosterol intake and, thereby, lower cholesterol absorption.

2.2 Micronutrients—Minerals and Vitamins

Seaweeds are very rich, more than terrestrial vegetables, in fundamental minerals such as iron, zinc, iodine, calcium, potassium, sodium, magnesium, copper and selenium and can contribute to a sustainable diet (both human and animal) (Table 3). Ash content is comparable to that of spinach, *H. elongata* has ten times more iron than lentils, and calcium is higher in *U. pinnatifida* and *C. crispus* than in milk, giving an alternative to lactose intolerant consumers. Calcium and potassium together have higher concentrations in seaweed than in apples, oranges, carrots or potatoes (Bocanegra et al. 2009, Pereira 2011).

Iodine and iron, however, are the most acknowledged seaweed minerals, especially with the severe worldwide deficiencies in these nutrients previously mentioned (FAO 2013a). Most of the seaweed supplements in the market exist to supply extra sources of iodine, derived from brown seaweed species like *S. japonica*, which has been used for centuries in China, its main producer, to prevent goitre. Other iodine sources include *U. pinnatifida, Fucus* spp. or *A. nodosum*. As for iron, Garcia-Casal et al. (2007) showed that this mineral is bioavailable in seaweed. *Porphyra* species can be an important source of selenium (Ródenas de la Rocha et al. 2009). As we've seen for other nutrients, mineral composition is also highly variable, depending on the species, mineralization processes, location, seasonality, physiological factors and others (Bocanegra et al. 2009, Larrea-Marin et al. 2010).

One of the current drivers for seaweed consumption is the possibility of using it for table salt replacement. As denoted in Table 4, several grams of seaweed are needed to completely replace the sodium found in one gram of table salt. However, apart from sodium, seaweed usage brings a combination of savoury flavours arising from potassium and glutamate. Using seaweed for the reduction of sodium intake with minimal adulteration in taste is therefore a complex issue. Nonetheless, any salt reduction is regarded as beneficial and a perfect cookery balance between seaweed's flavour and its salt content will probably eliminate the need for table salt.

Seaweeds are also an excellent source of vitamins. Red seaweeds, especially Nori species have the richest array of vitamins (Table 3). Among them, the vitamin B complex (thiamine, riboflavine), vitamin A and pro-vitamin A (betacarotene) and vitamin E. Vitamin C (Garcia-Casal et al. 2007, 2009) is also a key element, especially in green and brown species. Brown seaweeds also have the highest levels of vitamin E, namely in *Fucus* spp. and *A. nodosum* (Bocanegra et al. 2009). Other microelements include the polyphenols, namely the phlorotannins, extremely important for their antioxidant capacity and application in the health market. *A. nodosum* and *Fucus* species have the highest abundance of these compounds and are heavily commercially exploited for that.

More work is needed to assess the bioavailability of seaweed nutrients for humans and the effects of large-scale seaweed consumption (Urbano and Goñi 2002, Bocanegra et al. 2003), although recent *in vivo* experiments confirmed seaweeds (sea lettuce, nori and wakame) as excellent nutrient

Table 4. Relation between sodium (Na) content in edible seaweed species and table salt (adapted from Hotchkiss 2010).

Species	Na per g dry seaweed	g of dried seaweed required to match Na in 1 g table salt
A. nodosum	0.022	18
H. elongata	0.041	10
L. digitata	0.047	9
A. esculenta	0.04	10
S. latissima	0.025	16
P. palmata	0.025	16
Ulva spp.	0.037	11

sources (Taboada et al. 2011, 2013). Also, what happens when different cooking processes take place is rather unknown. The seaweeds *Himanthalia, Undaria, Palmaria and Porphyra* maintain a good nutritional profile, even after drying or canned preservation, with the exception being the lower fatty-acids levels (Sanchez-Machado et al. 2004); Garcia-Sartal et al. (2013), however, demonstrated that most seaweed trace elements are not bioavailable after cooking processes.

3 Contaminants and Notes on Legislation

As for any other natural food product, it is imperative to know the origin of the seaweed. For that reason, spontaneous/uncontrolled harvesting from the wild populations, when no information in water quality or resource management plans is known, should always be avoided. Besides accurate control of microbiological contaminants that is compulsory in food quality control processes, the presence and quantification of hazardous heavy metals should be monitored in edible seaweeds. Although no specific EU legislation exists, France has established the following limits in edible seaweeds: Lead (Pb) and Tin (Sn) <5, Cadmium (Cd) <0.5, Mercury (Hg) <0.1, and inorganic arsenic (As_{in}) <3 μg g-1, expressed on a dry weight basis (Garcia-Sartal et al. 2013).

The potential hazards for human and animals of arsenic present in seaweed have received increased attention. However, health hazards for live organisms are only expected from the exposure to the toxic forms of arsenic (inorganic As(III) and As(V)). Most seaweed species detoxify inorganic arsenic through an arseno-betaïne pathway and ultimately accumulate arsenosugars, a non-toxic form of arsenic (García-Sartal et al. 2012). The exception seems to be the Sargassaceae species, namely the popular Hijiki (*Sargassum fusiforme* (Harvey) Setchell) that accumulates inorganic arsenic up to 86% of its total arsenic content (Ichikawa et al. 2006, Besada et al. 2009). Hijiki consumption may therefore be of potential health hazard for human health (Yokoi and Konomi 2012). However more studies are needed where the impact of seaweed processing and cooking would be taken into account. Actually, several studies have demonstrated that arsenic speciation and bioavailability patterns change with a heat/cooking process (Almela et al. 2005, Laparra et al. 2004, Sartal et al. 2012).

4 So, How are We Eating Seaweed?

4.1 Sea Vegetables

The majority of the seaweed produced worldwide, as we have mentioned before, is for human consumption. Japanese are the biggest consumers *per capita* with 10–15% of their diet consisting of seaweed. There, as in China and Korea and some other countries (e.g., Indonesia and southern Chile), seaweeds are part of the traditional diet. They are appreciated for their aspect, texture and flavour like any other food item. Although this is still not the case for most western countries, in some specific locations (e.g., France, UK, Ireland, Canada, USA, Australia) it is starting to happen.

The consumption of seaweed is, for the majority of people in western countries, mostly correlated to the consumption of sushi (Fig. 1), a trend that has been greatly increasing (Jaspars and Holmer 2013). Also the nutritional richness and array of applications of the "sea veggies" is receiving greater recognition. An indicator of this is the rise in new culinary books published in recent years, some fully dedicated to seaweed recipes. According to the online retailer Amazon.co.uk, over the last five years sixteen new cookery books containing the word *seaweed* in the title and/or in the keywords were launched. This will help to fill in the gap recognized from the few seaweed consumption studies conducted so far (Edwards et al. 2012, Idéalg 2013), i.e., the lack of culinary knowledge on seaweed usage.

The most common uses for seaweed continue to be in salads, sushi and soups. Appealing and high quality products like the Pink, Green and Yellow Hana-Tsunomata™ (*C. crispus*) produced in the land-based farm of Acadian Seaplants in Canada are exclusively sold at very high prices to Japan, to be incorporated into a healthy salad or to decorate plates in high-end cuisine.

Figure 1. Sushi preparation in Portugal by chef Gingko SushiSashimi. © ALGAplus, Ltd.

Seaweed products originated from Asian large suppliers can be easily found in most food retailers, usually dried Wakame (*U. pinnatifida*), Nori (*Porphyra/Pyropia* sp.), Kombu (*Saccharina/Laminaria*), Arame (*Eisenia* Areschoug sp.) and Hijiki (*S. fusiforme*) in the whole or sheets format.

Small seaweed producers thriving in Europe, Australia and the US, however, tend to position their products in the organic and healthy-gourmet segments. Besides selling dehydrated seaweeds, some companies (especially in France) diversified into fresh seaweed preserved in salt or in brine and sell these products mainly to speciality restaurants. Current European seaweed companies often incorporate the whole production and packaging chain. This allows them to maximize the design and appeal of their product to demanding consumers that acknowledge the origin and quality of what they buy and are willing to pay a higher price for it. These consumers are young adults, well educated and environment and health-conscious (Idéalg 2013).

Apart from being used as a side vegetable, flakes and/or powder of the several seaweed species are also incorporated into innovative food products. These are developed many times in partnerships among different food type producers, providing a differentiated flavour to traditional meals. These include condiments (sea salt, sugar, savoury sprinkles, dried and oil dressings for salads), sauces (pesto, tartars, mustard, mayonnaise), premixes for soups and pasta sauces, baked goods (biscuits, bread, crackers) and pasta, as well as drinks (teas, wine and beer) or even pizza, sausages, hamburgers and chocolate.

The increasing search of consumers for organic products has also been a good opportunity for seaweed marketing.

4.2 Seaweed Snacks and Food Supplements

The consumption of snacks is increasing worldwide. According to a 2012 article, Australians may spend around 7.6 billion dollars a year, just for evening snacks (http://www.reuters.com/article/2012/05/18/idUS171175+18-May-2012+BW20120518). The association of quick and functional foods is indicated as a reason for this success.

Seaweeds are also making their way into the healthy snacks market (http://healthland.time.com/2013/06/13/what-to-eat-now-seaweed), becoming more popular everyday in the western countries, with several brands launching new products (e.g., http://www.gimmehealth.com). Snack bars, crackers, or simply stripes of nori sheets are used. Strong marketing and communication tools are applied, including for example, multi-language packaging, mention of the correct seaweed name instead of the generic "algae" or "seaweed" words used so far and direct approaches to children (SIA 2013). The usage of seaweed snacks in countries with a more traditional approach to seaweed consumption is also a reality, in order to diversify products, add value to the crops and also attract new consumers. Even large western brands like "Lays" and "Pringles" are now producing seaweed chips for Asian consumers.

Another way of consuming seaweed and benefiting from their nutraceutical properties is by taking food supplements. The definitions of these products given by the US Food and Drug Administration (USFDA) or the European Food Safety Authority (EFSA) are similar and basically state that a food supplement is a concentrated source of nutrients or dietary ingredients intended to supplement a normal diet and taken by mouth (pills, tablets, softgels, liquids, powders, etc.). As for snacks, the supplements market is growing largely in the US and Europe (SIA 2013). Most common European seaweed supplements are from *Fucus* species, normally associated with suppressing appetite (high fibre contents) and from kelp species (kombu and wakame) to fight iodine deficiency.

4.3 Polysaccharides as Natural Food Additives

The extraction of polysaccharides is the major transformation industry related to seaweed. These hydrocolloids are long-chain molecules extracted from red (agar and carrageenan) and brown (alginates) seaweed species, being mainly used in processed food industry as texturizing agents, thickeners, stabilizers and emulsifiers. The new demand for gluten-free products in the baking industry may lead to the application of seaweed polysaccharides as gluten substitutes (Jaspars and Holmer 2013). In Europe, the consumer can be aware of these ingredients by their specific codes: E400–E405 (alginates), E406 (agar) and E407 and E407a (respectively, refined and semi-refined carrageenan). Despite the typical instability in commodities markets, the polysaccharide industry has had positive, although slow, growth rates in the last years. Since it is hard to get updated information in this market, we based our information on the review produced by Bixler and Porse in 2010 (Table 4).

Agar was the first phycocolloid to be discovered in Japan during the mid 17th century (McHugh 2003). Global production of this polysaccharide in 2009 was around 9600 tons (Table 4). It is formed by two components, agarose and agaropectin. It is tasteless, not interfering with the flavours of foodstuffs and being mostly used as a thickening/gelling agent and stabilizer in jams and marmalades, fruit jellies, sauces and condiments, juices and baked goods. Agar use is widespread and it can be found in powder or strips within most supermarkets. Other innovative uses are also appearing, like colourful and edible cups made from this polysaccharide (Fig. 2).

The red species *Gracilaria* is currently the main raw material for agar extraction, being massively cultivated for that purpose in Indonesia, China

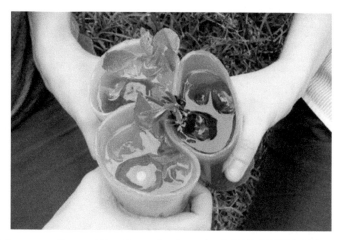

Figure 2. Agar cups. Source: http://www.toxel.com/tech/2010/07/25/edible-cups-made-from-jello/.

Color image of this figure appears in the color plate section at the end of the book.

and Chile. *Gelidium* J.V. Lamouroux species give the higher quality agar but its availability is scarce since its production relies solely on wild harvest, despite the various attempts to promote its cultivation (McHugh 2003). Agar extraction procedure is quite simple and relies on alkali pre-treatment (for *Gracilaria* species only) and hot water extractions. Microwave-assisted extraction (MAE) was recently applied to obtain agar from *Gracilaria* but for now, only at a laboratory scale. This environmental friendly procedure has lower energy and solvent consumptions, and higher recoveries while producing an agar with enhanced gelling properties when compared with the traditional extracted agar (Sousa et al. 2010, Sousa et al. 2012).

A British pharmacist firstly extracted carrageenan from *C. crispus* in 1862. This polysaccharide, also extracted from red seaweeds, is one of the most important hydrocolloids in the food industry, being a key element in dairy and meat produce. Another application re-gaining importance is its use as a beer clarifier, mainly in craft breweries (http://byo.com/component/k2/item/2838-clarification-of-beer-advanced-brewing). There are three types of carrageenan that differ in their gelling properties: kapa, iota and lambda. The first two types form gels and are mainly obtained from the seaweed *Kappaphycus alvarezii* (Doty) Doty ex. P.C. Silva (kappa) and *Eucheuma denticulatum* (N.L. Burman) F.S.Collins & Hervey (iota) massively cultivated in Indonesia, the Philippines and Tanzania. Carrageenans extracted from other species (Table 4) are not as pure but the hybrids produced can be modified into kappa and iota types during transformation processes (Bixler and Porse 2010, Pereira 2011).

Table 5. Information regarding 2009 on production values, price, major production areas and seaweed species used as raw material for agar, alginate and carrageenan. Source: Bixler and Porse 2010.

	Ingredient Label	Global Production (t)	Major producing countries*	Average Price (US$/kg)	Main seaweed genera
Agar	E406	9,600	Chile China Korea Indonesia Morocco Spain	18	*Gelidium* *Gracilaria* *Pterocladiella* B. Santelices & Hommersand
Alginate	E400 E401 E402 E403 E404 E405	30,000	Chile China France Japan Norway	12	*Laminaria* J.V. Lamouroux *Lessonia* Bory de Saint-Vincent *Macrocystis* C. Agardh *Durvillaea* Bory de Saint-Vincent *Ecklonia* Hornemann
Carrageenan	E407 E407a	50,000	China France Indonesia Philippines Spain	10.5	*Kappaphycus* Doty *Eucheuma* J. Agardh *Gigartina* Stackhouse *Chondracanthus* Kutzing *Sarcothalia* Kutzing *Chondrus* Stackhouse

*Countries not listed by production levels.

The term alginates encompasses alginic acid and the salts sodium, calcium, potassium and ammonium alginate. It was the last phycocolloid to be discovered (around 1930), but its extraction became an important industry in the period following World War II, namely in Europe, the USA and Japan. Alginate is extracted from brown seaweed species, mostly harvested from wild populations, although more and more kelp cultivated in China ends up in alginate factories. Global alginate production reaches 30,000 tons (Table 4). The extraction of high quality alginate was for a long time almost exclusive to western countries. Nowadays, China is gaining importance in this market and it is one of the main buyers of high quality seaweed in southern America (Bixler and Porse 2010). Alginates are widely used as thickening and gelling agents, emulsifier, stabilizer and texture-improvers, namely in products like beer and wine, canned and processed meat (Pereira 2011). They can also be used for the production of propylene glycol alginate (PGA) that acts as an emulsifier in ready-made sauces and salad dressings (Jaspars and Holmer 2013). Sodium and calcium alginate have gained a new importance in the new and trendy molecular cuisine, being an obligatory ingredient for the spherification process (Fig. 3).

Figure 3. Spherification of apple juice. By J. Lastras at http://flickr.com/photos/22662305@ N04/3104467425 (license Creative Commons by 3.0).

5 Conclusion

Marine macroalgae are now entering every household, at least in developed countries across the globe, in one way or another. From published research, empirical knowledge and media channels, consumers have realized the benefits of eating seaweed, a nutritionally balanced food. The reasons are many: seaweeds are pretty, they are a suitable food source for losing weight or getting fit, seaweeds can be eaten by everyone from meat lovers to vegetarians, vegans or gluten-free diets, they carry scientifically proven health promoters, from the basic nutrients to iodine, selenium or carotenoids; also, seaweeds grow faster than terrestrial crops and don't need to consume freshwater.

However, despite the acknowledgment of all these good reasons for eating seaweeds, most people in the western world still see them as an exotic food and/or associate them with specific diets (e.g., vegetarian, vegan, macrobiotic, etc.). Changing food habits can be a long journey. One way to overcome this mindset can be through good marketing and communication tools. The term "sea veggies" is making its way and motivating people to use fresh or dried seaweed as a healthy and nutritious ingredient. Seeking for flavours close to the traditional ones can also be a strategy for seaweed-based products development. On the other hand, in countries facing food deprivation with associated nutrient deficiencies, the introduction of seaweed may provide those people with a balanced and highly nutritious diet, as it is done in the most populated country in the world, China. The title of a new book by Ole Mouritsen says it all: Seaweeds, edible, available and sustainable.

Acknowledgments

The authors would like to acknowledge the editorial team for the invitation to participate in this project and the reviewers the text for improving the chapter.

References Cited

Abreu, M.H., R. Pereira, C. Yarish, A.H. Buschmann and I. Sousa-Pinto. 2011. IMTA with *Gracilaria vermiculophylla*: productivity and nutrient removal performance of the seaweed in a land-based pilot scale system. Aquaculture 312: 77–87.

Almela, C., J.M. Laparra, D. Velez, R. Barbera, R. Farre and R. Montoro. 2005. Arsenosugars in raw and cooked edible seaweed: characterization and bioaccessibility. Journal of Agricultural and Food Chemistry 53: 7344–7351.

Baik, I., M. Lee, N.-R. Jun and J.-Y. Lee. 2013. A healthy dietary pattern consisting of a variety of food choices is inversely associated with the development of metabolic syndrome. Nutrition Research and Practice 7: 233–241.

Besada, V., J.M. Andrade, F. Schultze and J.J. González. 2009. Heavy metals in edible seaweeds commercialised for human consumption. Journal of Marine Systems 75: 305–313.

Bixler, H.J. and H. Porse. 2010. A decade of change in the seaweed hydrocolloids industry. Journal of Applied Phycology 23: 321–335.

Bocanegra, A., A. Nieto, B. Blas and F.J. Sánchez-Muniz. 2003. Diets containing a high percentage of Nori or Konbu algae are well-accepted and efficiently utilised by growing rats but induce different degrees of histological changes in the liver and bowel. Food and Chemical Toxicology 41: 1473–1480.

Bocanegra, A., S. Bastida, J. Benedí, S. Ródenas and F.J. Sánchez-Muniz. 2009. Characteristics and nutritional and cardiovascular-health properties of seaweeds. Journal of Medicinal Food 12: 236–258.

Burlingame, B. and S. Dernini. 2010. Sustainable diets and biodiversity: directions and solutions for policy, research and action. *In*: Proc. of the International Scientific Symposium "Biodiversity and Sustainable Diets United against Hunger", 3–5 November 2010, FAO Headquarters, Rome. Rome, FAO and Biodiversity International.

Buschmann, A.H., D.A. Varela, M.C. Hernández-González and P. Huovinen. 2008. Opportunities and challenges for the development of an integrated seaweed based aquaculture activity in Chile: determining the physiological capabilities of Macrocystis and Gracilaria as biofilters. Journal of Applied Phycology 20: 571–577.

Chandini, S.K., P. Ganesan and N. Bhaskar. 2008. *In vitro* antioxidant activities of three selected brown seaweeds of India. Food Chem. 107: 707–713.

Chen, C.-Y. and H.-N. Chou. 2002. Screening of red algae filaments as a potential alternative source of eicosapentaenoic acid. Marine Biotechnology 4: 189–192.

Chopin, T. 2012. Seaweed aquaculture provided diversified products, key ecosystem functions. Part II. Recent Evolution of Seaweed Industry. Global Aquaculture Advocate: July/August 2012: 24–27.

Chopin, T., C. Yarish, R. Wilkes, E. Belyea, S. Lu and A. Mathieson. 1999. Developing Porphyra/salmon integrated aquaculture for bioremediation and diversification of the aquaculture industry. Journal of Applied Phycology 11: 463–472.

Chopin, T., S.M.C. Robinson, M. Troell, A. Neori, A.H. Buschmann and J. Fang. 2008. Multitrophic integration for sustainable marine aquaculture. pp. 2463–2475. *In*: Jørgensen, S.E. and B.D. Fath (eds.). Ecological Engineering: Encyclopedia of Ecology, 5 vols. Elsevier, Oxford.

Cofrades, S., I. López-lopes, L. Bravo, C. Ruiz-Capillas, S. Bastida, M.T. Larrea and F. Jimenez-Colmenero. 2010. Nutritional and antioxidant properties of different brown and red Spanish edible seaweeds. Food Science and Technology International 16: 361–370.

Corey, P., J.K. Kim, J. Duston, D.J. Garbary and B. Prithiviraj. 2013. Nutrient uptake by *Palmaria palmata* and *Chondrus crispus* at combined concentrations of nitrate and ammonium. Journal of Applied Phycology 215: 1349–1358.

D'Orazio, N., E. Gemello, M.A. Gammone, M. de Girolamo, C. Ficoneri and G. Riccioni. 2012. Fucoxantin: A treasure from the sea. Marine Drugs 10: 604–616.

Dawczynski, C., R. Schubert and G. Jahreis. 2007. Amino acids, fatty acids, and dietary fibre in edible seaweed products. Food Chemistry 103: 891–899.

Dillehay, T.D., C. Ramírez, M. Pino, M.B. Collins, J. Rossen and J.D. Pino-Navarro. 2008. Monte Verde: seaweed, food, medicine, and the peopling of South America. Science 320: 784–786.

Edwards, M.D. and S.L. Holdt. 2012. Algal eating habits of phycologists attending the ISAP Halifax Conference and members of the general public. Journal of Applied Phycology 24: 627–633.

FAO. 2013a. The State of Food and Agriculture. Food and Agriculture Organization of the United Nations. Rome, 2013, 114 pp.

FAO. 2013b. FishStatJ: software for fishery statistical time series. Data Set: 1950–2011. Version 2.1.0. Fisheries and Aquaculture Department of FAO.

Figueroa, F.L., L. Escassi, E. Pérez-Rodriguez, N. Korbee, A.D. Gelies and G. Johnsen. 2003. Effects of short-term irradiation on photoinhibition and accumulation of mycosporine-like amino acids in sun and shade species of the red algal genus Porphyra. Journal of Photochemistry and Photobiology B: Biology 69: 21–30.

Fleurence, J. 1999. Seaweed proteins: biochemical, nutritional aspects and potential uses. Trends in Food Science and Technology 10: 25–28.

Galland-Irmouli, A.V., J. Fleurence, R. Lamghari, M. Luçon, C. Rouxel, O. Barbaroux, J.P. Bronowicki, C. Villaume and J.L. Guéant. 1999. Nutritional value of proteins from edible seaweed Palmaria palmata (dulse). The Journal of Nutritional Biochemistry 10: 353–359.

García-Casal, M.N., A.C. Pereira, I. Leets, J. Ramírez and M.F. Quiroga. 2007. High iron content and bioavailability in humans from four species of marine algae. Journal of Nutrition 137: 2691–2695.

García-Casal, M.N., J. Ramírez, I. Leets, A.C. Pereira and M.F. Quiroga. 2009. Antioxidant capacity, polyphenol content and iron bioavailability from algae (*Ulva* sp., *Sargassum* sp. and *Porphyra* sp.) in human subjects. British Journal of Nutrition 101: 79–85.

García-Sartal, C.M., Barciela-Alonso and P. Bermejo-Barrera. 2012. Effect of the cooking procedure on the arsenic speciation in the bioavailable (dialyzable) fraction from seaweed. Microchemical Journal 105: 65–71.

Garcia-Sartal, C.M., C. Barciela-Alonso, A. Moreda-Piñeiro and P. Bermejo-Barrera. 2013. Study of cooking on the bioavailability of As, Co, Cr, Cu, Fe, Ni, Se and Zn from edible seaweed. Microchemical Journal 108: 92–99.

Guiry, M.D. and G.M. Guiry. 2013. AlgaeBase. World-wide electronic publication, National University of Ireland, Galway. http://www.algaebase.org; searched on 05 November 2013.

He, P.M., S.N. Xu, H.Y. Zhang, S.S. Wen, Y.J. Dai, S.J. Lin and C. Yarish. 2008. Bioremediation efficiency in the removal of dissolved inorganic nutrients by the red seaweed, Porphyra yezoensis, cultivated in the open sea. Water Research 42: 1281–1289.

Holdt, S.L. and S. Kraan. 2011. Bioactive compounds in seaweed: functional food applications and legislation. Journal of Applied Phycology 23: 543–593.

Hotchkiss, S. 2010. Investigation of the flavouring and taste compounds of Irish seaweeds. Industry-Led award final report. 24 pp.

Hotchkiss, S. and A. Trius. 2007. Seaweed: the most nutritious form of vegetation on the planet? Food Ingredients—Health and Nutrition, January/February, 22–33.

Ichikawa, S., M. Kamoshida, M.H. Ken'ichi Hanaoka, T. Maitani and T. Kaise. 2006. Decrease of arsenic in edible brown algae Hijikia fusiforme by the cooking process. Applied Organometallic Chemistry 20: 585–590.

Idéalg. 2013. Forum Idéalg, 23 October 2013. La consomattion des algues: enquête sur la France.

Jaspers, M. and F. Folmer. 2013. Sea Vegetables for Health. University of Aberdeen, 29 pp.

Kumari, P., M. Kumar, V. Gupta, C.R.K. Reddy and B. Jha. 2010. Tropical marine macroalgae as potential sources of nutritional important PUFAs. Food Chemistry 120: 749–757.

Laparra, J.M., D. Velez, R. Montoro, R. Barbera and R. Farre. 2004. Bioaccessibility of inorganic arsenic species in raw and cooked Hizikia fusiforme seaweed. Applied Organometallic Chemistry 18: 662–669.

Larrea-Marin, M.T., M.S. Pomares-Alfonso, M. Gómez-Juaristi, F.J. Sánchez-Muniz and S. Ródenas de la Rocha. 2010. Validation of an ICP-OES method for macro and trace element determination in Laminaria and Porphyra seaweeds from four different countries. Journal of Food Composition and Analysis 23: 814–820.

Lopes, G., C. Sousa, J. Bernardo, P.B. Andrade and P. Valentão. 2011. Journal of Phycology 47: 1210–1218.

MacArtain, P., R.G. Christophe, M. Brooks, R. Campbell and I.R. Rowland. 2007. Nutritional value of edible seaweeds. Nutrition Reviews 65: 535–543.

Madden, M., M. Mitra and D. Ruby. 2012. Seasonality of selected nutritional constituents of edible Delmarva seaweeds. Journal of Phycology 48: 1289–1298.

Matos, J., S. Costa, A. Rodrigues, R. Pereira and I. Sousa Pinto. 2006. Experimental integrated aquaculture of fish and red seaweeds in Northern Portugal. Aquaculture 252: 31–42.

McHugh, D.J. 2003. A guide to the seaweed industry. FAO Fisheries Technical Paper. No. 441. Rome, 105 pp.

Mithrill, C., L.O. Dragsted, C. Meyer, I. Tetens, A. Biltof-Jensen and A. Astrup. 2012. Dietary composition and nutrient content of the new Nordic diet. Public Health Nutrition 16: 777–785.

Msuya, F. and A. Neori. 2008. Effect of water aeration and nutrient load level on biomass yield, N uptake and protein content of the seaweed Ulva lactuca cultured in seawater tanks. Journal of Applied Phycology 20: 1021–1031.

Niwa, F., H. Furuita and Y. Aruga. 2003. Free amino acid contents of the gametophytic blades from the green mutant conchocelis and the heterozygous conchocelis in *Porphyra yezoensis* Ueda (Bangiales, Rhodophyta). Journal of Applied Phycology 15: 407–413.

Nwosu, F., J. Morris, V.A. Lund, D. Stewart, H.A. Ross and G.J. Mcdougall. 2011. Anti proliferative and potential anti-diabetic effects of phenolic-rich extracts from edible marine algae. Food Chemistry 126: 1006–1012.

Patarra, R.F., L. Paiva, A.I. Neto, E. Lima and J. Baptista. 2010. Nutritional value of selected macroalgae. Journal of Applied Phycology 23: 205–208.

Pereira, L. 2011. A review of the nutrient composition of selected edible seaweeds. pp. 15– 47. *In*: Pomin, V.H. (ed.). Seaweed: Ecology, Nutrient Composition and Medicinal Uses. Nova Science Publishers Inc., New York.

Pereira, R., C. Yarish and A.T. Critchley. 2012a. Seaweed aquaculture for human foods, land-based. pp. 9109–9128. *In*: Meyers, R.A. (ed.). Encyclopedia of Sustainability Science and Technology. Springer Science, New York.

Pereira, H., L. Barreira, F. Figueiredo, L. Custódio, C. Vizetto-Duarte, C. Polo, E. Resek, A. Engelen and J. Varela. 2012b. Polyunsaturated fatty acids of marine macroalgae: potential for nutritional and pharmaceutical applications. Marine Drugs 10: 1920–1935.

Ródenas de la Rocha, S., F.J. Sánchez-Muniz, M. Gómez-Juaristi and M.T. Larrea Marín. 2009. Trace elements determination in edible seaweeds by an optimized and validated ICP-MS methods. Journal of Food Composition and Analysis 22: 330–336.

Rouxel, C. and K. Crouan. 1995. Variations de la composition chimique de l'algue brune *Himanthalia elongata* (L.) Gray durant le printemps. Acta Botanica Gallica 142: 109–118.

Sanchez-Machado, D.I., J. López-Cervantes, J. López-Hernández and P. Paseiro-Losada. 2004. Fatty acids, total lipid, protein and ash contents of processed edible seaweeds. Food Chemistry 85: 439–444.

SIA. 2013. The world of seaweed cuisine: a study of snacks, salads, spices, spreads and other foods. Seaweed Market Reports series, July 2013, 82 pp.

Sousa, A.M.M., V.D. Alves, S. Morais, C. Delerue-Matos and M.P. Gonçalves. 2010. Agar extraction from integrated multitrophic aquacultured Gracilaria vermiculophylla: evaluation of a microwave-assisted process using response surface methodology. Bioresource Technology 101: 3258–3267.

Sousa, A.M.M., S. Morais, M.H. Abreu, R. Pereira, I. Sousa-Pinto, E.J. Cabrita, C. Delerue-Matos and M.P. Gonçalves. 2012. Structural, physical, and chemical modifications induced by microwave heating on native agar-like galactans. Journal of Agricultural and Food Chemistry 60: 4977–4985.

Stengel, D.B., S. Connan and Z.A. Popper. 2011. Algal chemodiversity and bioactivity: sources of natural variability and implications for commercial application. Biotechnology Advances 29: 483–501.

Tabarsa, M., M. Rezaei, Z. Ramezanpour, J.R. Waaland and R. Rabiei. 2012. Fatty acids, amino acids, mineral contents, and proximate composition of some brown seaweeds. Journal of Phycology 48: 285–292.

Taboada, C., R. Millán and M.I. Míguez. 2011. Evaluation of the Marine Alga Ulva rigida as a food supplement: effect of intake on intestinal, hepatic, and renal enzyme activities in rats. Journal of Medicinal Food 14: 161–166.

Taboada, C., R. Millán and M.I. Míguez. 2013. Nutritional value of the marine algae wakame (Undaria pinnatifida) and nori (Porphyra purpurea) as food supplements. Journal of Applied Phycology 25: 1271–1276.

Teas, J., S. Vena, D.L. Cone and M. Irhimeh. 2013. The consumption of seaweed as a protective factor in the etiology of breast cancer: proof of principle. Journal of Applied Phycology 25: 771–779.

Urbano, M.G. and I. Goñi. 2002. Bioavailability of nutrients in rats fed on edible seaweeds, nori (Porphyra tenera) and wakame (Undaria pinnatifida), as a source of dietary fibre. Food Chemistry 76: 281–286.

Yokoi, K. and A. Konomi. 2012. Toxicity of so-called *hijiki* seaweed (Sargassum fusiforme) containing inorganic arsenic. Regulatory Toxicology and Pharmacology 63: 291–297.

CHAPTER 10

Marine Macroalgae and Human Health

Sarah Hotchkiss[a], * and *Catherine Murphy*[b]

1 Introduction

There is a wealth of historical and anecdotal evidence for the health and nutrition benefits of eating marine macroalgae or using marine algal derived products. In recent years however, macroalgae have become a key focus area for the discovery, validation and development of new ingredients and natural products for use in the food, health and wellness and pharmaceutical industries.

Macroalgae have been eaten by many cultures worldwide for thousands of years. In Asia, they have always been, and remain, a staple ingredient in the diet. Records of their use as a food source and medicine date back to 600 BC. They have also been traditionally eaten in seaboard countries elsewhere in the world, such as those in the North Atlantic (Canada, Ireland, Scotland, Wales, England, France, Spain and Iceland) but to a much lesser extent.

Worldwide, the typical modern diet is moving towards one that is high in refined and processed products and lacks the levels of dietary nutrients that are necessary to keep us healthy. Many of the illnesses and diseases that plague our society—the so called "diseases of affluence or excess"—can be linked to poor diet. Current global food and health trends show that key industry players are seeking novel sources of nutritional components and marine algae have attracted much attention in this respect. Macroalgae are a rich source of key nutritional components such as fiber and other carbohydrates, proteins, antioxidants, polyunsaturated fatty acids, vitamins

CyberColloids Ltd, Unit 4A, Site 13, Carrigaline Industrial Estate, Carrigaline, Co. Cork, Ireland.
[a] Email: Sarah@cybercolloids.net
[b] Email: catherine_n_murphy@yahoo.co.uk
* Corresponding author

and minerals at levels similar to, or often greater than, traditional sources. In addition they also contain a diverse range of primary and secondary metabolites and thus represent a rich source of bioactive components. Macroalgal derived bioactives have been shown, for example, to modulate inflammation; reduce oxidative stress; modulate the metabolism of fats and sugars with beneficial implications for diabetes and weight management; improve cardiovascular function; improve brain and cognitive function; induce immunomodulatory and chemopreventative effects (Tables 4 to 8). A number of these properties are already exploited commercially by the health and wellness and pharmaceutical industries.

The aim of this review is twofold: (i) to summarize the key nutritional and structural components of marine macroalgae as potential sources of nutrients and bioactives and (ii) to present information on the potential benefits of these to human health.

An estimated 19 million tonnes (wet) of macroalgae were harvested globally in 2010, the majority (95.5%) from aquacultured stocks in Southeast Asia and Tanzania and the rest from wild biomass around the world (FAO 2012). Approximately 5% of the harvested biomass is used for the production of hydrocolloids (Bixler and Porse 2010) but the rest is essentially used for human consumption (FAO 2010a, FAO 2012). Small amounts of biomass are used for horticulture, animal feed, production of supplements and bioactives for human health and wellness, biotech and some industry. However, these amounts are globally insignificant when compared with the harvest used for human food. Asian cultures in particular have always accepted macroalgae as a nutritious wholefood with recognised health and medicinal potential. These cultures are also open to its innovative use in food and health products.

When considered as wholefoods, i.e., as edible seaweeds or sea vegetables, macroalgae represent a potentially rich and balanced source of key nutrients. This nutritional potential is well documented in the literature; see in particular Holdt and Kraan (2011) for the most recent summary. It is also reported that macroalgal nutrients can be found at levels similar to, or often greater than, terrestrial wholefood equivalents (MacArtain et al. 2007). The Japanese in particular consume large quantities of macroalgae as a whole food, approximately 1.6 kg (dry weight) per person per year (Chandini et al. 2008). It has long been proposed that the relatively healthy status of the Japanese population was due largely to diet, and to macroalgae as a key part of that diet. Indeed, epidemiological studies from Southeast Asia have shown a low prevalence of diseases such as coronary heart disease and diet-related cancers in countries with high seaweed consumption (Kono et al. 2004, Yang et al. 2010, Teas et al. 2011) and that a daily dietary intake of macroalgae, approximately 2 g dry weight or 30 g wet weight, is sufficient to reduce disease risk in populations under study (Brownlee et al. 2012).

Most research concerning the potential health promoting benefits of macroalgae has been conducted in Southeast Asia. A summary of recent observational studies relating to dietary macroalgal intake and health is provided by Brownlee et al. (2012); all seven studies were conducted in Japan or Korea. Although many other cultures have a history of eating macroalgae, it has not until recently featured strongly in the modern daily diet outside of Southeast Asia. In fact, in many cultures macroalgae are considered as "poor man's" food and are not accepted as a potentially healthy and nutritious foodstuff. However, this situation is changing.

Marine industries outside of Asia have traditionally harvested macroalgae for hydrocolloids production, animal fodder, horticultural products and to a lesser degree, human consumption. In general, these are all high volume—low value industries. It is only in recent years that research and industry have turned their attention to finding innovative and value added uses for this natural resource, which in many parts of the world, is considered to be underexploited. Research aimed at the discovery and validation of bioactives and natural products from macroalgae has been a key focus. In addition, macroalgae have not escaped the global drive to find alternatives to fossil fuels and in wake of this, have become a focus resource for the development of new biotechnologies to unlock the necessary feedstuffs for biorefineries and the like. Consequently, scientific support for the many health promoting properties of non-Asian species of macroalgae is building.

"Health and Wellness" is now a key driver across all sectors of the food and health industries. Consumers are demanding "natural", "free-from", "organic" products and manufacturers are being driven towards using clean label ingredients. In addition, health authorities worldwide are promoting preventative health care strategies that are based upon the general message to eat healthily. In all areas, macroalgae show great potential. Food products that contain "healthy" macroalgal ingredients and products for health and wellness that contain macroalgal derived bioactives (often referred to as neutraceutical products) are becoming more widely available on a commercial basis outside of Asia.

2 Macroalgae as a Source of Nutrients and Health Promoting Bioactives

Macroalgae mostly comprise carbohydrates (Table 1), approximately 40–75% of the dry weight (Inst. de Phytonutrition 2004, MacArtain et al. 2007, Holdt and Kraan 2011) and water (<70% wet weight). Red and green species can be high in protein with some red species comprising <50% of the dry weight (Inst. de Phytonutrition 2004, Holdt and Kraan 2011). All

Table 1. Macronutrient composition of key species of European macroalgae: illustrating relative composition of selected species; note that nutritional composition varies significantly with species, season, and locality.

Seaweed	Protein[12]	mg per 100 g dry weight			
		Carbohydrate[12]	Sugars[1]	Fiber[1]	Lipids[12]
Laminaria digitata	3.7–14	48–61	24	29.5–37.8	0.6–1.8
Saccharina latissima	4.3–23.2	59.4–61	25.5	27.3–30.5	0.5–2.9
Himanthalia elongata	4.9–23.5	40–61	28.5	19.9–48.2	0.5–8
Palmaria palmata	7.5–35.6	38–74	21.8	20.4–34.2	0.2–4
Chondrus crispus	1.6–26.5	55–68.3	24.9	30.3	0.5–5.2
Porphyra spp.	10.1–47	30–76	12.9	23.7–44.6	0.12–5.7
Ulva spp.	2.6–28.9	42–46	12.7	21.9–53.5	0.04–7.2
Data source	1. By permission of CEVA unpublished data, 2. Institute of Phytonutrition (2004)				

species are also rich in minerals and trace elements (Table 3), generally 8–40% dry weight, with some calcified forms of red macroalgae comprising <35% in calcium alone (Morrissey et al. 2001). Macroalgae are reported to contain a range of water soluble and fat soluble vitamins such thiamine (B1), riboflavin (B2), cobalamin (B12), ascorbic acid (C), Beta-carotene, vitamin A and vitamin E (Table 2). Trace amounts of other B vitamins such as niacin, biotin and folates have also been identified (Skrovankova 2011). The lipid content of macroalgae is actually quite low (Table 1), 1–4.5% of the total dry weight (Morrissey et al. 2001, Dawczynski et al. 2007, MacArtain et al. 2007); however more than half of this amount can comprise polyunsaturated fatty acids, including the essential omega (n)-3s LNA, EPA and DHA (*resp.* α-linoleic, Eicosapentaenoic and Docosahexaenoic acids) and omega-(n)-6 LA (linoleic acid).

It must be noted that composition is highly variable, both within and between species (Table 1), and is also dependant on a number of biological and environmental factors (Lobban and Harrison 2000). Macroalgae have the ability to store certain components such as minerals or carbohydrates during periods of excess and then utilize these when conditions are less favorable. Hence seasonal and spatial variation in their composition is also characteristic.

2.1 Carbohydrates

Macroalgae are renowned for the diversity of their polysaccharides which generally perform either a structural or storage function. The characteristic structural polysaccharides of brown algae are alginate and fucoidan (sometimes also referred to in the literature as fucoidin). Fucoidan is a

Table 2. Typical vitamin content for selected European species: illustrating relative composition of selected species, note that nutritional composition varies significantly with species, season and locality.

Seaweed	Per 100 g dry weight						
	B1 (mg)[12]	B2 (mg)[12]	B12 (µg)[12]	C (mg)[12]	B-carotene (µg)[12]	A (mg)[12]	E (mg)[12]
Laminaria digitata	0.06	0.06	0.003	17.2	n/d	9.8–20	0.09–1.6
Saccharina latissima	0.4	0.3	n/d	11.3	n/d	0.099	0.5–0.6
Himanthalia elongata	0.3	4.5	n/d	27–500	n/d	99–407	5.3–6.5
Palmaria palmata	0.15–63	0.48–0.53	9–10	17–83.9	1.6–10.9	3.7	2.2–13.9
Chondrus crispus	n/d	n/d	n/d	n/d	n/d	n/d	n/d
Porphyra spp.	0.16–0.5	1.9–2.3	29–43.9	1–100	n/d	0.8–4.4	4.6
Ulva spp.	0.06–0.6	0.03–0.3	10.7–63	10–100	186	0.1–0.3	1.6
Data source	1. By permission of CEVA unpublished data, 2. Institute of Phytonutrition (2004), n/d denotes not determined.						

generic term used to describe a class of sulphated fucan polysaccharides that are found in the cell walls of brown algae (Berteau and Molloy 2003).

Alginate is composed of 1–4 linked α-L-guluronic acid (G) and β-D-mannuronic acid (M) pyranose residues that form an unbranched chain. The G and M residues occur as homopolymeric MM or GG-blocks and also as heteropolymeric sequences in MG or GM blocks (Zhang et al. 2004, Sawabe et al. 1997), the arrangement and ratio of which vary with species, locality and growing conditions. The ratio of M:G blocks controls both the biological and chemical functionality of the alginate. Alginates are widely used in the food industry as stabilizers and in restructured fruit and meat products (INS Food additives No. 400—alginic acid, 401—sodium alginate and 404—calcium alginate).

Fucoidans are mainly found in species of Fucales and Laminariales but also in species of Chordariales, Dictyotales, Dictyosiphonales, Ectocarpales, and Scytosiphonales. The structure of fucoidan varies between algal species (Li et al. 2008, Ale et al. 2011, Jiao et al. 2011). The main component is a sulphated L-fucose backbone that is built either of (1→3)-linked α-L-fucopyranosyl or of alternating (1→3)- and (1→4)-linked α-L-fucopyranosyl residues. However, sulphated galactofucans with backbones built of (1→6)-β-D-galacto- and/or (1→2)-β-D-mannopyranosyl units with fucose or fuco-oligosaccharide branching, and/or glucuronic acid, xylose or glucose substitutions have also been identified in brown algae (Ale et al. 2011). Sulphated fucans are also found in some marine invertebrates but they are not referred to as "fucoidans" (Berteau and Molloy 2003). Sulphated fucans are defined as *"polysaccharides based mainly on sulphated L-fucose, with less than 10% other monosaccharides"* (Berteau and Molloy 2003).

Fucoidans exhibit a wide range of health promoting properties, including anticoagulant and antithrombotic, antivirus, antitumor and immunomodulatory, anti-inflammatory, reduction of blood lipids, antioxidant and anticomplementary properties. This activity has been comprehensively described in a number of review articles (Berteau and Molloy 2003, Li et al. 2008, Wijesekara et al. 2010, Ale et al. 2011, Fitton 2011, Jiao et al. 2011).

There is evidence to suggest that the bioactivity and bioavailability of fucoidans are related to the type of glycosidic linkages present, molecular weight and degree of sulphation. A number of commonalities in structure have been identified between fucoid derived fucoidans such those derived from *Ascophyllum nodosum* (Linnaeus) Le Jolis and species of *Fucus* Linnaeus and these appear to be different from non-fucoid derived fucoidans, however, there is still insufficient evidence to form any systematic structural relationship (Berteau and Molloy 2003, Li et al. 2008, Wijesekara et al. 2010, Ale et al. 2011, Jiao et al. 2011).

The storage polysaccharides of brown algae are laminaran and mannitol. Laminaran is a $\beta(1\rightarrow3)$-glucan with $\beta(1\rightarrow6)$-linkages that comprises soluble and insoluble molecules, some of which have mannitol attached. Mannitol is a monomeric sugar alcohol. It is one of the primary photosynthetic products in brown algae and along with amino acids, forms the bulk of translocated material in the larger brown species. Seasonal variation in laminaran and mannitol content is well documented (Lobban and Harrison 2000). Mannitol is used in the food industry (INS Food additive No. 421), as a sweetener, thickener, emulsifier and anti caking agent.

The structural polysaccharides of red species mostly comprise $\alpha(1\rightarrow3)$ and $\beta(1\rightarrow4)$ galactans with varying degrees of sulphation, methylation and pyruvation. The most important are agars and carrageenans which are commercial hydrocolloids. Agar is primarily found in species belonging to the Gracilariaceae and Gelidiaceae. It has a linear sugar skeleton consisting of alternating units of 1-4-linked 3-6-anhydro-α-l-galactose and 1-3-linked β-D-galactopyranose. Agar can be fractionated into two components, agarose and agaropectin (Osumi et al. 1998, Chen et al. 2004, Kazlowski et al. 2008). Agarose is the gelling fraction; it is a neutral linear molecule with no sulphation. Agaropectin is the non gelling fraction, it is slightly branched and is essentially a heterogeneous mixture of smaller molecules including some substituted galactose residues and some methylated or sulphated sugar units (Osumi et al. 1998, Chen et al. 2004, Kazlowski et al. 2008). The relative proportions of agarose and agaropectin vary with species, locality and environmental conditions (Osumi et al. 1998). Agarose itself comprises two disaccharide components, agarobiose and neoagarobiose and on hydrolysis, yields two types of agaro oligosaccharides depending on the site of cleavage (Chen et al. 2004, Hu et al. 2006, Kazlowski et al. 2008). Agar (INS Food additive no. 406) is used extensively in the food industry as a gelling agent, stabilizer and thickener.

Carrageenans are linear polymers of approximately 25,000 galactose units their structure varies with macroalgal source and is also dependant on extraction method. Three fractions of carrageenan are recognized (INS Food additive no. 407), based essentially on their solubility in potassium chloride (KCl) and degree of sulphation: lambda carrageenan has the strongest degree of sulphation, it is typical of *Chondrus crispus* Stackhouse and species of *Gigartina* Stackhouse; kappa carrageenan has the least degree of sulphation and is typical of *Kappaphycus alvarezii* (Doty) Doty ex P.C. Silva (formerly known as *Eucheuma Cottonii* Weber-van-Bosse); and iota carrageenan, typical of *Eucheuma denticulatum* (N.L. Burman) F.S. Collins & Hervey (formerly known as *E. spinosum* J. Agardh) is somewhat intermediate in sulphation. Essentially, the stronger the degree of sulphation, the lower the gel strength and each is used for specific functionality in foodstuffs. Red algae also contain a number of xylans such as porphyran from species of

Porphyra C. Agardh which comprise units of β-xylose. The primary storage polysaccharide is floridean starch which is a branched glucan similar to amylopectin (Lobban and Harrison 2000).

The structural polysaccharides of green algae are highly complex, sulphated heteropolysaccharides comprised of different sugars: glucuronic acid, xylose, rhamnose, arabinose and galactose (Michel and MacFarlane 1995), relatively little is known about these polymers and at present there is no commercial application. As in terrestrial plants, starch is the major storage polysaccharide in green algae. It comprises units of unbranched amylose, i.e., $\alpha(1\rightarrow3)$-D-galactose and branched amylopectin which also has $\alpha(1\rightarrow4)$ linkages (Lobban and Harrison 2000).

Macroalgal carbohydrates represent a source of soluble and insoluble fiber; many of the polysaccharides exhibit beneficial health properties and also act as a source of bioactives. The soluble and insoluble fiber content of macroalgae ranges from 17.2–58.6% and 4.7–25.6%, respectively, of the total biomass (Warrand 2006) and is thus primarily soluble (Jiminez-Escrig and Sanchez-Munez 2000). MacArtain et al. (2007) showed the fiber content of selected European species of macroalgae to similar to or even higher than the fiber content of comparable wholefoods such as rice, lentils and bananas. They went on to suggest that an 8 g serving of dried macroalgae could provide <12.5% of a person's daily fiber needs. The guideline daily amount (GDA) for dietary fiber is 24 g per day for an adult. Although macroalgae contain large amounts of polysaccharides that are resistant to degradation by human endogenous enzymes, most undergo some degree of fermentation by the gut microflora and thus have some potential to benefit gut health (Deville et al. 2004, 2007). This is covered in more detail in the following section.

2.2 Proteins, Peptides and Amino Acids

In comparison to carbohydrates, relatively little is known about the structure and biological activity of the proteinaceous components of macroalgae. However, macroalgae are considered to be a good source of nutritional protein and bioactive peptides with health promoting properties and also show potential for development as functional food ingredients (Harnedy and FitzGerald 2011). Most research has focused on quantifying total protein content and in determining amino acid profiles. There are large differences in the protein content and amino acid profile of macroalgae (Table 1), depending on species, season and environmental conditions (Fleurence 1999, Galland-Irmouli et al. 1999, Venugopal 2009). Red and green species typically contain 10–47% protein (dry weight) but brown species typically have a lower protein content on average of 3–15% dry weight (Fleurence 1999) although this can be higher (Table 1).

The amino acid profile has been determined for several species of macroalgae (Fleurence 1999a,b, Galland-Irmouli et al. 1999, Wong and Cheung 2000, Marrion et al. 2003, Sanchez-Machado et al. 2003, Dawczynski et al. 2007). Of the 20 naturally occurring amino acids (AAs), most are found commonly in macroalgae including all of the essential amino acids (EAAs): histidine, isoleucine, leucine, lysine, methionine, phenylalanine, threonine, tryptophan and valine (Cerna 2011, Holdt and Kraan 2011). Lysine, threonine, tryptophan and the sulphur amino acids can be the limiting amino acids in macroalgal proteins under certain environmental conditions (Galland-Irmouli et al. 1999).

The ratio of EAAs:AAs can be very favorable and some macroalgae can comprise <50% of the total AA content as EAAs (Cerna 2011). Such values are comparable with egg protein. Essential amino acid scores in the range of 60–90% have been reported which puts macroalgae above many vegetables and cereals in terms of the protein quality (Holdt and Kraan 2011); however, these scores are not corrected for protein digestibility. Although the amino acid profile of macroalgae is indicative of their nutritional potential, protein digestibility is also of key importance as it determines the availability of bound amino acids. In humans, the digestibility of native macroalgal protein is quite low because the cell wall encapsulates cytoplasmic proteins and limits bioavailability. In addition there are possible interactions between proteinsand soluble polysaccharide that prevent proteolysis during digestion (Gall and-Irmouli et al. 1999, Fleurence et al. 2012). However, digestibility is seen to improve with chemical or enzymatic treatment, e.g., alkali-soluble proteins or enzymatic extracts of proteins (Marrion et al. 2003, Marrion et al. 2005, Fleurence et al. 2012) and this is seen as a possible way to develop protein available ingredients for human consumption (Fleurence 1999, Fleurence et al. 2012).

The recommended daily protein intake for male and female adults is 0.83 g/kg per day, for proteins with a protein digestibility-corrected amino acid value of 1.0 (WHO 2002). Obtaining a balanced intake of dietary protein is however, not generally a problem for conscientious consumers, whether this is derived from animal or plant sources. More often concerns lie with obtaining an adequate intake of the essential amino acids and other nutrients like Vitamin B12, iron and retinol that are associated with animal proteins and thus may be lacking from vegetarian and vegan diets (FSA 2008).

When compared to plant and animal derived proteins, relatively little is known about the bioactivity of macroalgal proteins. Research has essentially focused on lectins, phycobiliproteins and bioactive amino acids (Harnedy and Fitzgerald 2011). Bioactive peptides have been isolated from macroalgae at laboratory scale; these tend to be from 2 to 20 amino acid residues in size and are normally inactive until released from the parent protein by hydrolysis (Harnedy and FitzGerald 2011). Demonstrated bioactive properties include:

antihypertensive, antioxidative, antithrombotic, hypocholesterolemic, opioid agonist and antagonist activity, mineral-binding, anti-appetizing, antimicrobial, immunomodulatory and cytomodulatory (Korhonen and Pihlanto 2006, Tan et al. 200, Silva et al. 2010).

2.3 Lipids

The lipid component of macroalgae essentially comprises fatty acids, phospholipids, glycolipids, pigments, sterols and fat soluble vitamins. Total lipid content is low, generally <5% dry weight but this varies considerably with a number of environmental factors (Table 1). This is discussed in detail by Holdt and Kraan (2012).

The fatty acid profile of many species has been determined (Dawczynski et al. 2007, van Ginnekin et al. 2011, Pereira et al. 2012); macro (and micro) algae are particularly rich sources of PUFAs (poly unsaturated fatty acids) including the essential omega (n)-3s LNA, EPA and DHA (respectively α-linoleic, Eicosapentaenoic and Docosahexaenoic acids) and omega-(n)-6 LA (linoleic acid). The n-3 and n-6 fatty acids have opposing physiological functions that require a balance for normal growth and development in humans and until recently an intake ratio has been advised. However, the WHO has now set recommended daily intakes for n-6 and n-3 fatty acids and providing that these are met, there is no longer a recommended n-6:n-3 intake ratio (FAO 2010b). European diets are generally over rich in n-6 (Dawczynski et al. 2007) which are mainly derived from vegetable oils and it is suggested that the addition of certain macroalgae to the diet could be highly beneficial on account of their high n-3 content (van Ginnekin et al. 2011). Brown and red seaweeds provide a good balance with ratios that are comparable with cold fish sources (Dawczynski et al. 2007, Pereira et al. 2012).

Fatty acids can comprise <50% of the total lipid content in some species (Dawcynzki 2007). Content varies significantly with environmental factors, in particular temperature and cold water species tend to contain higher levels of PUFAs (Holdt and Kraan 2011, van Ginnekin et al. 2011). The predominant fatty acid in most macroalgae is EPA (C20:5, n-3), this fatty acid can comprise <59% of the total fatty acid content of the red species *Palmaria palmata* (Linnaeus) Weber & Mohr (van Ginnekin et al. 2011). Other n-3 fatty acids also commonly found are C16:3 (isomer n-3), C16:4 (n-3), C18:3 (a-linolenic acid), C18:4 (stearidonic), and also C22:6 (DHA: docosahexaenoic acid) (Dawczynski et al. 2007, van Ginnekin et al. 2011). PUFA 18:4 n-3 is not found in other organisms (Holdt and Kraan 2011). N-6 fatty acids commonly found are C18:2 (linoleic acid), C18:3 (g-linolenic acid), C20:4 (arachidonic acid) and C22:5 (docosapentaenoic acid) (Dawczynski et al. 2007, van Ginnekin et al. 2011, Pereira et al. 2012).

In general, a balanced dietary intake of both n-3 and n-6 fatty acids from any source is shown to modulate inflammatory processes and other cell functions. Most studies relating to the health benefits of algal derived fatty acids have focused on microalgal species (see Bernstein et al. 2011 for a comprehensive meta-analysis). Information regarding the health benefits of macroalgal derived fatty acids, including bioavailability and bioactivity is generally lacking. Conquer and Holub (1996) showed supplementation with a commercially available DHA enriched encapsulated triglyceride oil (DHASCO ™—presumably from the brown algal species *Ascophyllum*) to have positive effects on the DHA status of certain vegetarian groups and to also moderate blood lipid levels.

Macroalgae contain a number of pigments. Those that have most beneficial potential for human health are the carotenoids. Carotenoids are the most widespread pigments in nature. They are C_{40} tetraterpenes; carotenoids comprise carotenes which are hydrocarbons and xanthophylls that contain one or more oxygen molecules (Lobban and Harrison 2000). They are found in all macroalgae and in wide variety. β-carotene is found in species of red, green and brown algae along with lutein in the reds and greens and various xanthins (Holdt and Kraan 2011). Brown macroalgae contain fucoxanthin which is a brown photosynthetic accessory pigment that belongs to the xanthophyll group of carotenoids and is probably the most abundant carotenoid in nature. Fucoxanthin is a potent antioxidant. The fucoxanthin content of macroalgae is generally low, approximately 3–6% of the total lipid content (TL) of brown species (TL = 1–10% of the dry weight of the macroalga) (Miyashita 2011). Other xanthophylls and xanthan pigments from red and brown species have been identified (Cornish and Garbery 2010).

Some carotenoids also function as vitamins (see relevant section below). β-carotene from red species like *Porphyra* and *Palmaria palmata* exhibits provitamin A activity. Its content ranges from 36 to 4,500 mg/kg dry weight and is seasonally variable (Holdt and Kraan 2011).

In addition to their antioxidant activity, carotenoids in general are known to have anti-inflammatory and anti-cancer activity (Maeda et al. 2008, Miyashita 2011); fucoxanthin also shows anti-diabetic and anti-obesity effects that are not found in the other carotenoids that lack the allenic bonding (Maeda et al. 2008, Miyashita 2011). The metabolic pathway of fucoxanthin is fairly well described (Miyashita 2011, Peng et al. 2011). It is absorbed and converted into fucoxanthinol and halocynthiaxanthin, which is further metabolized to amarouciaxanthin A. These metabolites are then concentrated in adipose tissues where they exert a number of key effects: (i) antioxidant effects; (ii) increased synthesis of DHA (docosahexaenoic acid) in the liver; (iii) upregulation of UCP-1 (uncoupling protein involved in heat generation through fat burn) in white adipose tissue and (iv) glucose-regulating effects in muscles. These effects all have implications for the management of diabetes, obesity and Syndrome X (Maeda et al. 2008, Miyashita 2011).

Fucoxanthin and fucoxanthin rich seaweed extracts are shown to have a number of beneficial effects on skin health including antioxidative, anti-inflammatory (Peng et al. 2011) and inhibition of key enzymes such as elastase, hyaluronidase and collagenase. Fucoxanthin also inhibits tyrosinase and melanogenesis and thus has skin whitening properties (Maeda et al. 2008, Peng et al. 2011).

2.4 Vitamins

The literature concerning macroalgal derived vitamins is not extensive. However, macroalgae are reported to contain a range of water soluble vitamins such as thiamine (B1), riboflavin (B2), cobalamin (B12) and ascorbic acid (C) and fat soluble (Beta-carotene, A and E) vitamins. Trace amounts of other B vitamins such as niacin, biotin and folates have also been identified (Skrovankova 2011). Macroalgae have been promoted as an excellent vegetarian source of vitamin B12 (Rauma et al. 1995, MacArtain et al. 2007, Skrovankova 2011). However, the situation is controversial. Studies have shown that the B12 found in macroalgae is actually an analogue form that is not nutritionally beneficial and that current methods used to estimate bioavailability are not adequate (Dagnelie et al. 1991, 1994). The situation requires clarification but at present, vegetarian and vegan societies are warning against the reliance on macroalgae to supply B12. Despite this, evidence suggests that the general vitamin content of macroalgae (Table 2) could contribute to a balanced nutritional intake (MacArtain et al. 2007, Skrovankova 2011).

As with minerals, the level of vitamins found in macroalgae varies on a seasonal and geographic basis. Brown macroalgae are typically higher in vitamin C than other vitamins (Table 2), as are many green species (Mabeau and Fleurence 1993). The red macroalgae *Palmaria palmata* and species of *Porphyra* contain high amounts of B1, B2 and provitamin A (Skrovankova 2011). The green species *Ulva* Linnaeus is also rich in B vitamins (Table 2).

Comprehensive studies on the bioavailability of macroalgal vitamins, other than B12, are lacking. In general, bioavailability and absorption is dependent on solubility, and for the fat soluble vitamins, is thus also dependent on whether the vitamins are consumed with food containing lipids (Skrovankova 2011). Bioavailability is also dependent on the presence of carbohydrates and fiber as vitamins are known to bind with these nutrients. Macroalgae are rich in carbohydrates and Suzuki et al. (1996) found binding of thiamine to be as high as 45% in some Japanese species.

2.5 Minerals and Trace Elements

Macroalgae can comprise greater than 50% of their dried biomass as minerals and trace elements (Holdt and Kraan 2011) but generally this falls within 8–40% (Ruperez 2002, Inst. de Phytonutrition 2004, Holdt and Kraan 2011). Some species of red macroalgae, e.g., *Lithothamnion* Heydrich and *Phymatolithon* Foslie (also known as Maerl) are coated with deposits of calcium and magnesium carbonates. The calcium and magnesium content of these deposits alone can comprise over 30% of their dry weight (Morrissey et al. 2001).

All of the essential minerals and trace elements that are required in the diet are found in macroalgae. The typical mineral and trace element content of selected European species is given in Table 3 below. For certain minerals, a few grams of dried macroalgae could represent a significant % of the daily RNI (MacArtain et al. 2007), however, information on the bioavailability of macroalgal derived minerals is not well documented and a number of factors could interfere with their uptake and absorption (see below).

The main minerals required in the diet are calcium (Ca), magnesium (Mg), phosphorous (P), potassium (K) and sodium (Na). Recommended dietary intake of these minerals, based on UK RNI (Reference Nutrient Intake) values, ranges from 270–3500 mg/day for adult males and non-pregnant or lactating females over the age of 19 (Table 3). Different RNI values exist for children and adolescents. Based on similar criteria, the recommended dietary intake of trace elements: copper (Cu), chromium (Cr), cobalt (Co), fluoride, iodine (I), manganese (Mn), molybdenum (Mo), selenium (Se) and zinc (Zn) is generally less than <100 mg/day and <100 µg/day in the case of selenium and iodine (Buttriss 2000). Non-meat eaters need to be especially aware of the need for iodine, selenium and zinc. Macroalgae contain selenium and zinc at levels equivalent to or in excess of most fruit and vegetables (MacArtain et al. 2007). Some macroalgae, in particular the large brown species, can bioaccumulate high levels of iodine which could be a problem depending on the iodine status of the population (Scientific Committee on Food 2002). However, extracts of brown species or "kelp extracts" are sold as iodine and mineral supplements.

2.6 Phenolics

Marine algae produce a range of secondary metabolites, many of which have been shown to have positive benefits to human health (Holdt and Kraan 2012). Of these, the phenolic substances are probably best associated with human health and are also commercially exploited. The polyphenolic substances that are found in brown macroalgae, known as phlorotannins, are potent antioxidants (Ragan and Glombitza 1986, Cornish and Garbery

Table 3. Typical mineral and trace element content for selected European species: illustrating relative composition of selected species, note that nutritional composition varies significantly with species, season and locality. NB: There is no RNI value for Mn.

Seaweed	mg per 100g dry weight										μg/100 g dw
	Ca	Mg	P	K	Na	Mn	Fe	Cu	Zn	I	Se
Laminaria digitata	700–3400	300–2000	120–2520	1200–20230	800–4490	700–4530	2.3–550	0.2–3.4	1.6–17	51–992	<0.01
Saccharina latissima	630–1350	380–2790	140–4310	3720–9300	2530–5060	380–2520	1.8–16.2	0.17–0.5	1.8–3.8	252–680	n/d
Himanthalia elongata	200–900	510–6960	70–140	3610–9000	2950–4070	0.08–4.8	0.2–32.8	0.1–0.4	2.5–7.9	7.3–44	n/d
Palmaria palmata	84–1750	84–830	178–569	2077–12441	178–3000	0.9–75.8	9.5–150	0.25–4.8	1.1–20	6.7–120	0.02
Chondrus crispus	370–2340	650–3750	8	2610–4010	3430–9150	1.2	3.5–44.4	0.2–1.8	4–21.8	19.1–54.5	n/d
Porphyra spp.	3–797	4–1792	188–699	161–330	108–5877	1.8–9	5.6–380	0.3–2.3	0.8–9.6	0.5–20.6	0.8–70
Ulva spp.	270–5120	150–6630	30–460	150–5480	90–5530	0.4–9.1	4.1–512	0.2–5.5	0.5–18	1.1–26.5	0.9–48.5
UK RNI for adults in mg/day, except * = μg/day	700	270–300	550	1600	3500	-	87–148	12	70–95	140*	60–70*
Data source	1. By permission of CEVA unpublished data, 2. Institute of Phytonutrition (2004), n/d denotes not determined.										

2010). Phlorotannins are unique to species of brown macroalgae and are the only polyphenols found in brown macroalgae (unlike terrestrial plants that typically contain a range of different polyphenols). Phlorotannins are exclusively polymers of phloroglucinol (1,3,5-trihydroxybenzene) and are structurally less complex than other polyphenols. In brown macroalgae, phlorotannins can comprise <25% of dry weight but this varies across location and season (Targett et al. 1992).

Red and green macroalgae contain low levels of phenolic compounds when compared to brown species and these are more analogous with terrestrial plant polyphenols. Red and green species contain flavonoids, catechins and epicatechins amongst others in the range of 4–11 mg/g dry weight (Holdt and Kraan 2012).

Phlorotannins are stored in physodes, which are membrane bound cell organelles in the cytoplasm. This form of phlorotannin is referred to as soluble phlorotannin and is easily extracted. Phlorotannins are also present in bound form as the physodes fuse to cell walls and the phlorotannins complex with the structural polysaccharides, alginate in particular. This form of phlorotannin is referred to as bound or CWB (cell wall bound) phlorotannin and only a few studies have attempted to extract and quantify them. It is soluble phlorotannin that is generally estimated as total phlorotannin (Targett et al. 1992, Targett and Arnold 1998, Koivikko et al. 2005, Koivikko 2008).

Phlorotannins typically have a molecular weight (MW) range of 126 Da to 650 kDa but more commonly fall in the range of 10 to 100 kDa. In ecological terms, high molecular weight (HMW) phlorotannins (≥10 kDa) are most bioactive whereas in human systems, it is most likely that low MW phlorotannins (<10 kDa) are bioactive (Targett et al. 1992, Targett and Arnold 1998, Koivikko et al. 2005, Koivikko 2008). However, little is known about the bioavailability and metabolic fate of macroalgal derived polyphenols.

The documented evidence for the beneficial effects of phlorotannins is extensive. Phlorotannins are shown to modulate inflammation; reduce oxidative stress; modulate the metabolism of fats and sugars with beneficial implications for diabetes and weight management; improve cardiovascular function; improve brain and cognitive function; induce immunomodulatory and chemopreventative effects; promote skin repair and skin whitening processes. Cornish and Garbary (2010) provide an excellent review. A number of phlorotannin rich extracts are commercially available; these are sold directly or as functional components in a range of health and wellness products. Phlorotannin rich extracts are primarily marketed on the basis of their antioxidant and anti-inflammatory benefits and for weight management.

2.7 Nutrient Bioavailability

Despite the potential nutritional composition of macroalgae being well documented, information regarding the bioavailability of macroalgal derived nutrients is limited. A few studies have focussed on minerals and trace elements, including heavy metals (Dominquez-Gonzalez et al. 2010, Moreda-Pinero et al. 2011, Romaris-Hortas et al. 2011, Garcia-Sartal et al. 2011, 2012, Nakamura et al. 2012).

Bioavailability is defined as the amount of nutrient digested, absorbed and metabolized by the organism. It can also be defined as the amount of nutrient, after absorption, used in biological processes. There are two major components of bioavailability: (i) the absorption phase, in which nutrients are transferred through the intestine cell wall and (ii) the metabolism phase, during which the nutrients are metabolized. The term "bioaccessibility" is often used synonymously in the literature with "bioavailability" but the two are different. Bioaccessibility refers to the fraction of the nutrient which is released from the food matrix and is thus available for intestinal absorption; bioavailability describes the fraction that is absorbed.

In the majority of studies, *in vitro* methods have been used to simulate gastric and intestinal digestion of the macroalga to bring about the release of the nutrient from the matrix; the fraction of the nutrient that is available for absorption is then quantified. Many variations on the simple model exist and all have their pros and cons. Modifications of the basic system tend to focus on pH adjustment during digestion, buffer used in dialysis step and molecular weight cut off (Dominquez-Gonzalez et al. 2010).

A few studies have looked at the bioavailability of key trace elements and heavy metals from wild harvested European species using *in vitro* models (Dominquez-Gonzalez et al. 2010, Garcia-Sartal et al. 2011, 2012, Moreda-Pinero et al. 2011, Romaris-Hortas et al. 2011). A few commercially available prepared food products have also been investigated. The studies show considerable variation in the bioavailability of minerals and trace elements between species. Manganese seems to be highly available in many species including *Palmaria palmata* and species of *Porphyra* (Dominquez-Gonzalez et al. 2010). Bioavailability of a range of minerals and trace elements was consistently high (average of 30–74.7%) in seven different macroalgae, including *Palmaria palmata* and species of *Porphyra* and *Ulva* (Moreda-Pinero et al. 2011). Nakamura et al. (2012) investigated the bioavailability of magnesium from several commercially available Japanese species and found this to vary between species; bioavailability was high in the brown species Kombu (*Saccharina japonica* (Areschoug) C.E. Lane, C. Mayes, Druehl & G.W. Saunders) but much lower in the red and green species studies. The authors report that these findings were in conflict with other similar studies on Japanese species and therefore conclude that bioavailability appears not to be dependent on division of macroalgae (Nakamura et al. 2012).

Bioavailability of magnesium has been investigated *in vivo* using Sprague-Dawley rats. Magnesium levels in plasma and bone were measured after 28 days of controlled feeding using three different Japanese macroalgae. The results showed that differences in magnesium content of the macroalgae, absorption rate and solubility affect the bioavailability so that species with varying magnesium content can have similar bioavailability (Nakamura et al. 2012).

The effects of cooking on mineral and trace element bioavailability have been investigated but only in as far as the sea vegetables have been cooked in water and the amount of leaching into the cooking water has been quantified (Garcia-Sartal et al. 2010, 2012a,b). This is a common consideration with regard to heavy metals. Macroalgae are renowned for the bioaccumulation of all minerals, including heavy metals and leaching of certain species, in particular inorganic arsenic and iodine; their cultivation is seen as a method to reduce potentially toxic levels (Romaris-Hortas et al. 2011, Garcia-Sartal et al. 2012a,b). Cooking in water has variable effects on the leaching of minerals from the macroalgal matrix, not all are released. In the case of *Porphyra,* zinc in particular was retained in the matrix (Garcia-Sartal et al. 2012a).

The presence of macro nutrients such as carbohydrates, protein and lipids can affect mineral bioavailability; however the effects appear to be highly variable. Garcia-Sartal et al. (2012) found a highly positive correlation between the presence of protein and the bioavailability of Cr, Fe, Co, Ni and Cu, whereas Moreda-Pinero et al. (2011) found a highly negative correlation between protein and a similar suite of minerals. Both studies confirmed a negative correlation between mineral bioavailability and the presence of lipids and carbohydrates. Although not investigated, the interaction of different cell wall polysaccharides was proposed as a major influence on the bioavailability of magnesium in Japanese species (Nakamura et al. 2012).

Preliminary findings from an *in vivo* study aimed at investigating the absorption and metabolism of macroalgal derived phlorotannins in human volunteers indicate that absorption and metabolism is occurring (Corona et al. 2012). The documented evidence for the health promoting effects of phlorotannins is extensive and is addressed in the following section.

3 Health Promoting Benefits of Macroalgae

In recent years, a number of commercial products that contain macroalgal derived bioactives have become available on the wider global market. Such products are commonly found in Southeast Asia and have been for a number of years. These products are primarily marketed for their benefits to humans in the following areas: reduction of oxidative stress (antioxidants): modulation of inflammation; modulation of fat and sugar metabolism; improved digestive health and improved cardiovascular health. The most commonly exploited bioactives are: fucoidans; fucoxanthin, brown macroalgal phlorotannins; polysaccharides and oligosaccharides. This section of the review will primarily focus on these commercially relevant bioactives and their health promoting benefits to humans. The potential health promoting benefits of other bioactives and metabolites is discussed further in Chapter 6.

In general, the commercial products that are on the market are derived from species of brown macroalgae such as *Fucus* and *Ascophyllum*. This is due in part to the commercial availability of these species. A far wider range of extracts and species is used in the cosmetics industry, however this is outside the scope of this review. It must be noted, that the extracts used in studies to investigate the health potential of macroalgae can be in the form of a relatively pure fraction, a "bioactive rich" extract or even a crude extract depending on whether suitable methodology for the extraction and separation is available. In so far as it is possible, information on the type and nature of the extracts used in the studies reported here, are given in Tables 4 to 8 below.

3.1 Reduction of Oxidative Stress

Oxidative stress results from an imbalance in the body's equilibrium between the formation of potentially damaging RS (reactive species) like hydrogen peroxide, hydroxyl radicals and superoxides and the endogenous antioxidant defense mechanisms that exist. Cells possess a variety of defense mechanisms including antioxidant enzymes (such as superoxide dismutases SODs, catalases and peroxidases) and endogeneous antioxidants such as glutathionine, vitamins C and E which can remove RS from body. ROS (reactive oxygen) and RNS (reactive nitrogen) species have been implicated in over 150 human disorders (Cornish and Garbery 2010). Macroalagal extracts and bioactives are shown to mitigate the effects of oxidative stress (Table 4). The key antioxidants in macroalgae include many of the fat soluble vitamins: vitamin E, carotenoids, vitamin C and vitamin B1 (Skrovankova 2011), pigments, polyphenolic substances and polysaccharides (Cornish and Garbery 2010, Holdt and Kraan 2011).

Various fucoidans and/or fucoidan rich extracts of brown seaweeds have shown antioxidant ability in terms of radical scavenging ability (hydroxyl and superoxide radicals), reducing power and ferrous ion II chelating abilities and also the ability to inhibit DNA damage (Table 4). LMW

Table 4. Selected studies showing the potential of macroalgal derived bioactives and extracts to reduce oxidative stress.

Detail including extraction methods, bioactivity and test procedures	Reference
Phloroglucinol from *Ecklonia cava* Kjellman: Inhibitory effect of (i) intracellular ROS generation by hydrogen peroxide using DCFCH-DA assay; (ii) membrane protein oxidation; (iii) radical mediated DNA damage and increased production of GSH in HT1080 human fibrisarcoma cells	Kim and Kim 2010
Ethanolic extract of *Ascophyllum nodosum:* Antioxidant activity NBT assay	Zhang et al. 2007
Methanolic extract of *Padina antillarum* (Kutzing) Piccone: Antioxidant activity, DPPH, FRAP, ferric Ion chelation assay, Beta carotene bleaching assay	Chew et al. 2008
Acetone extracts of *Fucus vesiculosus* Linnaeus, *F. serratus* Linnaeus, *Ascophyllum nodosum, Laminaria hyperborea* (Gunnerus) Foslie, *L. digitata* Lamouroux, *L. saccharina, Alaria esculenta* (Linnaeus) Greville: Antioxidant activity, DPPH, ORAC	Wang et al. 2009
Water: ethanol extracts of *Undaria pinnatifida* (Harvey) Suringar, *Saccharina japonica* & *Hizikia fusiformis* (Harvey) Okamura: Antioxidant activity, DPPH, Beta carotene bleaching assay	Ismail and Hong 2002
Ethanol: water extracts of *Fucus vesiculosus, F. serratus* & *Ascophyllum nodosum*: Antioxidant activity DPPH, cyclic voltammetry	Keyrouz et al. 2010
Solvent extracts of *Eisenia bicyclis* (Kjellman) Setchell, *Kjellmaniella crassifolia* Miyabi, *Alaria crassifolia* Kjellman, *Sargassum horneri* (Turner) C. Agardh & *Cystoseira hakodatensis* (Yendo) Fensholt: Antioxidant capacity synergistic with fucoxanthin (DPPH, peroxyl radical, ABTS, nitric oxide & activity in a liposome system)	Wijesekara et al. 2010
Methanolic extraction & fractionation of *Ascophyllum nodosum*: Antioxidant activity, ABTS, DPPH and cyclic voltammetry	Blanc et al. 2011
Various solvent extractions & dialysis of *Ascophyllum nodosum*: Antioxidant activity, DPPH	Breton et al. 2011
Methanolic extracts of *Ascophyllum nodosum, Pelvetia canaliculata* (Linnaeus) Decaisne & Thuret, *Fucus serratus, F. vesiculosus* & *L. digitata*: different species showed (i) High FRAP activity, (ii) Effective radical scavenging (DPPH) and prevention of b-carotene bleaching, (iii) increased glutathione (GSH) content, (iv) decreased DNA damage, (v) significant protection against H_2O_2-mediated SOD reduction	O'Sullivan et al. 2011
Trifucodiphlorethol, fucotriphlorethol & trifucotriphlorethol from *Fucus vesiculosus*: Antioxidant activity, DPPH, ORAC, Inhibition of cytochrome P450 1A	Parys et al. 2010
Himanthalia Lyngbye, *Undaria, Fucus, Sargassum* C. Agardh: Improved antioxidant potential in food products to improve quality and shelf life (meat, fish, cereals, pasta, oils)	Gupta and Abu-Ghannam (2010) Ngo et al. 2010

Table 4. contd....

Table 4. contd.

Detail including extraction methods, bioactivity and test procedures	Reference
Enzyme assisted extracts from species of *Ecklonia* Hornemann, *Ischige* Yendo C. Agardh, *Sargassum* & *Scytosiphon* C. Agardh: Antioxidant capacity and radical scavenging ability (DPPH, superoxide anion Hydroxyl radical and hydrogen peroxide scavenging assay), inhibition of DNA damage (comet assay)	Heo et al. 2005
Methanolic extracts of *Bifurcaria bifurcata* R. Ross, *Cystoseira tamariscifolia* (Hudson) Papenfuss, *Fucus ceranoides* Linnaeus and *Halidrys siliquosa* (Linnaeus) Lyngbye, fractionated using solid phase extraction: Antioxidant activity DPPH, reducing activity and b-carotene–linoleic acid system	Zubia et al. 2009
Hot water extraction & ethanol precipitation of *Laminaria japonica* extract Antioxidant capacity and radical scavenging ability (superoxide radical assay, hydroxyl radical assay & reducing power assay)	Zhang et al. 2010
Enzyme-digested extracts from *Cladosiphon novae-caledoniae* Kylin: Amelioration of oxidative stress in human fibrosarcoma HT1080 cells—levels of intracellular and released hydrogen peroxide greatly repressed	Ye et al. 2005
Proteolytic & solvent based extracts of *Sargassum filipendula*: Antioxidant capacity (total antioxidant capacity, scavenging hydroxyl and superoxide radicals, reducing power and ferrous ion II chelating)	Costa et al. 2011
Acetone extracted fractions of *Lobophora variegata* (J.V. Lamoouroux) Womersley ex E.C. Oliveira: Antioxidant capacity and radical scavenging ability (superoxide radical assay, hydroxyl radical assay)	Paiva et al. 2011
Sequential extraction using chloroform, ethyl acetate, acetone & ethanol *of Hijikia fusiformis, Undaria pinnatifida* & *Sargassum fulvellum* (Turner) C. Agardh: DPPH radical scavenging assay	Yan et al. 1999
Extracts of different brown seaweeds using methanol, ethanol, acetone, chloroform, ethyl acetate, and n-hexane: DPPH & AAPH radical scavenging assays showed synergistic antioxidant activity of phlorotannins and fucoxanthin	Widjaja-Adhi Airanthi et al. 2011
Agaro-oligosaccharides of varying DP prepared via acid hydrolysis and gel permeation separation: Antioxidant effects measured via DPPH	Chen and Yan 2005
Agaro-oligosaccharides prepared by acid hydrolysis of agar: Antioxidant effects in human hepatocyte L-02 measured using fluorescence and 2', 7'-dichlorofluorescin diacetate probe	Chen et al. 2006

fucoidan is generally more potent than HMW fucoidan; this is believed to be due to the LMW forms being able to donate protons more effectively than HMW forms (Wijesekara et al. 2010).

Oligosaccharides derived by hydrolysis of agar (agaro oligosaccharides) are shown to have a range of bioactivities including antioxidant activity (Chen et al. 2005) and this appears to be linked to degree of polymerisation.

Table 5. Selected studies showing the potential of macroalgal derived bioactives and extracts to modulate inflammation.

Detail including extraction methods, bioactivity and test procedures	Reference
Phloroglucinol from *Ecklonia cava*: Inhibitory effect on (i) production of inflammatory mediators TNF-α, IL-1β, IL-6 and PGE2 in RAW264.7 cells; (ii) activity of MMPs in HT1080 cells; (iii) expression of inflammatory proteins in RAW264.7 cells and HT1080 cells	Kim and Kim 2010
Isolates of *Eisenia arborea* Areschoug: Inhibition of enzyme activity cyclooxygenase (COX)-2, lipoxygenase (LOX), phospholipase A$_2$ (PLA$_2$) and hyaluronidase (HA)	Sigiura et al. 2011
Ecklonia cava: Inhibition of MMP activity	Wijesekara et al. 2010
Acetone extracted fractions of *Lobophora variegata*: Anti-inflammatory activity in acute zymosan-induced arthritis in rats	Paiva et al. 2011
Purified fucoidan from *Fucus vesiculosus* from Sigma Aldrich.: Inhibition of (i) production of nitric oxide (NO); (ii) prostaglandin E$_2$ (PGE$_2$) in LPS-stimulated BV2 microglia. Attenuated expression of (i) inducible nitric oxide synthase (iNOS), (ii) cyclooxygenase (COX)-2, (iii) monocyte chemoattractant protein-1 (MCP-1), and (iv) pro-inflammatory cytokines, including interleukin-1β (IL-1β) and tumor necrosis factor (TNF)-α. Suppression of nuclear factor-kappa B (NF-κB) activation and down-regulation of (i) extracellular signal-regulated kinase (ERK), (ii) c-Jun N-terminal kinase (JNK), (iii) p38 mitogen-activated protein kinase (MAPK), and (iv) AKT pathways	Park et al. 2011
Purified fucoidan from *Fucus vesiculosus* from Sigma Aldrich & Ascophyllan from *Ascophyllum nodosum*—hot acidic extraction with alginate lyase.: Potential to induce NO and cytokine production (tumor necrosis factor-α (TNF-α) and granulocyte colony-stimulating factor (G-CSF)) in murine RAW264.7 cells	Jiang et al. 2011
Fuoidan rich extracts from 7 species including *Ascophyllum nodosum, Fucus, Laminaria* & *Cladosiphon*, calcium chloride extract followed by precipitation of acidic polysaccharides with Cetavlon: Anti-inflammatory activity through (i) inhibition of neutrophil extravasation & (ii) reduction of polymorphonuclear leucocytes (PMNs) extravasation via P-selectin dependant mechanism in rat peritonitis model	Cumashi et al. 2007
High, medium and low molecular weight fucoidans from acid hydrolysis of *Undaria pinnatifida* (Bion Co. Ltd).: Low molecular weight fucoidans inhibited factors involved in the progression of collagen-induced arthritis in mice. High molecular weight fucoidans were pro-inflammatory	Park et al. 2010
Methanol extract of *Ishige okamurae* Yendo, subsequent partitioning with chloroform & methanol.: Reduction in levels of pro-inflammatory mediators including NO, PGE2, IL-1β, TNF-α, and IL-6 via the inhibition of NF-κB activation and the suppression of MAPK phosphorylation in RAW 264.7 cells	Kim et al. 2010

Table 5. contd....

Table 5. contd.

Detail including extraction methods, bioactivity and test procedures	Reference
Methanol extract of numerous seaweeds including *Myagropsis myagroides* (Mertens ex Turner) Fensholt, subsequent partitioning with chloroform & methanol.: Strongest effects seen in extracts of *Myagropsis myagroides*—reduction in prostaglandin E2 (PGE2) production; inhibition of inducible nitric oxide synthase (iNOS) and cyclooxygenase 2 (COX-2) protein expressions; suppressed expression of iNOS and COX-2 mRNA; reduction in release of tumor necrosis factor-a (TNF-a), interleukin-1b (IL-1b), and interleukin-6 (IL-6). General inhibitory effect of nitric oxide (NO) production in lipopolysaccharide (LPS) induced RAW 264.7 macrophage cells	Heo et al. 2010
Fucoxanthin: *In vitro* assessment of T cell differentiation— suppression of interleukin-17 secretion from CD4+ T cells under IL-17-producing T (Th17) cell development conditions	Kawashima 2011

In particular, a number of studies have shown agaro oligosaccharides to exert a hepatoprotective effect against ROS (reactive oxygen species) in *in vitro* and *in vivo* (rats) models (Chen et al. 2006).

Carotenoids in general are known to have antioxidant properties and fucoxanthin rich extracts are shown to exert radical scavenging effects (Yan et al. 1999, Gupta and Abu-Ghannam 2011, Pangestuti and Kim 2011, Peng et al. 2011). A number of studies have shown fucoxanthin to exert stronger antioxidant activity than other carotenoids present. Pangestuti and Kim (2011) propose that this stronger activity is related to the presence of the allenic bonds.

Phlorotannins have been shown to be active against the effects of reactive species by playing an inhibitory role in their generation (Table 4). Phlorotannins are also shown to prevent DNA and membrane damage and to stimulate the production of glutathionine in affected cells (Table 4).

3.2 Modulation of Inflammation

Acute inflammation is a part of the body's natural defense system against injury and disease. However, chronic inflammation can ensue if the natural response gets out of control and the inflammatory state is sustained. Chronic inflammation is considered to be an underlying factor in many of the chronic diseases affecting our society such as cancer, diabetes mellitus, neurodegenerative, cardiovascular and inflammatory diseases.

Fucoidans are recognised inhibitors and ameliorators of inflammatory disorders such as arthritis, osteoarthritis, psoriasis and inflammatory bowel disease (Table 5). Fucoidans act in a number of ways to modulate the inflammatory response: (i) by inhibiting the production of pro-inflammatory mediators including interleukin-1β (IL-1β) and tumor necrosis factor

Table 6. Summary of studies showing the potential of macroalgal derived bioactives and extracts to modulate fat and sugar metabolism.

Detail including extraction methods, bioactivity and test procedures	Reference.
Ethanolic extract of *Ascophyllum nodosum:* Anti-diabetic activity—inhibition of rat α-glucosidase.	Zhang et al. 2007
Extracts of *Alaria esculenta* & *Ascophyllum nodosum*: Inhibition of α-amylase and α-glucosidase.	Nwosu et al. 2011
Ecklonia cava: Inhibition of α-amylase and α-glucosidase.	Wijesekara et al. 2010
V InSea2™—commercial extract of *Ascophyllum nodosum* and *Fucus vesiculosus* Significant reduction of post prandial levels of gucose and insulin in blood, *in vivo* Wistar Rat model.	Roy et al. 2011
Fucofuroeckol and dioxinodehydroeckol from *Eisenia bicyclis:* Potent inhibition of α-amylase and α-glucosidase.	Eom et al. 2012
Fucoxanthin-P1 from Oryza Oil & Fat Chemical Co., Ltd: Reduced fat mass, body weight and BMI in Japanese human volunteers taking 1–3 mg fucoxanthin daily.	Oryza Oil and Fat Company 2010
Solvent extraction of lipid fraction of *Undaria pinnatifida*: Reduced WAT (white adipose tissue) abdominal fat in rats and mice and significant reduction in body weight, through expression of uncoupling protein 1 (UCP1); Decrease in blood glucose and plasma insulin in mice; Increased levels of hepatic fatty acid DHA (docosahexaenoic acid) in mice model.	Maeda et al. 2008
78% Fucoxanthin extract from lipid extract of *Undaria pinnatifida*: Reduced WAT (white adipose tissue) abdominal fat in obese mouse models and in obese human volunteers through induced expression of UCP1 in the WAT, leading to heat production.	Miyashita 2011
Fucoxanthin rich extract of *Undaria pinnatifida*: Anti-obesity effect through UCP1 expression in white adipose tissues of obese mice	Maeda et al. 2005
Fucoxanthin and fucoxanthinol: Inhibition of lipase activity in the gastrointestinal lumen and suppression of triglyceride absorption in model rat system.	Matsumoto et al. 2010
Commercially available powders of different European seaweeds used to supplement feed of Large White pigs: high viscosity fibers had impact on intestinal absorption of glucose and post prandial insulin response.	Vaugelade et al. 2000
Sodium alginate supplement to diet assessed in a randomized, controlled, 2 way cross over intervention in humans: daily ingestion shown to have significant effect on energy intake and appetite control with significant reductions in daily fat, sugar, carbohydrate intake.	Pamnan et al. 2008

(TNF)-α, inducible nitric oxide synthase (iNOS), (ii) cyclooxygenase (COX)-2; and (iii) by suppression of inflammatory regulators such as nuclear factor-kappa B (NF-κB), extracellular signal-regulated kinase (ERK) and

Table 7. Selected studies showing the potential of macroalgal derived bioactives and extracts to improve digestive health.

Detail including extraction methods, bioactivity and test procedures	Reference
Alginate oligosaccharides prepared by enzymatic hydrolysis: stimulated growth of bifidobacteria and lactobacilli *in vivo* and *in vitro* when compared with FOS control.	Wang et al. 2006
Crude extracts of 22 edible brown species containing variable laminaran, fucoidan and alginates used to supplement diet of Wisar rats and inoculated with human faecal bacteria *in vitro:* evidence of increased faecal weight and increase in bifidobacteria.	Kuda et al. 1998a
Laminaran and depolymerised alginate used to supplement diet of Wistar rats and inoculated with human fecal bacteria: dose dependent effects *in vitro* and *in vivo* on increased bifidobacteria, and lowering of plasma cholesterol & triglycerides. In some cases, high doses led to diarrhoea.	Kuda et al. 1998a
Purified and crude extracts of *Himanthalia elongata, Laminaria digitata* and *Undaria pinnatifida* assessed *in vitro* using human fecal microflora: 60–90% of fibers digested but not completely metabolized to SCFA (47–85%) with laminaran being most utilized. Sulphated fucans were not degraded.	Michel et al. 1996
Neoagaro-oligosaccharides prepared by enzymatic hydrolysis of agar: *in vitro* and *in vivo* prebiotic effects studied in anaerobic medium and in mice (respectively). Evidence for prebiotic effects *in vitro* and *in vivo*.	Hu et al. 2006
Fucoidan extracted from *Cladosiphon okamuranus* Tokida: promotion of growth factors basic fibroblast growth factor (bFGF) and epidermal growth factor (EGF) that are recognized to promote ulcer healing.	Juffrie et al. 2006
Phloroglucinol shown to relieve symptoms of IBS.	Chassany et al. 2007
10 low molecular weight agars and alginates prepared by hydrolysis using acid or hydrogen peroxide supplemented to pH, temperature controlled anaerobic batch cultures inoculated with human feces: fermentation properties and prebiotic potential assessed. Some of the LMW agars and alginates promoted growth of blactobacilli and bifidobacteria, and also significantly increased the production of SCFAs.	Ramnani et al. 2011
Commercially available and laboratory extracted laminaran assessed in *in vitro* anaerobic batch culture fermenters using human microflora and *in vivo* in Wistar rats: <90% of laminaran degraded *in vitro* although SCFA production was promoted, laminaran did not promote selected growth of bifidobacteria and lactobacilli. Evidence for effects of intestinal metabolic modulaltion in rats.	Deville et al. 2007

c-Jun N-terminal kinase (JNK). Fucoidans also inhibit a number of tissue degradative enzymes such as heparanase, elastases and hyaluronidase that can be characteristic of inflammatory disorders (Fitton et al. 2007).

Table 8. Selected studies showing the potential of macroalgal derived bioactives and extracts to improve cardiovascular health.

Detail including extraction methods, bioactivity and test procedures	Reference
Free radical degradation of HMW fucoidan from *Ascophyllum nodosum*: Anticoagulant and antithrombotic activity (APTT—activated partial thromboplastin time.	Nardella et al. 1996
Free radical degradation and hydrolysis of fucoidan from *Ascophyllum nodosum*: Anticoagulant activity assessed vs. structure NMR structural determination .	Chevolot 1999
Ecklonia cava: Inhibition of ACE (Angiotensin-I-converting enzyme) and potential inhibitory effect against growth of human breast cancer cells MCF-7 with induction of apoptosis.	Wijesinghe et al. 2011
Phlorotannins, phloroglucinol, triphlorethol-A, eckol, dieckol and eckstolonol isolates from *Ecklonia cava*: demonstrated ACE (Angiotensin-I-converting enzyme) inhibition in endothelial cell line EAhy926.	Wijesekara et al. 2011
Dietary study amongst Japanese school children who have daily intake of seaweed: demonstrated correlation between high seaweed intake and lower blood pressure.	Wada et al. 2011
Commercial extract of *Fucus vesiculosus* Healsea: *in vivo* studies in rats and mice to demonstrate antioxidant and anti-inflammatory efficacy.	Besnard et al. 2008

Fucoidans also block the presence of leukocytes (white blood cells) and impair the action of selectins (Table 5). As part of the immune response process, leukocytes are moved through the blood vessels to the site of tissue damage or infection. The leukocytes bind with selectins which slows their progress and allows them to pass out of the blood vessels into the site of tissue damage or infection. This process is known as leukocyte extravasation.

A recent review by Fitton (2011) highlights a small number of studies that show the beneficial effects of fucoidans extracted from *Fucus vesiculosus* Linnaeus, *Macrocystis pyrifera* (Linnaeus) C. Agardh, *Saccharina japonica* and *Undaria pinnatifida* (Harvey) Suringar on osteoarthritis. A commercial extract containing a blend of fucoidan from *Fucus vesiculosus* (85% w/w), *Macrocystis pyrifera* (10% w/w) and *Saccharina japonica* (5% w/w) plus vitamin B6, zinc and manganese has been assessed in a human clinical study. The extract was shown to reduce osteoarthritis symptoms by up to 52% in a dose dependant manner when taken orally over 12 weeks (Myers et al. 2010).

As with other bioactives, molecular weight is important. A study by Park et al. (2010) highlights the differential effects of molecular weight on the bioactivity of fucoidan. The study focuses on the inflammatory processes involved in collagen induced arthritis in a mouse model. Low molecular weight fucoidans (approx 1 kDa) were anti-inflammatory whereas the high molecular weight fucoidans (approx 100 kDa) were pro-inflammatory.

Fucoxanthin has been shown to modulate the production of pro-inflammatory mediators such as nitric oxide (NO), prostaglandin E2 (PGE2), pro-inflammatory cytokines [tumor necrosis factor-a (TNF-a), interleukin-6 (IL-6) and interleukin-1b (IL-1b)] (Chen et al. 2005, 2006, Kim et al. 2010, Pangestuti and Kim 2011).

Phlorotannins have been shown to be effective in controlling inflammation via several pathways (Table 5). Phlorotannins can have an inhibitory effect on pro-inflammatory mediators. These mediators include nitric oxide (NO), prostaglandin E2 (PGE2) and the pro-inflammatory cytokines TNF-α (tumor necrosis factor) and interleukins (IL)-1β and IL-6. Pro-inflammatory mediators trigger the body's initial inflammatory response by increasing blood flow and permeability of the blood vessels. This results in swelling, heat and redness, i.e., the visible symptoms of inflammation.

Phlorotannins are also active against other enzymes such as cyclooxygenase (COX) –2, lipoxygenase (LOX), phospholipase A_2 (PLA$_2$) and hyaluronidase (HA) that play a role in promoting the inflammatory response (Table 5). The production of pro-inflammatory mediators is under the control of regulatory enzymes such as nuclear factor-kappa B (NF-kB) and AP-1 transcription factors. These enzymes play a critical role in modulating the inflammatory response through regulation of the genes that encode the pro-inflammatory mediators. Matrix metalloproteinase (MMP) enzymes play an important role in the digestion of extra cellular components and are closely associated with chronic inflammation, wrinkle formation, arthritis, osteoporosis and tumor formation. MMPs can also facilitate the recruitment of inflammatory cells such as eosinophils and neutrophils to sites of injury or infection. MMP genes are also mediated by inflammatory regulators such as NF-kB and AP-1 transcription factors. Phlorotannins are shown to inhibit the activity of regulatory enzymes (Table 5) and thus the production of pro-inflammatory mediators, pro-inflammatory enzymes and MMPs.

3.3 Modulation of Fat and Sugar Metabolism

The modulatory effects of macroalgal isolates, e.g., polysaccharides, polyphenols and fucoxanthin, on the uptake and metabolism of dietary fat and sugar are well documented (Table 6). A number of mechanisms have been identified with wider implications for the management of dietary related conditions such as obesity, diabetes and cardiovascular disease. A recent report by Myers (2012) lists over 70 companies that are selling fucoxanthin based products and ingredients for weight management.

Macroalgal fibers that form viscous gels in the digestive tract, such as alginate, produce hypocholesterolemic and hypolipidemic responses due to decreased absorption of cholesterol (Vaugelade et al. 2000, Brownlee et al. 2005, Paxman et al. 2008a,b). Gel formation of alginate in the stomach and small intestine restricts the absorption of bile acids by the body, these compounds bind to the alginate gel and are excreted. In turn, this forces the liver to scavenge cholesterol from the blood to synthesize and replace the bile, thus lowering cholesterol levels in the blood (Smit 2004, Warrand 2006). Alginates have also been linked to significantly lower post-prandial rises in blood glucose, serum insulin and plasma C-peptide in studies including pigs and humans (Vaugelade et al. 2000, O'Sullivan et al. 2010).

Phlorotannins can have a significant impact on the digestive process of complex dietary carbohydrates, through inhibiting the activity of important enzymes such as α-glucosidase and α-amylase. Inhibition of these enzymes can retard the digestion of oligosaccharides and disaccharides delay the glucose absorption in the small intestine and reduce glucose levels in plasma. Extracts of *Ascophyllum nodosum* (Ismail and Hong 2002, Zhang et al. 2007, Nwosu et al. 2011), *Ecklonia cava* Kjellman (Wijesekara and Kim 2010), *Eisenia bicyclis* (Kjellman) Setchell (Eom et al. 2012) and *Alaria esculenta* (Linnaeus) Greville (Ismail and Hong 2002, Nwosu et al. 2011) all inhibit the activity of α-glucosidase and α-amylase. A phlorotannin rich extract of *Ascophyllum nodosum* and *Fucus* is commercially available in Europe. The scientific support for this extract shows a significant reduction in post prandial levels of glucose and insulin in blood of Wistar rats (Roy et al. 2011).

The excessive accumulation of body fat, in particular, white adipose tissue (WAT), leads to obesity and the disruption of cytokine secretion in the adipose tissue. This in turn results in an increased risk of a number of diseases such as type II diabetes, hyperlipidaemia, hypertension and cardiovascular disease (Peng et al. 2011). The anti-diabetic and weight management properties of fucoxanthin are well documented (Table 6) and have been the key focus in the development of a number of commercial products.

Fucoxanthin is an effective modulator of body weight via a number of different pathways: (i) thermogenesis via the stimulation of key proteins, e.g., Uncoupling protein-1 (UCP-1) that cause the oxidation of fat and its conversion to energy or heat; (ii) reduction of fat uptake; (iii) reduction of fatty acid synthesis in the liver and adipose tissues; (iv) regulation of glucose uptake. Peng et al. (2011) provide a comprehensive review. Humans posses both white adipose tissue (WAT) and brown adipose tissue (BAT). Fat degradation through thermogenesis occurs in the BAT where the UCP-1 protein is present. However, excess fat is stored in the WAT. Animal studies have shown that UCP-1 activity can be induced in the white adipose tissue

(WAT) of fucoxanthin fed obese mice (Maeda et al. 2008) and thus induce weight loss.

Reduction of WAT also has implications for diabetes. Adipose tissues secrete leptin which is a key regulatory hormone in the energy metabolism and appetite control. Fucoxanthin fed mice are shown to have lower levels of blood glucose, serum insulin and serum leptin (Oryza Fat and Oil Company 2008).

3.4 Improved Digestive Health

Macroalgal derived fiber can have positive effects on gut health and a number of studies have looked for potential prebiotic activity (Michel and MacFarlane 1996, Warrand 2006, O'Sullivan et al. 2010). There is still however, very little known about the chemical, physiochemical and fermentation characteristics of macroalgal fiber in the human gut or the individual species of bacteria that are responsible for these beneficial activities but macroalgal carbohydrates do appear to have chemical, physicochemical and fermentation characteristics that differ from higher plant carbohydrates (O'Sullivan et al. 2010). To date, most studies have used *in vitro* assessments or small animal models (albeit by introducing human gut microflora) to determine such things (Table 7). In contrast, information regarding potential prebiotic activity in pigs and cattle is building (O'Sullivan 2010).

Alginate is shown to alter the human gut flora and affects the production of short chain fatty acids (Michel and MacFarlane 1996, Brownlee et al. 2005, Wang et al. 2006). There is emerging evidence to show that lower molecular weight polysaccharides and oligosaccharides can also act as a source of soluble fiber and may have prebiotic activity. As little as a 2.5% (weight per volume) addition of alginate oligosaccharides to the diet of rats and humans significantly increased the levels of bifidobacteria and lactobacilli and to reduce levels of putrefaction in the gut (Wang et al. 2006). Ramnani et al. (2011) investigated the fermentation properties and prebiotic potential of ten low molecular weight polysaccharides derived from *Ascophyllum nodosum* and species of *Gracilaria* Greville and *Gelidium* J.V. Lamouroux using *in vitro* batch screening methods. A number of the polysaccharides were fermented by the gut microflora and increased the production of SCFA. In particular, a low molecular weight *Gelidium* polysaccharide showed significant prebiotic potential by increasing bifidobacterial numbers and the production of acetate and propionate.

The fact that alginate appears resistant to total fermentation in the GI may benefit extended fermentation throughout the colon; alginate has been used to coat beads of resistant starch as a protective strategy that allows the starch to be delivered to the distal parts of the colon where it and the alginate are fermented (Rose et al. 2007).

A number of studies have also shown laminaran to have beneficial effects on gut health. Laminaran is not hydrolyzed by human endogenous enzymes; is rapidly and extensively fermented (<90%) by intestinal bacteria (Kuda et al. 1998a,b, Deville et al. 2007, Kuda et al. 2005) and produces high levels of SCFAs, in particular butyrate and propionate (Deville et al. 2007). Generation of high levels of butyrate is common to glucose containing polymers that undergo fermentation (Kuda et al. 1998).

Neoagaro oligosaccharides (NAOS) prepared from agar using enzyme hydrolysis have demonstrated prebiotic potential. In *in vitro* studies, NAOS were resistant to hydrolysis by upper GI tract enzymes (Hu et al. 2006). Supplementation of NAOS positively influenced the proliferation of colonic beneficial bacteria (Bifidobacteria and Lactobacilli) with no side effect both *in vivo* and *in vitro*, and their effects were better (P=0.05) than FOS. There were no proliferation effects on undesirable bacteria such as *Escherichia coli* or *E. faecal*. In this study, two NAOS fractions were prepared using different β-agarases, the first fraction comprised 51% neoagarotetraose and 49% neoagarohexaose. The second fraction comprised 43.1% neoagarooctaose, 37.4% neoagarodecaose and 1905% dodecaose. The NAOS fraction with the larger total DP showed better prebiotic response.

Some studies suggest that certain alginates may enhance the repair of mucosal damage in the gut (Warrand 2006) and can be beneficial to overall gut function in the following ways; reduced intestinal absorption, increased satiety, reduced damaging potential of GI luminal contents, modulation of colonic microflora, and elevation of colonic barrier function (Brownlee et al. 2005).

A few studies have investigated the benefits of fucoidan with regard to ulcers and gastritis (Fitton 2011). Binding of *Helicobacter pylori* to the gastric mucosa is inhibited by fucoidan *in vitro* and *in vivo*. Infection by *Helicobacter pylori* can lead to peptic and duodenal ulcers and also gastritis. Fucoidan is also shown to promote the healing process in ulcers by inducing the production of basic fibroblast growth factor (bFGF) and epidermal growth factor (EGF) that are recognized to promote ulcer healing (Juffrie et al. 2006).

3.5 Improved Cardiovascular Function

There are many risk factors associated with cardiovascular diseases. Oxidative stress and inflammation are both key factors and modulation of both by macroalgal derived isolates and bioactives are considered to be beneficial to overall cardiovascular health (Bocanegra et al. 2009). A commercial extract of *Fucus vesiculosus* is specifically marketed for its benefits to cardiovascular health (Besnard et al. 2008). Oxidative stress and inflammation can lead to conditions such as atherosclerosis (abnormal thickening and hardening of the inner linings of arteries) and stenosis (narrowing of blood vessels). Clinical studies have shown the antioxidant benefits of the *Fucus* extract (primarily from the phlorotannins), to play a preventative role in the oxidation of cholesterol lipoproteins, the development of atherosclerosis and also to act as vaso relaxants (Besnard et al. 2008).

High blood pressure (hypertension) is the single biggest risk factor for cardiovascular health; it is modifiable and can be treated. A few studies in animals have demonstrated a link between dietary intake of macroalgae and lowered blood pressure but studies in humans have been inconclusive (see Bocanegra et al. 2009, Wada et al. 2011). Wada et al. (2011) demonstrated a negative correlation between dietary intake of macroalgae and hypertension in Japanese school children although the underlying mechanisms were not determined.

Inhibition of Angiotensin-I-converting enzyme (ACE) is probably the most common mechanism that underlies the lowering of blood pressure. ACE inhibitory activity has been demonstrated for a number of different macroalgal isolates and extracts including phlorotannin rich isolates of *Ecklonia cava* (Wijesinghe et al. 2011). Suetsuna and co-workers found tetrapeptides and dipeptides from *Undaria pinnatifida* to have ACE inhibitory activity *in vitro* (Suetsuna and Nakano 2000) and in hypertensive rats (Suetsuna et al. 2004).

Abnormal blood lipid levels (i.e., high cholesterol, high total cholesterol, high levels of triglycerides, high levels of low-density lipoprotein or low levels of high-density lipoprotein (HDL) cholesterol) are also modifiable risk factors. The potential benefits of macroalgal derived bioactives and isolates were detailed above and selected studies summarized in Table 6.

The anticoagulant and antithrombotic properties of fucoidans are probably the most well documented health benefits of fucoidans (Bocanegra et al. 2009, Wijesekara et al. 2010b, Ale et al. 2011, Fitton 2011). Fucoidans derived from *Ascophyllum nodosum* and *Fucus vesiculosus* are exploited commercially for this purpose (Smit 2004). Fucoidans have shown *in vivo* and *in vitro* heparin-like antithrombotic and anticoagulant activities that result from the direct interaction of fucoidan-thrombin. A number of structural characteristics have been shown to influence this activity, including sugar composition, molecular weight, sulphation level and the position of sulphate groups on the sugar backbone (Jiao et al. 2011).

4 Conclusion

Consumer and commercial attitudes to macroalgae as both a wholefood and functional ingredient are changing. This offers great potential for the development of new products and ingredients from what is still considered to be an interesting and innovative foodstuff. However, there are still gaps in our current knowledge that need to be addressed if we are to understand the true benefits of macroalgae to human health.

The available information on the nutritional composition of macroalgae would suggest them, in general, to be a nutritious wholefood. Indeed, historical and anecdotal evidence would support this. However, data from epidemiological and dietary intervention studies is limited and relatively little is known about the bioavailability of key macronutrients, vitamins, minerals and trace elements. Evidence suggests the bioavailability of certain nutrients in the raw state to be hampered by the complex carbohydrate and protein matrix but virtually nothing is known of the fate of these nutrients when the macroalgae are incorporated into food products as ingredients. In addition, very little is known of the effects of cooking and processing on bioavailability.

The number of commercially available products for the health and wellness market that contain extracts or isolates of macroalgae are on the increase outside of Southeast Asia. Whilst the mechanisms underlying their efficacy have been generally demonstrated *in vitro* and *in vivo* in small animal studies, relatively few studies have been conducted on humans. If the full potential of macroalgal derived extracts and bioactives for human health is to be realized, then this must become a primary focus for future research.

Acknowledgments

The authors would like to thank the Centre d'Étude et de Valorisation des Algues (CEVA), France for the use of their nutritional data on edible macroalgae.

Keywords: Nutrition, human health, bioactives, "health & wellness", bioavailability

References Cited

Ale, M.T., J.D. Mikkelsen and A.S. Meyer. 2011. Important determinants for fucoidan bioactivity: A critical review of structure-function relations and extraction methods for fucose-containing sulfated polysaccharides from brown seaweeds. Marine Drugs 9: 2106–2130.

Bernstein, A.M., E.L. Ding, W.C. Willett and E.B. Rimm. 2012. A meta-analysis shows that docosahexaenoic acid from algal oil reduces serum triglycerides and increases HDL-cholesterol and LDL-cholesterol in persons without coronary heart disease. Journal of Nutrition 142: 99–104.

Berteau, O. and B. Mulloy. 2003. Review. Sulphated fucans, fresh perspectives: structures, functions and biological properties of sulphated fucans and an overview of enzymes active towards this class of polysaccharide. Glycobiology 13(6): 29–40.

Besnard, M., D. Megard, C. Inisan, I. Rousseau, Y. Lerat, P. Grandieu, M.-T. Mitjavila and P. Simonetti. 2008. Algae Extract Containing Polyphenols. European Patent # FR2914186.

Bocanegra, A., S. Bastida, J. Benedi, S. Rodenas and F.J. Sanchez-Muniz. 2009. Characteristics and nutritional and cardiovascular-health properties of seaweeds. Journal of Medicinal Food 12(2): 236–258.

Brownlee, I.A., A. Allen, J.P. Pearson, P.W. Dettmar, M.E. Havler, M.R. Atherton and E. Onsøyen. 2005. Alginate as a source of dietary fiber. Critical Reviews in Food Science and Nutrition 45(6): 497–510.

Brownlee, I., A. Fairclough, A. Hall and J. Paxman. 2012. The potential health benefits of seaweed and seaweed extract. pp. 119–136. *In*: POMIN, H. Vitor (ed.). Seaweed: Ecology, Nutrient Composition and Medicinal Uses. Marine Biology: Earth Sciences in the 21st Century. Nova Science Publishers, Hauppauge, New York.

Buttriss, J. 2000. Nutrient requirements and optimisation of Intakes. British Medical Bulletin 56(1): 18–33.

Cerna, M. 2011. Seaweed proteins and amino acids as nutraceuticals. Advances in Food and Nutrition Research 64: 297–312.

Chandinia, S.K., P. Ganesana and N. Bhaskar. 2008. *In vitro* antioxidant activities of three selected brown seaweeds of India. Food Chemistry 107: 707–713.

Chassany, O., B. Bonaz, S. Bruley des Varannes, G. Cargill, B. Coffin, P. Ducrotte and V. Grange. 2007. Acute exacerbation of pain in irritable bowel syndrome: efficacy of phloroglucinol trimethylphloroglucinol—a randomized, double-blind, placebo-controlled study. Aliment Pharmacology and Therapeutics 25: 1115–1123.

Chen, H. and X.-J. Yan. 2005. Antioxidant activities of agaro-oligosaccharides with different degrees of polymerization in cell-based system. Biochimica et Biophysica Acta. 1722: 103–111.

Chen, H., X. Yan, P. Zhu and J. Lin. 2006. Antioxidant activity and hepatoprotective potential of agaro-oligosaccharides *in vitro* and *in vivo*. Nutrition Journal 5: 31.

Chen, H.-M., Z. Li, W. Lin and X.-J. Yan. 2004. Product monitoring and quantitation of oligosaccharides composition in agar hydrolysates by precolumn labeling HPLC. Talanta 64: 773–777.

Conquer, J.A. and B.J. Holub. 1996. Supplementation with an algae source of docosahexaenoic acid increases (n-3) fatty acid status and alters selected risk factors for heart disease in vegetarian subjects. Journal of Nutrition 126: 3032–3039.

Cornish, L. and D. Garbery. 2010. Antioxidants from macroalgae: potential applications in human health and nutrition. Algae 25(4): 155–171.

Dagnelie, P.C. and W.A. van Staveren. 1994. Macrobiotic nutrition and child health: results of a population-based, mixed-longitudinal cohort study in the Netherlands. The American Journal of Clinical Nutrition 59(suppl): 1187S–96S.

Dagnelie, P.C., W.A. van Staveren and H. van den Berg. 1991. Vitamin B-12 from algae appears not to be bioavailable. The American Journal of Clinical Nutrition 53: 695–697.

Dawczynski, C., R. Schubert and G. Jahreis. 2007. Amino acids, fatty acids and dietary fibre in edible seaweed products. Food Chemistry 103: 891–899.

Deville, C., J. Damas, P. Forget, G. Dandrifosse and O. Peulen. 2004. Laminarin the dietary fibre concept. Journal of the Science of Food and Agriculture 84: 1030–1038.

Deville, C., M. Gharbi, G. Dandrifosse and O. Peulen. 2007. Study on the effects of laminarin, a polysaccharide from seaweed, on gut characteristics. Journal of the Science of Food and Agriculture 87: 1717–1725.

Domínguez-González, R., V. Romarís-Hortas, C. García-Sartal, A. Moreda-Piñeiro, M.C. Barciela-Alonso and P. Bermejo-Barrera. 2010. Evaluation of an *in vitro* method to estimate trace elements bioavailability in edible seaweeds. Talanta 82(5): 1668–1673.

Eom, S.H., S.H. Lee, N.Y. Yoon, W.K. Jung, Y.J. Jeon, S.K. Kim, M.S. Lee and Y.M. Kim. 2012. Inhibitory activities of phlorotannins from *Eisenia bicyclis*. Journal of the Science of Food and Agriculture 92(10): 2084–2090.

[FAO] Food and Agriculture Organisation of the United Nations. 2010a. The State of World Fisheries and Aquaculture. Fisheries and Aquaculture Department. Food and Agriculture Organisation of the United Nations. Rome.

[FAO] Food and Agriculture Organisation of the United Nations. 2010b. Fats and Fatty Acids in Human Nutrition. Report of an expert consultation. FAO Food and Nutrition Paper 91. Rome.

[FAO] Food and Agriculture Organisation of the United Nations. 2012. The State of World Fisheries and Aquaculture. Fisheries and Aquaculture Department. Food and Agriculture Organisation of the United Nations. Rome.

Fitton, H. 2011. Therapies from fucoidan; multifunctional marine polymers. Mar. Drugs 9: 1731–1760.

Fitton, J., M. Irhimeh and N. Falk. 2007. Macroalgal fucoidan extracts: a new opportunity for marine cosmetics. Cosmetics and Toiletries 125: 55–64.

Fleurence, J. 1999a. Seaweed proteins: biochemical, nutritional aspects and potential uses. Trends in Food Science and Technology 10: 26–29.

Fleurence, J. 1999b. The enzymatic degradation of algal cell walls: A useful approach for improving protein accessibility. Journal of Applied Phycology 11: 313–314.

Fleurence, J., M. Morancais, J. Dumay, P. Decottignies, V. Turpin, M. Munier, N. Garcia-Bueno and P. Jaouen. 2012. What are the prospects for using seaweed in human nutrition and for marine animals raised through aquaculture? Trends in Food Science and Technology 21(1): 57–61.

[FSA] Food Standards Agency. 2008. The FSA Manual of Nutrition, 11th Edition. TBC, UK, 188 pp.

Galland-Irmouli, A.V., J. Fleurence, R. Lamghari, M. Lucon, C. Rouxel, O. Barbaroux, J.P. Bronowicki, C. Villaume and J.L. Gueant. 1999. Nutritional value of proteins from edible seaweed Palmaria palmata (Dulse). Journal of Nutritional Biochemistry 10: 353–359.

García-Sartal, C., V. Romarís-Hortas, C. Barciela-Alonso Mdel, A. Moreda-Piñeiro, R. Dominguez-Gonzalez and P. Bermejo-Barrera. 2011. Use of an *in vitro* digestion method to evaluate the bioaccessibility of arsenic in edible seaweed by inductively coupled plasma-mass spectrometry. Microchemical Journal 98: 91–96.

Garcia-Sartal, C., S. Tabunpakul, E. Stokes, C. Barciela-Alonso Mdel, P. Bermejo-Barrera and H. Goenaga-Infante. 2012. Two-dimensional HPLC coupled to ICP-MS and electrospray ionisation (ESI) MS/MS for investigating the bioaccessibility of arsenic species from edible algae. Analytical and Bioanalytical Chemistry 402: 3359–3369.

Gupta, S. and N. Abu-Ghannam. 2011. Recent developments in the application of seaweeds or seaweed extracts as a means for enhancing the safety and quality attributes of foods. Innovative Food Science and Emerging Technologies 12(4): 600–609.

Harnedy, P.A. and R.J. FitzGerald. 2011. Bioactive proteins, peptides, and amino acids from macroalgae. Journal of Phycology 47(2): 218–232.

Holdt, S.L. and S. Kraan. 2011. Bioactive compounds in seaweed; functional food applications and legislation. Journal of Applied Phycology 23: 543–597.

Hu, B., Q. Gong, Y. Wang, Y. Ma, J. Li and W. Yu. 2006. Prebiotic effects of neoagaro-oligosaccharides prepared by enzymatic hydrolysis of agarose. Anaerobe 12: 260–266.

Institut de Phytonutrition. 2004. Functional health and therapeutic effects of algae and seaweed. Electronic database Version 1.5. Bolsoleil, France.

Ismail, A. and T.-S. Hong. 2002. Antioxidant activity of selected commercial seaweeds. Malaysian Journal of Nutrition 8(2): 167–177.

Jiao, G., G. Yu, J. Zhang and S. Ewart. 2011. Chemical structures and bioactivities of sulfated polysaccharides from marine algae. Marine Drugs 9: 196–223.

Jiminez-Escrig, A. and F.J. Sanchez-Munez. 2000. Dietary fibre from edible seaweeds: chemical structure, physicochemical properties and effects on cholesterol metabolism. Nutrition Research 20: 585–598.

Juffrie, M., I. Rosalina, A. Rosalina, W. Damayanti, A. Djumhana and H. Ahmad. 2006. The efficacy of fucoidan on gastric ulcer. Indones. Journal of Biotechnology 11: 908–913.

Kazłowski, B., C.-L. Pan and Y.-T. Ko. 2008. Separation and quantification of neoagaro- and agaro-oligosaccharide products generated from agarose digestion by b-agarase and HCl in liquid chromatography systems. Carbohydrate Research 343: 2443–2450.

Kim, K.-M., S.-J. Heo, W.-J. Yoon, S.-M. Kang, G.-N. Ahn, T.-H. Yi and Y.-J. Jeon. 2010. Fucoxanthin inhibits the inflammatory response by suppressing the activation of NF-κB and MAPKs in lipopolysaccharide-induced RAW 264.7 macrophages. European Journal of Pharmacology 649: 369–375.

Koivikko, R. 2008. Brown Algal Phlorotannins: improving and applying chemical methods. SARJA - SER. A I OSA - TOM. 381Astronomica - Chemica - Physica – Mathematica.

Koivikko, R., J. Loponen, T. Honkanen and V. Jormalainen. 2005. Contents of soluble, cell-wall-bound and exuded phlorotannins in the brown alga *Fucus vesiculosus*, with implications on their ecological aspects. Journal of Chemical Ecology 31: 195–212.

Kono, S., K. Toyomura, G. Yin, J. Nagano and T. Mizoue. 2004. A case-control study of colorectal cancer in relation to lifestyle factors and genetic polymorphisms: design and conduct of the fukuoka colorectal cancer study. Asian Pacific Journal of Cancer Prevention 5(4): 393–400.

Korhonen, H. and A. Pihlanto. 2006. Bioactive peptides: production and functionality. International Dairy Journal 16: 945–960.

Li, B., F. Lu, X. Wei and R. Zhao. 2008. Fucoidan: structure and bioactivity. Molecules 13: 1671–1695.

Lobban, C.S. and P.J. Harrison. 2000. Seaweed Ecology and Physiology. Cambridge University Press, UK, 366 pp.

Mabeau, S., J. Fleurence and C. Dagorn-Scaviner. 1993. Seaweed in food products: biochemical and nutritional aspects. Trends in Food Science and Technology 4: 103–107.

MacArtain, P., C.I. Gill, M. Brooks, R. Campbell and I.R. Rowland. 2007. Special article: nutritional value of edible seaweeds. Nutrition Reviews 65(12): 535–543.

Maeda, H., T. Tsukui, T. Sashima, M. Hosokawa and K. Miyashita. 2008. Seaweed carotenoid, fucoxanthin, as a multi-functional nutrient. Asia Pacific Journal of Clinical Nutrition 17(S1): 196–199.

Marrion, O., J. Fleurence, A. Schwertz, J.L. Guéant, C. Villaume, L. Mamelouk and J. Ksouri. 2005. Evaluation of protein *in vitro* digestibility of *Palmaria palmata* and *Gracilaria verrucosa*. Journal of Applied Phycology 17: 99–102.

Michel, C. and G.T. MacFarlane. 1995. Digestive fates from soluble polysaccharides from marine macroalgae: involvement of the colonic microflora and physiological consequences for the host. Journal of Applied Bacteriology 80: 349–369.

Michel, C., M. Lahaye, C. Bonnet, S. Mabeau and J.L. Barry. 1996. *In vitro* fermentation by human faecal bacteria of total and purified dietary fibres from brown seaweeds. British Journal of Nutrition 75(2): 263–280.

Miyashita, K. 2011. Beneficial Health Effects of Seaweed Bio-actives. Proceedings, XI Asian Congress of Nutrition (can), Singapore, July 2011.

Moreda-Piñeiro, A., V. Romarís-Hortas, R. Domínguez-González, E. Alonso-Rodríguez, P. López-Mahía, S. Muniategui-Lorenzo, D. Prada-Rodríguez and P. Bermejo-Barrera. 2012. Trace metals in marine foodstuff: bioavailability estimation and effect of major food constituents. Food Chemistry 134(1): 339–345.

Morrissey, J., S. Kraan and M.D. Guiry. 2001. A guide to commercially important seaweeds on the Irish coast. Published by Bord Iascaigh Mhara, Dun Laoghaire, Co. Dublin, 66 pp.

Myers, S. 2012. Weight Management Ingredients Buyers Guidebook. Natural Products Insider. February 2012. pp. 49.

Myers, S.P., J. O'Connor, J.H. Fitton, L. Brooks, M. Rolfe, P. Connellan, H. Wohlmuth, P.A. Cheras and C. Morris. 2010. A combined phase I and II open label study on the effects of a seaweed extract nutrient complex on osteoarthritis. Biologics: Targets and Therapy 4: 33–44.

Nakamura, N., H. Tai, Y. Uozumi, K. Nakagawa and T. Matsu. 2012. Magnesium absorption from mineral water decreases with increasing quantities of magnesium per serving in rats. Journal of the Science of Food and Agriculture 92: 2305–2309.

Nwosu, F., J. Morris, V.A. Lund, D. Stewart, H.A. Ross and G.J. McDougall. 2011. Antiproliferative and potential anti-diabetic effects of phenolic-rich extracts from edible marine algae. Food Chemistry 126: 1006–1012.

O'Sullivan, L., P. Murphy, P. McLoughlin, P. Duggan, P.G. Lawlor, H. Hughes and G.E. Gardiner. 2010. Prebiotics from Marine Macroalgae for Human and Animal Health Applications. Marine Drugs 8: 2038–2064.

Oryza Oil and Fat Chemical Company, Ltd. 2008. Fucoxanthin: dietary ingredient for prevention of metabolic syndrome, antioxidation and cosmetics. Version 1.

Osumi, Y., M. Kawai, H. Amano and H. Noda. 1998. Effect of oligosaccharides from porphyran on *in vitro* digestion, utilizations by various intestinal bacteria, and levels of serum lipids in mice. Nippon Suisan Gakkaishi 64: 98–104.

Pangestuti, R. and S.-K. Kim. 2011. Biological activities and health benefit effects of natural pigments derived from marine algae. Journal of Functional Foods 3: 255–266.

Park, S.-B., K.-R. Chun, J.K. Kim, K. Suk, Y.-M. Jung and W.-H. Lee. 2010. The differential effect of high and low molecular weight fucoidans on the severity of collagen-induced arthritis in mice. Phytother. Res. 24: 1384–1391.

Paxman, J.R., J.C. Richardson, P.W. Dettmar and B.M. Corfe. 2008a. Alginate reduces the increased uptake of cholesterol and glucose in overweight male subjects: a pilot study. Nutrition Research 28: 501–505.

Paxman, J.R., J.C. Richardson, P.W. Dettmar and B.M. Corfe. 2008b. Daily ingestion of alginate reduces energy intake in free-living subjects. Appetite 51: 713–719.

Peng, J., J.-P. Yuan, C.-F. Wu and J.-H. Wang. 2011. Fucoxanthin, a marine carotenoid present in brown seaweeds and diatoms: metabolism and bioactivities relevant to human health. Marine Drugs 9: 1806–1828.

Pereira, H., L. Barreira, F. Figueiredo, L. Custódio, C. Vizetto-Duarte, C. Polo, E. Rešek, A. Engelen and J. Varela. 2012. Polyunsaturated fatty acids of marine macroalgae: potential for nutritional and pharmaceutical applications. Marine Drugs 10: 1920–1935.

Ragan, M.A. and K.W. Glombitza. 1986. Phlorotannins, brown algal polyphenols. pp. 129–241. *In*: Round, F.E. and D.J. Chapman (eds.). Progress in Phycological Research. Biopress Ltd., Bristol, UK.

Rauma, A.L., R. Törrönen, O. Hänninen and H. Mykkänen. 1995. Vitamin B-12 status of long-term adherents of a strict uncooked vegan diet ("living food diet") is compromised. Journal of Nutrition 125: 2511–2515.

Reilly, P., J.V. O'Doherty, K.M. Pierce, J.J Callan, J.T. O'Sullivan and T. Sweeney. 2008. The effects of seaweed extract inclusion on gut morphology, selected intestinal microbiota, nutrient digestibility, volatile fatty acid concentrations and the immune status of the weaned pig. Animal 2(10): 1465–1473.

Romarís–Hortas, V., C. García-Sartal, M.C. Barciela-Alonso, R. Domínguez-González, A. Moreda-Piñeiro and P. Bermejo-Barrera. 2011. Bioavailability study using an *in vitro* method of iodine and bromine in edible seaweed. Food Chemistry 124(4): 1747–1752.

Rose, D.J., M. Venkatachalam, J. Patterson, A. Keshavarzian and B. Hamaker. 2007. *In vitro* fecal fermentation of alginate-starch microspheres shows slow fermentation rate and increased production of butyrate. The FASEB Journal 21: 853.2.

Roy, M.-C., R. Anguenot, C. Fillion, M. Beaulieu, J. Bérubé and D. Richard. 2011. Effect of a commercially-available algal phlorotannins extract on digestive enzymes and carbohydrate absorption *in vivo*. Food Research International 44(9): 3026–3029.

Ruperez, P. 2002. Mineral content of edible marine seaweeds. Food Chemistry 79: 23–26.

Sánchez-Machado, D.I., J. López-Hernández, P. Paseiro-Losada and L. Simal-Lozano. 2003. High-performance liquid chromatographic analysis of amino acids in edible seaweeds after derivatization with phenyl isothiocyanate. Chromatographia 58(3-4): 159–163.

Sawabe, T., M. Ohtsuka and Y. Ezura. 1997. Novel alginate lyases from marine bacterium *Alteromonas* sp. strain H-4. Carbohydrate Research 304: 69-76.

[SCF] Scientific Committee on Food. 2002. Opinion of the Scientific Committee on Food on the Tolerable Upper Intake Level of Iodine. European Commission, SCF/CS/NUT/ UPPLEV/26 Final, Brussels.

Singh, I.P. and S.B. Bharate. 2006. Phloroglucinol compounds of natural origin. Nat. Prod. Rep. 23: 558–591.

Skrovankova, S. 2011. Seaweed vitamins as neutraceuticals. Advances in Food and Nutrition Research 64: 357–369.

Smit, A.J. 2004. Medicinal and pharmaceutical uses of seaweed natural products: A review. Journal of Applied Phycology 16: 245–262.

Targett, N.M. and T.M. Arnold. 1998. Predicting the effects of brown algal phlorotannins on marine herbivores in tropical and temperate oceans. Journal of Phycology 34: 195–205.

Targett, N.M., L.D. Coen, A.A. Boettcher and C.E. Tanner. 1992. Biogeographic comparisons of marine algal polyphenolics—evidence against a latitudinal trend. Oecologia 89: 464–470.

Teas, J., M. Irhimeh, S. Druker, T. Hurley, J. Hébert, T. Savarese, T. and M. Kurzer. 2011. Serum IGF-1 concentrations change with soy and seaweed supplements in healthy postmenopausal American women. Nutrition and Cancer 63(5): 743–748.

The Martin Chapman Water Structure and Science website, London South Bank University (http://www1.lsbu.ac.uk/water/index2.html) Accessed March 2013.

van Ginneken, V.J.T., J.P.F.G. Helsper, W. de Visser, H. van Keulen and W.A. Brandenburg. 2011. Polyunsaturated fatty acids in various macroalgal species from north Atlantic and tropical seas. Lipids in Health and Disease 10: 104.

Vaugelade, P., C. Hoebler, F. Bernard, F. Guillon, M. Lahaye, P.-H. Duee and B. Dracy-Vrillon. 2000. Non-starch polysaccharides extractsed from seaweed can modulate intestinal absorption of glucose and insulin response in the pig. Reproduction Nutrition Development 40: 33–47.

Venugopal, V. 2009. Marine Products for Healthcare: Functional and Bioactive Nutraceutical Compounds from the Ocean. Boca Raton. CRC Press.

Wada, K., K. Nakamura, Y. Tamai, M. Tsuji, Y. Sahashi, K. Watanabe and S. Ohtsuchi. 2011. Seaweed intake and blood pressure levels in healthy pre-school Japanese children. Nutrition Journal 10: 83.

Wang, Y., F. Han, B. Hu, J. Li. and W. Yu. 2006. *In vivo* prebiotic properties of alginate oligosaccharides prepared through enzymatic hydrolysis of alginate. Nutrition Research 26: 597–603.

Warrand, J. 2006. Healthy polysaccharides. Food Technology 44(3): 355–370.

[WHO] World Health Organisation. 2002. Protein and Amino Acid Requirements in Human Nutrition: Report of a Joint FAO/WHO/UNU Expert Consultation on Protein and Amino Acid Requirements in Human Nutrition. Geneva, Switzerland.

Wijesekara, I., R. Pangestuti and S.-K. Kim. 2010. Biological activities and potential health benefits of sulfated polysaccharides derived from marine algae. Carbohydrate Polymers 84: 14–21.

Wong, K.H. and P.C.K. Cheung. 2000. Nutritional evaluation of some subtropical red and green seaweeds: Part I: proximate composition, amino acid profiles and some physicochemical properties. Food Chemistry 71(4): 475–482.

Wong, K.H. and P.C.K. Cheung. 2001. Nutritional evaluation of some subtropical red and green seaweeds Part II: *In vitro* protein digestibility and amino acid profiles of protein concentrates. Food Chemistry 72(1): 11–17.

Yan, X., Y. Chuda, M. Suzuki and T. Nagata. 1999. Fucoxanthin as the major antioxidant in Hijikia fusiformis, a common edible seaweed. Bioscience Biotechnology and Biochemistry 63: 605–607.

Yang, Y.J., S.-J. Nam, G. Kong and M.K. Kim. 2010. A case-control study on seaweed consumption and the risk of breast cancer. The British Journal of Nutrition 103(9): 1345–1353.

Zhang, J., C. Tiller, J. Shen, C. Wang, G.S. Girouard, D. Dennis, C.J. Barrow, M. Miao and H.S. Ewart. 2007. Antidiabetic properties of polysaccharide- and polyphenolic-enriched fractions from the brown seaweed *Ascophyllum nodosum*. Canadian Journal of Physiology and Pharmacology 85(11): 1116–1123.

Zhang, Z., G. Yu, H. Guan, X. Zhao, Y. Du and X. Jiang. 2004. Preparation and structure elucidation of alginate oligosaccharides degraded by alginate lyase from *Vibro* sp. 510. Carbohydrate Research 339: 1475–1481.

CHAPTER 11

Internet Information Resources for Marine Algae

Michael D. Guiry[1,]* and *Liam Morrison*[1,2]

1 Introduction

An information revolution, perhaps only matched by the development of writing (around 3500 BC), and by the invention of moveable type by Johannes Gutenberg in Germany (about AD 1439), has swept our planet in the last 30 years. Loosely referred to as the "Internet", this revolutionary communication and data-exchange system had its origins in the development of a TCP/IP (Transmission Control Protocol/Internet Protocol) network called CSNET linking academic and governmental computers in the USA in the early 1980s, its extension to the so-called "developed" world, and the commercialization of the system in the 1990s.

Today, the Internet is a global system of networked networks linking computers of all kinds that communicate freely and effectively with each other by simple protocols at speeds that would have been inconceivable in the days of the telegraph or telex. It is estimated that over 3 billion people are potentially in instant communication by various computer driven devices including personal computers, tablets and smartphones. Not only has the Internet brought about a revolution in the way we work and communicate but, in recent years, it has also been instrumental in political revolutions.

The process of obtaining information on the taxonomy and nomenclature of algae prior to the development of the Internet was a matter of laboriously

---

[1] AlgaeBase & Ryan Institute for Environmental, Marine and Energy Research, National University of Ireland, University Road, Galway, Ireland.

[2] Earth and Ocean Sciences, School of Natural Sciences, National University of Ireland, University Road, Galway Ireland.

* Corresponding author: michael.guiry@algaebase.org

consulting journals and books. The aspiring taxonomist or nomenclaturist used to have little choice but to become an obsessive-compulsive book and reprint collector, and had to visit major libraries in large cities on a regular basis to consult rare and obscure books and journals that survived in only in a few of the larger repositories. Universities, research institutes and museums in the capital cities of the Old World (mainly the ex-colonial powers of western Europe) and in the New World (mainly North America) were, to be candid, at a major advantage over those in the rest of the world. European wars in the 19th and 20th century were not helpful as travel became impossible to these important centers, and many essential resources in libraries and herbaria were destroyed or severely damaged at this time, both by political activity and by warfare.

Where once the algae nomenclaturist or taxonomist could be found cocooned by books and reprint boxes, spending long hours perspiring profusely over piping-hot photocopiers making personal copies of critical works that were otherwise not available, his or her modern equivalent can store all that is needed electronically, and this information can be retrieved almost instantaneously, even from quite remote locations. Data storage has become so cheap and universally connected that all this information can easily be carried around or accessed. No longer is the budding nomenclaturist or taxonomist confined to institutions with great libraries; also, the "developed" world no longer has an unfair advantage over the rest of the world, at least in this respect.

So, is everything perfect in this new electronic wonder-age? Alas, as with everything else, there are disadvantages as well as advantages. Initially, the costs of the networks that make up the Internet were borne by government agencies for strategic reasons, as the primary networks still are; but as the Internet became more pervasive, it had of necessity to become commercial to allow the networks to extend ubiquitously. Unfortunately (or fortunately, depending on which side of the fence you are on), it has had to become subscription, advertising or sales-driven, or various combinations of these, and this is becoming more common as the resources mature. With this commerciality came the necessity to reassure those who were accessing it that they were getting their money's worth. This led in turn to systems for counting the accesses to information but also to systems for indexing the information so that potential customers could get to the information or goods quickly and efficiently. By far the most successful, and now completely dominant, "indexer" of the Internet is Google (http://www. google.com). So pervasive is this company now that "to Google" has become a universally understood term for searching as in, "Let me Google that for you!", and such Googling can now be done virtually anywhere.

As part of its activities Google started in 2004, in collaboration with a number of libraries in the USA, the Google Print Library Project (http://

www.google.com/googlebooks/about), the aim of which was to scan all the books of the world and to make them available on the Internet. More and more libraries joined the project, particularly in the USA, and Google was able to announce in April 2013 (http://en.wikipedia.org/wiki/Google_Books#2013), that over 30 million books had been scanned of an estimated 130 million. To get an idea of the enormity of the Google project, the Library of Congress in the USA presently includes over 32 million cataloged books and other print materials in 470 languages.

This Google project and other scanning initiatives, while appearing very laudable at first sight, also have disadvantages. A commercial fact of life is that authors and publishers are only able to publish books if they can sell them. Unless copyright protection exists and is enforceable, authors cannot survive and publishers cannot make money, particularly in today's market where a single copy of a book can now be printed almost instantly from a PDF in countries with less-than-comprehensive copyright laws, and sold well below a publisher's costs and the levels of remuneration upon which authors can survive. Worse than that, these books can be imported into countries with more stringent laws and sold well below the selling price in that country. Nevertheless, because of copyright law in the USA and the EU, many of Google's scans are not available or not entirely online, and this will only change when a universal agreement can be negotiated so that all parties, including Google, can be compensated appropriately. As there are ongoing legal proceedings connected with the Google project (a settlement of a class action known as "The Google Books Settlement," was subsequently rejected by a court the USA in 2011), only time will tell what the outcome will be, and it may be that this matter is not for the courts to decide but is a matter for international agreements and legislation to ensure the protection of all the parties' rights.

No more than other groups in society, those working on all aspects of algae have been strongly affected by this information revolution. In the last 20 years, more and more algal information resources have become available, and we provide here a guide to the main sources of information and other resources for phycologists.

2 Brief Notes on Some Internet Conventions and Problems

It is not universally appreciated that Internet addresses (including mail addresses) can be in uppercase or lowercase letters or any mixture of the two. So, links typed as http://www.AlgaeBase.org. work just as well as http://www.algaebase.org and http://WWW.ALGAEBASE.ORG. Most links are now given entirely in lower case, probably because it looks neater and less complicated. Such hypertext addresses do not tolerate spaces so that http://www. algaebase.org will give a "Not found" message (HTTP

404 error). Secure connections are clearly indicated by "https://" rather than "http://" and will not work unless a security certificate is provided by the server. Domains are names such as "algaebase.org" or "algaebase.info" and subdomains can be **www**.algaebase.org or **info**.algaebase.org and these are kept on proxy servers which translate these friendly, generally easy-to-remember names, into Internet Protocol (IP), which are currently 32-bit numbers (IPv4), such as 172.16.254.1, but 128-bit IP numbers (IPv6) are gradually making a very necessary appearance as the world is running out of 32-bit IP numbers. It is these IP addresses that computers use to label packages of data for routing.

A particularly irritating aspect of the Internet, the fault of those who structure websites rather than of the designers of the Internet protocols, is the use of poorly constructed links that are difficult to type, and/or use multiple nested folders, and/or unnecessarily complex file names. It should be noted that a huge and complex company like Google uses a simple and memorable architecture:

http://www.google.com/citations or http://www.google.com/analytics

Contrast this with this sort of error prone and impossible to remember architecture:

http://www.fakewebsite.com/folder/subfolder/
anothersubfolder/yetanothersubfolder/
aVerySillyfileNamewithlotsofjunklettersandnumberslike3Jk98.html

Even more irritating are those who change with wild abandon, file structure or names of folders or files without any consideration of users and their needs. The Internet thrives on fixed links and those who make websites should remember this and only change addresses, folder names and locations as a last resort.

3 Main Taxonomic and Nomenclatural Sources

These are in alphabetical order; brief notes are given, often as quotations from the sites.

AlgaeBase
http://www.algaebase.org

AlgaeBase was made available on line in early 1996 by Michael Guiry as a taxonomic source of information on seaweeds, and has gradually been expanded over the years to include all algae. Some 135,000 names of species and infraspecific names are included, of which 40,000 (May 2014) are species. Some 7,000 genus names are listed and about 2,000 names of taxa above the level of genus. Nearly 17,000 images are available, about half of which are seaweeds and the rest are terrestrial and freshwater algae. Some 236,000 distributional records are included from 250 states, territories and countries, mainly extracted from major check-lists and monographs; the scale of distributional data is intended to be large and verified by reference to published information. Generic descriptions from the unpublished *Encyclopedia of Algal Genera* compiled in the 1990s as part of a Phycological Society of America project coordinated by Bruce Parker are included. The project, now curated by Michael & Wendy Guiry with the indispensable help of Salvador Miranda (Spain), David Garbary (Canada) and Anders Langangen (Norway), is currently sponsored by Ocean Harvest Ltd. (http://www.oceanharvest.ie), and the National University of Ireland, Galway (http://www.nuigalway.ie). AlgaeBase is intended to complement *Index Nominum Algarum* (see below), and is presently run along *pro bono publico* (for the public good) lines.

AlgaTerra
http://www.algaterra.net

Online since 1986 and curated by Regine Jahn and Wolf-Henning Kusber of the Department of Biodiversity Informatics and Laboratories, Botanic Garden and Botanical Museum Berlin-Dahlem, Germany, AlgaTerra offers "evaluated information on their synonyms and their concepts. In addition, it is providing taxonomic, molecular and ecological data of mainly non-marine micro algae. The focus of this pilot version is on type collections —mainly Ehrenberg, Hustedt [and] Lange-Bertalot—and living reference collections, therefore linking historical information to modern research. Since morphology and identity of microalgae depend on their visualization, many pictures are included." The site also provides lists of diatom names published by the extraordinarily prolific diatomist Horst Lange-Bertalot and his co-workers.

CAS: Catalogue of Diatom Names
http://researcharchive.calacademy.org/research/diatoms/names/about.html

Compiled by Elisabeth Fourtanier and J. Patrick Kociolek, the "... Catalogue of Diatom Names is a compilation of names of diatom genera, species and taxa at infraspecific ranks (62,000 names). It has been assembled during the past 12 years by staff at the California Academy of Sciences. It includes all scientific names of diatom genera, species, and taxa at infraspecific ranks, with authorship, date, place of publication, page of description, basionym or replaced name (if applicable), status (valid or invalid), and occasionally type information. The catalogue of diatom names includes a publication table. The publication table (12,600 entries) not only includes publications where new diatom taxa are described and new names are created, but other publications related to diatoms, in the fields of taxonomy, ecology, paleontology, etc. The table of genus names (1,160 entries) contains nomenclatural information such as genus name, authorship, reference, page, type and author of typification (if applicable), nomenclatural status, and for some records, reference to recent revisions and monographs." The whole catalogue is gradually being published formally (e.g., Fourtanier and Kociolek 1999).

CEDiT; Centre of Excellence for Dinophyte Taxonomy
http://www.dinophyta.org

This site is a source of information on dinoflagellates and the "... centre [provides] a complete check-list of all described living dinoflagellate genera and species with exact bibliographic details. In addition, we intend to provide all original descriptions electronically. Next to this core database it is planned to make available information of general interest for the dinoflagellate research community, e.g., references, bibliographies, important links and images. In the future one main task will be to archive reference material from cultures or field samples of dinoflagellate blooms. The centre [was] founded in July 2005 by Dr. Elbrächter, Prof. Dr. Martínez, Dr. Türkay, and Prof. Dr. Steininger and is now in the process of development."

Cyanophyta.DB
http://www.cyanodb.cz

Cyanophyta.DB provides comprehensive information on cyanobacteria, also known as blue-green algae and cyanophytes. "The project of the taxonomic database of cyanobacterial genera was started in order to provide as sufficient information about cyanobacteria as possible to all people interested. Either from scientific community or from outside it, e.g., schools, environment monitoring services, public health protecting services, etc." The project, compiled by Jiří Komárek & Tomáš Hauer, is supported by the Faculty of Science, University of South Bohemia, České Budějovice and by [the] Institute of Botany, Academy of Sciences of the Czech Republic, Třeboň, both in the Czech Republic.

Index Nominum Algarum (INA)
http://ucjeps.berkeley.edu/cpd/index.html

"The INA is a card file [containing] nearly 200,000 names of algae (in the broad sense). The BPU [*Bibliographia Phycologica Universalis*] is a card file containing bibliographic references pertaining to algal taxonomy." Both are part of the Silva Center for Phycological Documentation located at

The University and Jepson Herbaria of the University of California at Berkeley, USA. The Silva Center was established in 2004 by means of a generous endowment gift from the eminent phycologist Paul Silva. "The primary mission of the Center is to develop, maintain, and promulgate documentation in the field of algal systematics." The Centre's work and resources include: the California Seaweed Digitization Project, funded by the [US] National Science Foundation in 2011; DeCew's *Guide to the Seaweeds of British Columbia, Washington, Oregon, and Northern California* [most of the major brown and green seaweeds were online by August 2013]; additions and modifications to California seaweed species list since *Marine Algae of California*; type specimens of algae in UC (working list); images of type specimens of marine algae from the Galapagos; an Index to the *Phycotheca Boreali-Americana*, an important exsiccata; and an online version of the *Indian Ocean Catalogue* (Silva et al. 1996). A comprehensive account of the development of INA is given by Silva and Moe (1999).

4 General Sites that Include Information Useful for Phycologists

Index Herbariorum (IH)
http://sciweb.nybg.org/science2/IndexHerbariorum.asp

"There are approximately 3,400 herbaria in the world today, with approximately 10,000 associated curators and biodiversity specialists. Collectively the world's herbaria contain an estimated 350,000,000 specimens that document the earth's vegetation for the past 400 years. *Index Herbariorum* is a guide to this crucial resource for biodiversity science and conservation. The *Index Herbariorum* (IH) entry for an herbarium includes its physical location, Web address, contents (e.g., number and type of specimens), history, and names, contact information and areas of expertise of associated staff. Only those collections that are permanent scientific repositories are included in IH."

Index Nominum Genericorum (ING)
http://botany.si.edu/ing/

ING is a "…compilation of generic names published for organisms covered by the *ICN: International Code of Nomenclature for Algae, Fungi, and Plants.*" The database is searchable online and provides details of publication and the status of the types of the generic names. It is particularly critical that those proposing new names of genera check here that a name has not been previously published.

International Code of Nomenclature for algae, fungi, and plants (ICNafp)
http://www.iapt-taxon.org/nomen/main.php

The most recent ICNafp, the Melbourne Code adopted by the Eighteenth International Botanical Congress Melbourne, Australia, July 2011, is online and searchable. It is particularly gratifying that the title of the Code at last acknowledges algae and fungi.

The International Plant Names Index (IPNI)

IPNI " ... is a database of the names and associated basic bibliographical details of seed plants, ferns and lycophytes. Its goal is to eliminate the need for repeated reference to primary sources for basic bibliographic information about plant names. The data are freely available and are gradually being standardized and checked. IPNI will be a dynamic resource, depending on direct contributions by all members of the botanical community. IPNI is the product of a collaboration between The Royal Botanic Gardens, Kew, The Harvard University Herbaria, and the Australian National Herbarium." While IPNI does not currently include algal names, it is a very useful source of the names of authors of algal taxa (http://www.ipni.org/ipni/authorsearchpage.do), which includes a list of algal authorities compiled by Paul Silva and others, referred to as the "Berkeley list".

5 Main Online Libraries

Only major digital libraries with free, unfettered access to books and journals are included below. Google, as mentioned above, has scanned many works (http://books.google.com), but those in copyright are not currently fully available. Some digital libraries, for example, the Hathi Trust Digital Library (http://www.hathitrust.org), require an institutional subscription. The quality of scans is very variable also, some sources providing scans that are essentially images and are not searchable, and others that do not scan all the material essential to establish evidence for dates of publication of works issued in parts. In many cases, larger illustrative plates have not been scanned; for example, AlgaeBase has not been able to discover a satisfactory scan of Postels & Ruprecht's *Illustrationes Algarum* issued in 1840 in double-elephant-folio similar in size to Audubon's *The Birds of America*, probably because scanners of this size are even rarer than the *Illustrationes* or the *Birds* themselves!

Biodiversity Heritage Library
http://www.biodiversitylibrary.org

"The Biodiversity Heritage Library (BHL) is a consortium of natural history and botanical libraries that cooperate to digitize and make accessible the legacy literature of biodiversity held in their collections and to make that literature available for open access and responsible use as a part of a global 'biodiversity commons'. The BHL consortium works with the international taxonomic community, rights holders, and other interested parties to ensure that this biodiversity heritage is made available to a global audience through open access principles. In partnership with the Internet Archive and through local digitization efforts, the BHL has digitized millions of pages of taxonomic literature, representing tens of thousands of titles and over 100,000 volumes."

Botanicus
http://www.botanicus.org

Even though this site, like the IPNI site below, is primarily designed for "higher-plant" botanists, it has digitized many of the older botanical works, many of which are not freely available elsewhere. "To improve access to scientific literature, we have created *Botanicus*, a freely accessible, Web-based encyclopedia of digitized historic botanical literature from the Missouri Botanical Garden Library. We have been digitizing materials from our library since 1995, focusing primarily on beautifully illustrated volumes from our rare book collection."

Madrid Digital Library
http://bibdigital.rjb.csic.es

"The digital library of the Royal Botanic Garden (CSIC) was set up because of two factors: an extraordinary wealth of documents and active research. The Royal Botanic Garden, on account of its rich and lengthy history, has a magnificent collection of antiquarian botanical books. Apart from the intrinsic value of this historic and scientific heritage, the collection is constantly consulted by researchers investigating the organization and distribution of organisms, or the relationship between scientific names and the organisms they are applied to. Globalization of communications now provides the means to make the collection available for anyone who is interested in it. With this objective in mind, work began on digitalizing the antiquarian books in 2003. ... It was finished in 2005, precisely the year the Royal Botanic Garden celebrated its 250th anniversary. We feel very proud to be able to offer this library to Internet users, and have great hopes for its future which will be as open as the demands of users, and technology and resources permit."

Gallica
http://gallica.bnf.fr

"*Gallica* is the digital library of the *Bibliothèque nationale de France* [BNF] and its partners. Online since 1997, it is updated every week [with] thousands of new [works] and [it] now offers access to over 2 million documents. Written evidence of French heritage and its influence in Europe and in the world, the documents used by the BNF were selected to provide a reasoned and encyclopedic library, representative of the great French writers and current research and reflection through the centuries. Consisting of rare or difficult to access documents, this selection is supplemented by documents placing these works in their intellectual context (contemporary memoires, dictionaries, bibliographies). … Different types of media are represented: print (books, periodicals and press) as images and text [together with], manuscripts, sound recordings, music scores, graphic materials, maps mode." [Translated by the authors].

6 Major Phycological Society Websites

Millar (2011) gives a comprehensive list of the world's phycological societies based on a survey, and only the larger societies are treated here.

Australasian Society for Phycology and Aquatic Botany (ASPAB)
http://www.aspab.org/

"The Australasian Society of Phycology and Aquatic Botany (ASPAB) is a professional scientific society, formally established in May 1980, that aims to promote, develop and assist the study of, or an interest in, phycology (the study of macro- and micro-algae such as seaweeds and phytoplankton) and aquatic botany (the study of aquatic plants) within Australasia and elsewhere. Additionally the society aims to establish and maintain communication between people interested in phycology and aquatic botany."

To assist in promoting these aims ASPAB holds an annual conference (usually in November), produces a newsletter biannually and maintains an email discussion list for members (ASPAB-list).

British Phycological Society
http://www.brphycsoc.org

"The Society, a charity devoted to the study of algae founded in 1952, was one of the first to be established in the world, and is the largest in Europe. The British Phycological Society publishes the *European Journal of Phycology*, and organizes an Annual Meeting each year at locations throughout Britain and Ireland. It provides funding for students and others to attend meetings and field courses and organizes and funds annual field courses."

Czech Phycological Society (Česká Algologická Společnost)
http://www.czechphycology.cz

The "Czech Phycological Society (TIME) brings together experts who study algae and cyanobacteria from the theoretical and applied aspects. Work is a company focused organization determinative courses for water professionals, organizing annual conferences and publishing a journal *Fottea* (formerly *Czech Phycology*). TIME was founded in 2002 as a contiguous successor to the Phycological Section of the Czech Botanical Society (1955–2001)." [Translated by the authors using Google Translate.]

Federation of European Phycological Societies (FEPS)
http://www.feps-algae.org/

"The Federation of European Phycological Societies (FEPS) was founded by charter in July 2007 to promote the study, conservation and application of algae, to support educational activities and lobbying at the European level on matters involving algae, their present and projected roles. The Federation already includes, as full members, the national Phycological Societies/Algal Groups of: Belgium, Croatia, Czech, Germany, Greece, Ireland, Italy, Macedonia (FYROM), The Netherlands, Poland, Spain and the UK, representing more than 1000 scientists in these countries." FEPS organizes the European Phycological Congresses, held at various European locations at 4-yearly intervals.

International Phycological Society
http://www.intphycsoc.org/

"The Society was founded in 1960 ... and is dedicated to the development of phycology; the distribution of phycological information; and, international cooperation among phycologists and phycological organizations. The Society organizes the International Phycological Congresses at 4-yearly intervals." The IPS also publishes the journal *Phycologia*, which appears six times a year.

Hellenic Phycological Society
http://www.phycology.gr/en/index.php

"Hellenic Phycological Society (HPS) was founded in 1991, by a group of phycologists from the Universities of Thessalonica and Athens, as well as from the Hellenic Center for Marine Research (HCMR). The constitution of the Society was approved by the Court on 15 April 1991, when it began also to function officially."

International Society for Applied Phycology
http://www.appliedphycologysoc.org/

"The ISAP is a non-profit organization that aims to promote research, preservation of algal genotypes, and the dissemination of knowledge concerning the utilization of algae. In order to meet these objectives, the ISAP promotes the following activities: Organizes triennial meetings, encouraging wide participation from academia and industry; Forms an informed source of expertise among members who could provide an accurate, disinterested opinion concerning algal products or relating to environmental impact; Forms links and affiliations with existing national and regional organizations of applied phycology; Promotes exchange of students and researchers; Applies for special grants and donations from international bodies, governments and industry to support activities of the Society; Issues an electronic newsletter to publicise activities; Organizes workshops and training programs; Supports culture collections of algae; Organizes exhibitions of commercial algal products and similar pertinent exhibits along [with] the triennial meetings."

International Society for Diatom Research (ISDR)
http://www.isdr.org/

"Surprisingly, in view of the long history of research, there was never an International Society serving all diatomists whatever their interest until the Bristol [England] meeting in 1985 where the ISDR was formed and its first officers elected. At this meeting it was also decided that the biennial business meeting of the society would take place at the biennial symposia which would from that time be organized through the society. At Bristol it was also decided that the society would support its own journal which would be floated and published in two parts annually by Biopress with Frank Round as the executive editor. The 23rd volume has now appeared and *Diatom Research* now provides a focus for the continuing development of all fields of research on diatoms. It also draws together much work which until now has been scattered, often in relatively inaccessible publications."

The Japanese Society of Diatomology (JSD)
http://diatomology.org/eng/

Founded in 1980 as the Japanese Society of Diatomists, "… the name was changed in 1982 to the present title. The first volume of [the journal] DIATOM was published in 1985. The aims of JSD are to promote all aspects of the study of diatoms, to distribute the results of the research, to facilitate the exchange of information concerning diatoms, and to promote friendship among diatomists."

The Japanese Society of Phycology (JSP; Sôrui)
http://www.sourui.org

"The Japanese Society of Phycology (JSP) was founded in 1952, to promote all researches that are related to algae and phycology, and to become a central hub of people who are interested in phycology. Membership is open to any individual and group who is interested in phycology. The society currently has a worldwide membership of about 900 members and subscribers." The JSP publishes the journal *Phycological Research* (see below).

The Korean Society of Phycology (KSP)
http://www.algae.or.kr

The Korean Society of Phycology (KSP) was founded on Aug 19th, 1986 at Seoul National University. Its Annual Meeting is usually held in March each year. A KSP workshop is held during the summer or winter vacation. KSP publishes *Algae* (formerly the Korean Journal of Phycology, written in English and Korean) and a Newsletter written in Korean. Its website is currently only in Korean.

Phycological Society of America
http://www.psaalgae.org

"The Phycological Society of America (PSA) was founded in 1946 to promote research and teaching in all fields of Phycology. The society publishes the *Journal of Phycology* and the *Phycological Newsletter*. Annual meetings are held, often jointly with other national or international societies of mutual member interest. Phycological Society of America student awards include the Bold Award for best student paper at the annual meeting, the new Student Poster Award for the best student poster at the annual meeting. Other PSA honorary awards include the Provasoli Award for outstanding papers published in the Journal of Phycology, and the Prescott Award for the best Phycology book published within the previous two years. Phycological Society of America student awards include the Bold Award for best student paper at the annual meeting, the new Student Poster Award for the best student poster at the annual meeting. Other PSA honorary awards include the Provasoli Award for outstanding papers published in the Journal of Phycology, and the Prescott Award for the best Phycology book published within the previous two years. The society provides grants and fellowships to graduate student members through Croasdale Fellowships for enrollment in phycology courses at biological stations, Hoshaw Travel Awards for travel to the annual society meeting, and Grants-In-Aid for supporting research."

Phycological Society of South Africa
http://www.botany.uwc.ac.za/pssa/index2.htm

"The objectives of the Society is [sic] to promote interest in Phycology (the study of algae), and to establish and maintain communication between persons interested in the algae of southern Africa."

Polish Phycological Society (Polskie Towarzystwo Fykologiczne)
http://www.staff.amu.edu.pl/~fykolog/en/

"The idea to establish the Polish Phycological Society was born in September 2004, during a session of the Phycological Section, Polish Botanic Society in Toruń and Bydgoszcz. A year later, on 20th May 2005, when the PBS Phycological Section met in Krynica Morska, during a General Meeting a resolution was taken to form the Polish Phycological Society, an association of Polish phycologists from scientific establishments, universities and field sites, which conduct research into the taxonomy and ecology of algae and blue-green algae." The Society holds an International Conference of Polish Phycologists each year.

Sociedad Chilena de Ficología
http://www.ficologia.cl

"Chilean Society of Phycology (SOCHIFICO) is a scientific society, bringing together practitioners, researchers, and/or natural persons of technical careers or professionals in the field of Marine Biology, Aquaculture and related matters, interested in studying and promoting the development of phycology or science dedicated to the study of algae (micro and macroalgae) in Chile." [Translated by the authors].

Sociedade de Ficologia do Brasil (SBFic)
http://www.sbfic.org.br

The Brazilian Phycological Society was conceived in 1980 at a meeting held at the Department of Botany at USP in São Paulo. The Society holds regular meetings.

Sociedad Española de Ficología
http://www.sefalgas.org

"The Spanish Society of Phycology (SEF) was established in 1988. Since then it has tried to become a forum for meeting national and foreign professionals interested in the world of algae. Its current members work in a wide range of research fields, many of them of great public interest." [Translated by the authors.] It publishes the online bulletin *Algas. Boletin de la Sociedad Española de Ficología.*

7 Journals and Journal Websites

Αλγοσ (algos) is a Greek neuter noun literally meaning "pain". Unfortunately, many journals with the word "algology" in the title are medical journals, which creates confusion and leads to many tiresome after-dinner-speech jokes about algology being a "painful" subject. So, for example, the *Journal of the Turkish Society of Algology* is a publication on pain, and not a phycological journal. Even more confusing is the tendency of search-engines automatically to convert "phycology" into "psychology" on the assumption that you are an idiot. Try typing "phycological journals online" into Google and it will insist initially that you mean "psychological journals online" and deliver these in painfully suffocating numbers. The *Journal of Phycology* and the *Journal of Applied Phycology* must suffer considerable corporate pain from this unfortunate similarity in spelling.

Many of the journals below publish non-algal papers as well as those on algae. For example, the journals with "protist" in the title publish papers on non-photoautotropic Protozoa, and *Botanica marina* publishes papers on sea-grasses and marine fungi, although a majority of papers are on algae.

Journals are treated in alphabetical order and only extant journals are included. Most of the major journals have PDF versions available online of all papers published in the journal; some make these available immediately, some make them available after a number of years, and some require payment for all PDFs. In many cases, no PDFs are available online.

Algae
http://www.e-algae.kr/about/about.asp

"Algae is published by the Korean Society of Phycology and provides prompt publication of original works on phycology. ALGAE publishes articles on all aspects of phylogenetics and taxonomy, ecology and population biology, physiology and biochemistry, cell and molecular biology, and biotechnology and applied phycology. Checklists or equivalent manuscripts may be considered for publication only if they contribute original information on taxonomy (e.g., new combinations), ecology or biogeography of more than just local relevance. Contributions may take the form of Original Research Articles, Research Notes, Review Articles and Book Reviews."

Algologia
http://www.botany.kiev.ua/journal_en.htm

Algologia has been published since 2004 by the M.G. Kholodny Institute of Botany of the National Academy of Sciences of Ukraine and publishes a wide range of articles on taxonomy and ecology of algae.

Algological Studies

http://www.schweizerbart.de/journals/algol_stud

"Algological Studies publishes peer reviewed, original scientific papers of international significance from the entire field of phycology: taxonomy, systematics and floristics, physiology, biochemistry, genetic studies, hydrobiology, phytogeography, cultivation. This includes contributions to algal biotechnology and applied phycology. Algological Studies is an ideal platform to collect and publish special issues from conferences, covering all fields of phycology." It was published beginning in 1968 as a supplement to *Archiv für Hydrobiologie* but since 2007 it has been issued as a separate journal.

Bibliotheca Phycologica

http://www.schweizerbart.de/publications/list/series/bibl_phycol

Bibliotheca Phycologica publishes monographs on studies of algae in the widest sense. This includes papers on the physiology, taxonomy, morphology, ecological aspects, and biochemical aspects. Up to August 2013, 117 volumes had been published.

Botanica Marina

http://www.degruyter.com/view/j/botm

"The journal publishes contributions from all of the disciplines of marine botany at all levels of biological organization from subcellular to ecosystem. Subject areas are: marine algal and marine angiosperm systematics, floristics, biogeography, biochemistry, molecular biology, genetics, chemistry, industrial processes and utilization; marine mycology and marine microbiology. Original knowledge is disseminated to provide synopses of global or interdisciplinary interest, and to stress aspects of utilization. Applied science papers are especially welcome, when they illustrate the application of emerging conceptual issues or promote developing technologies. Checklists or equivalent manuscripts may be considered for publication only if they contribute new information on taxonomy (e.g., new combinations), ecology or biogeography of more than just local relevance. Checklists should be focused to highlight original information."

Cryptogamie: Algologie

http://www.cryptogamie.com/pagint_en/editeur/revues.php

Crypotogamie: Algologie continues *Revue algologique* "founded in 1924 by Pierre Allorge (1891–1944) and Gontran Hamel (1883–1944). Two series were published: the first series from 1924–1941 (vols. 1–12), and the second from 1954–1979 (vols. 1–14). "*Cryptogamie* is a French journal of international scope publishing in several European languages. It accepts original papers and review articles on the systematics, biology and ecology ...." The journal is published by *Association des Amis des Cryptogames* and has recently been made available online at http://www.bioone.org/loi/crya.

Diatom
http://diatomology.org/eng/diatom/

Initially published in 1985, *Diatom* is the official journal of the Japanese Society of Diatomology, "… publishing papers on all aspects of basic and applied research on diatoms."

Diatom Research
http://www.isdr.org/index.php/publications/diatom-research/

Diatom Research is the journal of the International Society for Diatom Research; manuscripts on any aspect of freshwater and marine diatom biology, ecology and palaeoecology are considered for publication.

European Journal of Phycology
http://www.tandfonline.com/action/aboutThisJournal?journalCode=te jp20#.UdwDIhbEjgo

"*European Journal of Phycology* is an important focus for the activities of phycologists all over the world. The Editors-in-Chief are assisted by an international team of Associate Editors who are experts in the following fields: macroalgal ecology, microalgal ecology, physiology and biochemistry, cell biology, molecular biology, macroalgal and microalgal systematics, applied phycology and biotechnology. *European Journal of Phycology* publishes papers on all aspects of algae, including cyanobacteria. Articles may be in the form of primary research papers, miniviews, which present the author's viewpoint on important recent developments, or full length reviews of topical subjects." The journal was previously published as the *British Phycological Journal* (1969–1992) and the *British Phycological Bulletin* (1952–1968).

European Journal of Protistology
http://www.elsevier.com/journals/european-journal-of-protistology/0932-4739#

"Articles deal with protists, unicellular organisms encountered free-living in various habitats or as parasites or used in basic research or applications. The *European Journal of Protistology* covers topics such as the structure and systematics of protists, their development, ecology, molecular biology and physiology. Beside publishing original articles the journal offers a forum for announcing scientific meetings. Reviews of recently published books are included as well. With its diversity of topics, the *European Journal of Protistology* is an essential source of information for every active protistologist and for biologists of various fields."

Fottea
http://fottea.czechphycology.cz

Fottea is a journal of Czech Phycological Society (formerly *"Czech Phycology"*), and "publishes papers on all aspects of the ecology, physiology, biochemistry, cell biology, molecular biology, systematics and uses of algae (including cyanobacteria)". The first volume was published in 2007, but as volume 7, continuing the volume sequence of its predecessor, *Czech Phycology*.

International Journal on Algae
http://www.begellhouse.com/journals/journal-on-algae.html

Published since 1999, the "... quarterly *International Journal on Algae* ... publishes selected papers translated from the first Russian language phycological journal, *Algologia*, founded in 1991 in the former Soviet Union. The aim of *Algologia* is to present recent advances in algology. The journal covers both fundamental and applied aspects in algology, including papers based on the results of a wide range of field and experimental studies, as well as reviews and surveys and procedure papers. The journal is intended for specialists in theoretical, experimental, and applied algology, hydrobiology, microbiology, all scientists using algae as a model organisms for research, and all those interested in general problems of biology. The aim and scope of IJA is to inform the western scientific community, especially algologists, about original studies by scientists of the former Soviet Union and Eastern Europe in the following subjects: General Problems of Algology; Morphology, Anatomy, Cytology; Reproduction and Life Cycles of Algae; Genetics; Physiology, Biochemistry and Biophysics; Ecology, Cenology and Conservation of Algae and their Role in Nature; Flora and Geography; Fossil Algae; Systematics, Phylogeny and Problems of Evolution of Algae; New Taxa and Noteworthy Records; and Applied Algology."

Journal of Applied Phycology
http://www.springer.com/life+sciences/plant+sciences/journal/10811

Published since 1989 "The *Journal of Applied Phycology* publishes work in the rapidly expanding field of the commercial use of algae. Coverage includes fundamental research and development of techniques and practical applications in such areas as algal and cyanobacterial biotechnology and genetic engineering, tissues culture, culture collections, commercially useful micro-algae and their products, mariculture, algalization and soil fertility, pollution and fouling, monitoring, toxicity tests, toxic compounds, antibiotics and other biologically active compounds. Each issue of the Journal also includes a short section for brief notes and general information on new products, patents and company news."

Journal of Phycology
http://onlinelibrary.wiley.com/journal/10.1111/(ISSN)1529-8817

Published since 1965, "all aspects of basic and applied research on algae are included to provide a common medium for the ecologist, physiologist, cell biologist, molecular biologist, morphologist, oceanographer, acquaculturist, systematist, geneticist, and biochemist. The Journal also welcomes research that emphasizes algal interactions with other organisms and the roles of algae as components of natural ecosystems."

Nova Hedwigia and Nova Hedwigia Supplements
http://www.schweizerbart.de/journals/nova_hedwigia

Published since 1959, "Nova Hedwigia is an international journal publishing original, peer-reviewed papers on current issues of taxonomy, morphology, ultrastructure and ecology of all groups of cryptogamic plants, including cyanophytes/cyanobacteria and fungi. The half-tone plates in *Nova Hedwigia* are known for their high quality, which makes them especially suitable for the reproduction of photomicrographs and scanning and transmission electron micrographs. As the length of papers published in Nova Hedwigia main volumes is limited, *Nova Hedwigia* publishes Supplement Volumes as *Beihefte zur Nova Hedwigia* in which more extensive monographic works (e.g., revisions of genera) are published." (http://www.schweizerbart.de/publications/list/series/nova_suppl).

Phycologia
http://www.phycologia.org

Published since 1961, "*Phycologia* is published bimonthly by the International Phycological Society and serves as a publishing medium for information about any aspect of phycology, basic or applied, including biochemistry, cell biology, developmental biology, ecology, evolution, genetics, molecular biology, physiology, and systematics."

Phycological Research
http://onlinelibrary.wiley.com/journal/10.1111/(ISSN)1440-1835

Published from 1952, initially as the *Japanese Journal of Phycology* (Sôrui; 1952–1995), the journal is published by the Japanese Society of Phycology. "The Journal publishes international research dealing with all aspects of phycology to facilitate the international exchange of ideas. The Journal publishes peer-reviewed research on all aspects of phycology."

Protist
http://www.elsevier.com/journals/protist/1434-4610#

Formerly *Archiv für Protistenkunde*, "*Protist* is the international forum for reporting substantial and novel findings in any area of research on protists. The criteria for acceptance of manuscripts are scientific excellence, significance, and interest for a broad readership. Suitable subject areas include: molecular, cell and developmental biology, biochemistry, systematics and phylogeny, and ecology of protists. Both autotrophic and heterotrophic protists as well as parasites are covered. The journal publishes original papers, short historical perspectives and includes a news and views section."

Protistology
http://protistology.ifmo.ru

"*Protistology* publishes original papers (experimental and theoretical contributions), full-size reviews, short topical reviews (which are supposed to be somewhat "provocative" for setting up new hypotheses), rapid short communications, book reviews, symposia materials, historical materials, obituary notices on famous scientists, letters to the Editor, comments on and replies to published papers. Chronicles will present information about past and future scientific meetings, conferences, etc."

8 List Server Lists

List server lists are private or public e-mail lists run in accordance with listserv protocols (http://www.lsoft.com/products/listserv.asp). Private lists (e.g., BPS-L run by the British Phycological Society, and PSA-L run by the Phycological Society of America) are those to which it is not possible to subscribe but are run for the benefit of members of particular groups including societies. Public lists are those to which anybody can subscribe.

ALGAE-L

This list was established in May 1995 by Michael Guiry. Currently, it has over 2600 subscribers and is monitored by the present authors. Information, including subscription details is at http://www.seaweed.ie/algae-l/ and an archive is available at https://listserv.heanet.ie/cgi-bin/wa?A0=ALGAE-L. ALGAE-L is currently run jointly by the present authors.

References Cited

Fourtanier, E. and J.P. Kociolek. 1999. Catalogue of the diatom genera. Diatom Research 14: 1–190.
Millar, A.J.K. 2011. Phycological Societies. Phycologia 50: 507–510.
Silva, P.C. and R.L. Moe. 1999. The Index Nominum Algarum. Taxon 48: 351–353.
Silva, P.C., P.W. Basson and R.L. Moe. 1996. Catalogue of the benthic marine algae of the Indian Ocean. University of California Publications in Botany 79: 1–1259.

Index

Color Plate Section

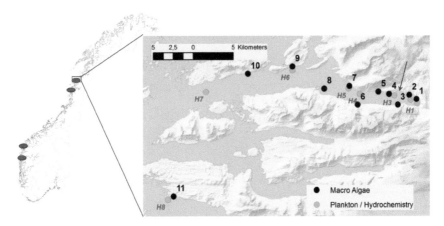

Figure 2. Stations for macro algal collections. Four stations (red dots) on the left figure, indicates areas of data collection for development of the RSLA-index. Map on right shows where macro alga (black dots) and hydrograhical/hydrochemitry sampling (blue dots) were carried out in Glomfjord and the reference station in Tjongsfjorden. The arrow indicates the discharge site for nitrogen and phosphorous compounds from a fertilizer plant.

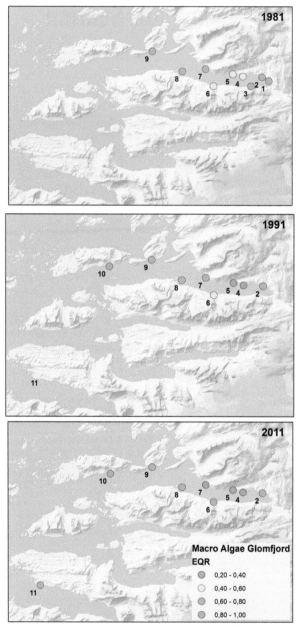

Figure 4. Status classes as EQR values according to definitions in WFD. Orange is "bad", yellow is "moderate", green is "good" and blue is "high". Note that not all stations were monitored among the three periods and reference station 11 was only registered in 2011.

Chapter 4

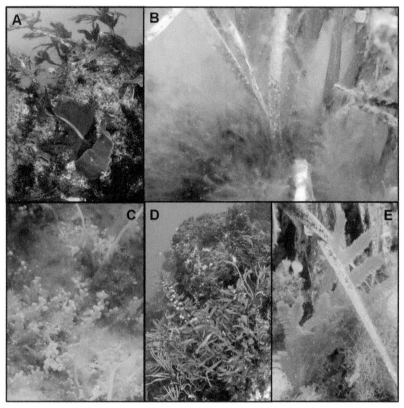

Figure 1. (A) Wakame *Undaria pinnatifida* (Heterokontophyta), native to Japan, [c] David Villegas; (B) Red macroalga *Lophocladia lallemandii*, introduced into the Mediterranean through the Suez Canal [c] Miguel Cabanellas-Reboredo; (C) *Caulerpa racemosa* var. *cylindracea* (Chlorophyta), native to SW Australia, [c] Miguel Cabanellas-Reboredo; (D) Japanese wireweed *Sargassum muticum* (Heterokontophyta), native to SE Asia, [c] David Villegas; (E) *Caulerpa taxifolia* (Chlorophyta), native to the tropical seas, [c] Miguel Cabanellas-Reboredo.

Chapter 7

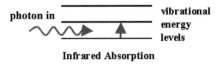

Infrared Absorption

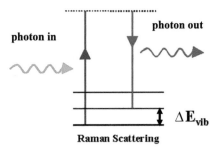

Raman Scattering

Figure 1. Diagram that illustrates infrared absorption and Raman scattering (Adapted from Pereira 2006).

Chapter 8

Figure 2. (a-e) Cultivation methods in Brazil, South America (Photos courtesy of RP Reis and RR Loureiro): (a) PVC raft; (b) tubular net; (c-d) stuffing of seedling into the tubular net; (e) tubular net with *Kappaphycus* 'seedlings'.

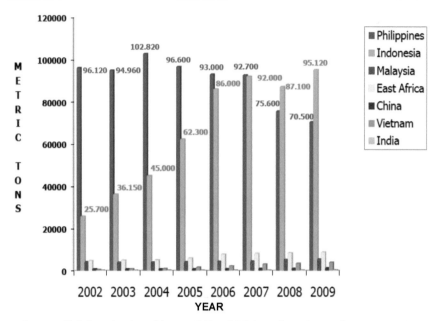

Figure 4. Global production of 'cottonii' (dwt, MT) from the major producing countries.

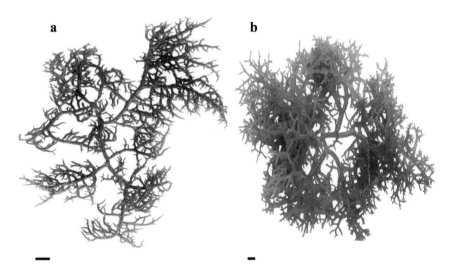

Figure 5. Two major species of *Kappaphycus*, 'cottonii' type (a) *alvarezii* and (b) *striatum*, commonly used in cultivation (bar = 1 cm, Photos courtesy of AQ Hurtado).

Figure 6. 'spinosum' type *Eucheuma denticulatum* (bar = 1 cm, Photo courtesy of AQ Hurtado).

Figure 7. A branch of *Kappaphycus* with an 'ice ice' (bar = 1 cm, Photo courtesy of AQ Hurtado).

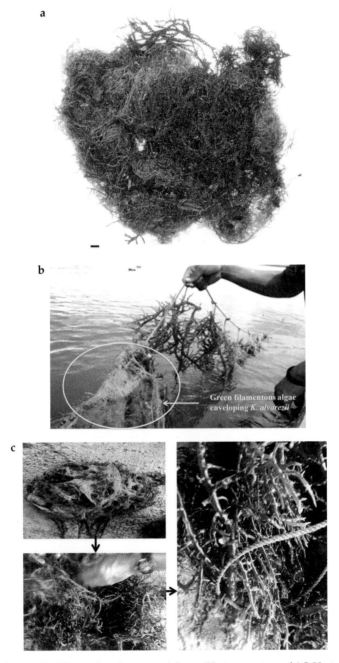

Figure 8. (a-c) Examples of macro-epiphytes (Photos courtesy of AQ Hurtado).

Figure 14. (a-d) Plantlet regenerants using tissue culture at NSTDC Cabid-an Sorsogon, Philippines (bar = 1 cm, Photos courtesy of IT Capacio).

Figure 16. Shoots in *K. alvarezii* treated with AMPEP (Photo courtesy of AQ Hurtado).

Chapter 9

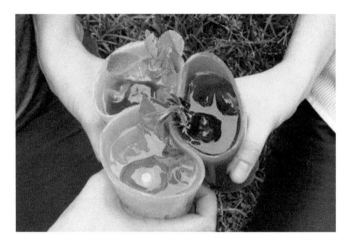

Figure 2. Agar cups. Source: http://www.toxel.com/tech/2010/07/25/edible-cups-made-from-jello/.